Global Warming: Understanding Climate Change

Global Warming:
Understanding Climate Change

Editor: Daisy Mathews

R CALLISTO REFERENCE

www.callistoreference.com

Callisto Reference,
118-35 Queens Blvd., Suite 400,
Forest Hills, NY 11375, USA

Visit us on the World Wide Web at:
www.callistoreference.com

ISBN: 978-1-64116-055-1 (Hardback)

Cataloging-in-Publication Data

Global warming : understanding climate change / edited by Daisy Mathews.
     p. cm.
Includes bibliographical references and index.
ISBN 978-1-64116-055-1
1. Global warming. 2. Climatic changes. 3. Global environmental change. I. Mathews, Daisy.
QC981.8.G56 G56 2019
363.738 74--dc23

Table of Contents

Preface

Decades of industrial growth and deforestation has resulted in many hazardous environmental effects like ozone depletion, global warming, etc. Climate change is a result of increasing global warming levels. One of the major causes of global warming is the emission of greenhouse gases into the atmosphere. Global warming can cause significant impacts such as expansion of deserts, rising sea levels, retreat of glaciers, increase in global temperatures, etc. Extreme weather phenomena such as droughts, floods, heat waves, heavy snowfall, etc. can also occur as a result of global warming. This may result in ocean acidification, species extinction, threat to food security, biodiversity loss, among others. This book aims to present the causes and effects of global warming with respect to climate change in a detailed manner while also highlighting the modern methods to control its impacts. In this book, using case studies and examples, constant effort has been made to make the understanding of difficult concepts of global warming as easy and informative as possible for the reader. Researchers and students in this field will be assisted by this book.

This book unites the global concepts and researches in an organized manner for a comprehensive understanding of the subject. It is a ripe text for all researchers, students, scientists or anyone else who is interested in acquiring a better knowledge of this dynamic field.

I extend my sincere thanks to the contributors for such eloquent research chapters. Finally, I thank my family for being a source of support and help.

Editor

Social Learning and the Mitigation of Transport CO$_2$ Emissions

Maha AlSabbagh

Environmental Management Programme, Arabian Gulf University, Manama 26671, Bahrain;
mahamw@agu.edu.bh

Academic Editor: Yang Zhang

Abstract: Social learning, a key factor in fostering behavioural change and improving decision making, is considered necessary for achieving substantial CO$_2$ emission reductions. However, no empirical evidence exists on how it contributes to mitigation of transport CO$_2$ emissions, or the extent of its influence on decision making. This paper presents evidence addressing these knowledge gaps. Social learning-oriented workshops were conducted to gather the views and preferences of participants from the general public in Bahrain on selected transport CO$_2$ mitigation measures. Social preferences were inputted into a deliberative decision-making model and then compared to a previously prepared participative model. An analysis of the results revealed that social learning could contribute to changes in views, preferences and acceptance regarding mitigation measures, and these changes were statistically significant at an alpha level of 0.1. Thus, while social learning evidently plays an important role in the decision-making process, the impacts of using other participatory techniques should also be explored.

Keywords: Bahrain; climate change mitigation; deliberative decision making; social learning; transport CO$_2$ emissions

---

1. Introduction

The twenty-first session of the Conference of Parties (COP21) achieved consensus among the 196 participating countries to limit the rise in the global average temperature to below 2 °C above pre-industrial levels, with efforts focusing on a target of 1.5 °C above those levels. Unlike the Kyoto protocol, developing countries will commit to reducing emissions by 2020, as indicated in their Intended Nationally Determined Contributions. A total of 186 parties submitted these intended contributions to emission reductions from different energy consuming sectors prior to the COP21. The road transport sector appears to be a priority among these sectors for developing countries. This is because emissions from the transport sector in developing countries are expected to increase rapidly over the coming years [1,2]. Reductions in transport CO$_2$ emissions can be achieved through technological advances. However, achieving substantial reductions requires also a change in the public's behaviour [3]. Targeting behavioural change is necessary, especially in the Gulf Cooperation Council (GCC) countries, where the per capita CO$_2$ emissions rank among the highest globally [4].

Social learning has been identified as a key factor for fostering behavioural changes relating to climate change mitigation. It has been variously defined, and several publications have explored its theoretical aspects (e.g., [5–10]). Two main perspectives have been observed with respect to the concept of social learning. The first perspective focuses on learning process to build a joint framework, whereas the second perspective devotes more attention to reaching convergence among different stakeholders to solve a given problem [11]. In this paper, taking into account the different concepts and applications of social learning in the different research areas, I applied the following definition, developed in the context of natural resource management, which is close to the climate change mitigation field:

"a process of social change in which people learn from each other in ways that can benefit wider social-ecological systems" [7]. Within the literature, social learning has been applied at two levels. The first entails learning at an individual level [12], while the second extends to the wider group or even to the societal level [7,11,13]. Recent researches have either adopted the group level (e.g., [11]) or combined both levels (e.g., [14]), and the later approach is applied here.

Outcomes of social learning vary within empirical literature; however, the outcomes can be classified into four main dimensions: the cognitive dimension, the moral dimension, the relational dimension and the agreement dimension (see Table 1). Types of process dynamics can also vary. Literature on social learning categorises these into three loops: single-loop learning, double-loop learning and triple-loop learning. In the first type of process dynamics, single-loop learning, participants express their views and preferences. At the double-loop learning level, the participants discuss their views, question them and change them as appropriate. Finally, when participants prioritise their preferences and reach consensus about decisions, this indicates the triple-loop learning level [13,15–17].

Table 1. Classification of potential outcomes for social learning as identified in relevant literature. (Adapted from: [12,14,18–21])

Main Dimensions	Cognitive Dimension	Moral Dimension	Relational Dimension	Agreement Dimension
Sub-dimensions	- Acquisition of new knowledge - Knowledge about preferences of other participants - Change of views - Change in understanding	- Understanding of others' perceptive - Understanding concerns of others - Respecting others viewpoints	- Building new relationships or improving existing ones - Willingness to cooperate - Showing commitment - Showing interest in other participants - Being interested in common good - Developing trust between participants	- Reaching consensus about decisions - Satisfaction about the final decisions

Applications of social learning in the environmental arena are relatively new [11]. This arena includes natural resource management [11,22–25], environmental education [15,26], ecological economics [15], sustainable societies [17] and climate change adaptation [27–30]. However, few studies have focused on climate change mitigation, and, specifically on the transport sector (e.g., [31]). This paper addresses this gap, presenting empirical data on the role of social learning in mitigating transport CO_2 emissions. This is its first contribution.

Social learning is thought to generally contribute to improving decision-making processes and, specifically, as a crucial complement to environmental decision-making [28]. However, the empirical literature on social learning is still evolving and remains limited [5,18,32,33], particularly on the extent to which social learning contributes to decision making [32], and assessing transport CO_2 mitigation measures [1]. This is the study's second contribution.

Accordingly, the study has two objectives. First, it highlights the outcomes of social learning in the transport CO_2 emissions mitigation arena. Second, it examines the extent to which social learning contributes to decision making in the same field. It applies a case study methodology in line with the prevailing empirical research on social learning [21]. Bahrain, an oil-exporting GCC country, was chosen as the study's location because of its depleting oil resources. Thus, reducing the country's energy demand (and consequently CO_2 emissions) is imperative. Additionally, energy and carbon intensity indicators are highest in Bahrain compared with other GCC countries [4], indicating the need to adopt energy-efficient measures.

2. Methods

To achieve the study's objectives, workshops targeting general public participants were conducted. Pre- and post-workshop surveys were implemented to gather participants' views and preferences regarding selected transport CO_2 mitigation measures. The compiled data were used to investigate how social learning could contribute to the mitigation of transport CO_2 emissions. Subsequently, the extent to which social learning influenced the decision-making process was assessed. Data were input into a deliberative decision-making model and results were compared with those obtained from a previously developed participative model [34] that lacked a social learning objective. Details on the applied methodology are provided in Figure 1.

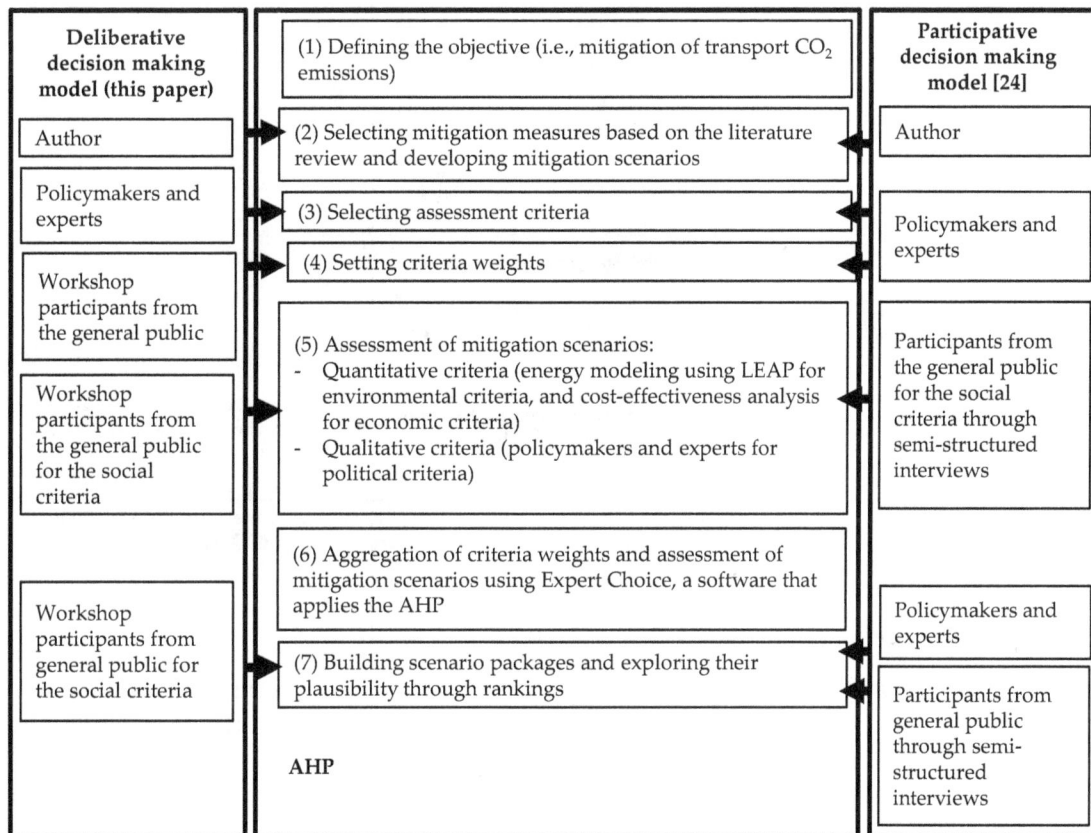

Figure 1. Overview of the applied methodology. Notes: CO_2 = carbon dioxide; AHP = Analytic Hierarchy Process; LEAP = Long Range Energy Alternatives Planning System.

2.1. Transport CO2 Mitigation Measures

Five key mitigation measures were investigated. These were: introducing annual vehicle registration fees based on CO_2 emissions, setting fuel economy standards, market penetration of hybrid cars, market penetration of dedicated Compressed Natural Gas (CNG) cars, and further development of the public transport system. Several assumptions were used to build mitigation scenarios. A total of nine mitigation scenarios were explored (Table 2) [35].

Table 2. Assumptions used to build mitigation scenarios (adapted from [35]).

Mitigation Measure	Scenario	Assumptions	Economic Inputs
Hybrid gasoline cars	Low penetration, low fuel economy	1%, 17.7 km/L	Average cost difference per car = USD 6250
	High penetration, low fuel economy	40%, 17.7 km/L	
Compressed natural gas cars	Low penetration	1%, 13.2 km/L	Cost of fuel station USD 1.5 million (a station per 1000 cars), difference in maintenance cost = USD 1033 every 5 years, difference in car price: USD 7000 for new car and USD 2000 for retrofitting.
Fuel economy standards (by 2030)	Low	15.4 km/L (the USA target for 2015)	USD 716
	High	23.5 km/L (the USA target for 2025)	USD 2067
Registration fees (RF) (using price elasticity of demand of –0.4)	Original RF	- The CO_2 limits are not tightened over time (starting from < 141 g CO_2/km till > 300 g CO_2/km, with 20 g CO_2/km intervals) - Fees start from 0 up to USD 600	
	RF 190	- Fees start from 0 up to USD 190	
	RF 100	- Fees start from 0 up to USD 100	
Public transport		- Introducing light rail transit (LRT) system and improving the current bus rapid transit (BRT) system. - 2.8 billion veh-km is saved	- Total capital cost: USD 5.3 billion - Total maintenance cost: USD 513 million

Notes: A discount rate of 3.3% was used to calculate the costs. This rate is the average for the period 2000–2010 for Bahrain. Registration fees are set here based on grams of CO_2 emissions per kilometre (gCO_2/km). Savings from improving the public transport system are measured in vehicle-kilometres (veh-km).

2.2. Social Learning Outcomes

In this paper, I identified outcomes resulting from social learning covering the four dimensions: cognitive, moral, relational and agreement (Table 1). Evidence and discussion of these outcomes is presented in Section 3.

2.3. Questionnaire Forms

Almost identical questionnaires that were used for the participative decision-making model applied in earlier studies [34,35] were applied for this study. This enabled comparisons to be made between the results of the deliberative and participative models. The questionnaires consisted of six sections: information on selected transport CO_2 mitigation measures, participants' views regarding these measures, preferences elicited through pair-wise comparisons, weights assigned to evaluation criteria, ranking of mitigation scenario packages and socio-economic information on the participants.

2.4. Sampling

Convenience sampling was used for this study. Three workshops, each with 14 participants from the general public, were conducted. Approximately 50% of participants were Bahrainis, in addition to the experts invited to attend the workshops. The sample from the general public was not intended to be statistically representative. Rather, it aimed to provide a societal cross section or "snapshot" [21,36–38].

2.5. Workshop Organisation

Participants at each of the workshops were divided into two small heterogeneous groups. The workshops were organised as three sessions: presentations, discussions and completion of individual questionnaires (Figure 2). The researcher's role during the workshops was that of a learning agent [15]. She delivered the presentations, acted as a facilitator, raised discussion questions and adjusted these questions where necessary. Observers were recruited to take notes, report responses and observe how group discussions developed. The workshops were also filmed and prior written consent was sought from participants and anonymity was guaranteed by not distributing the films.

The participants' responses were gathered prior to presenting the results of previously conducted environmental and economic assessments [35]. Responses were again sought at the end of the workshops to explore the impacts of the deliberations and social learning on how participants perceived transport CO_2 mitigation measures. The workshops also included feedback sessions. These were held after the groups' discussions to provide further opportunities for participants to learn about the views of those with whom they had not interacted.

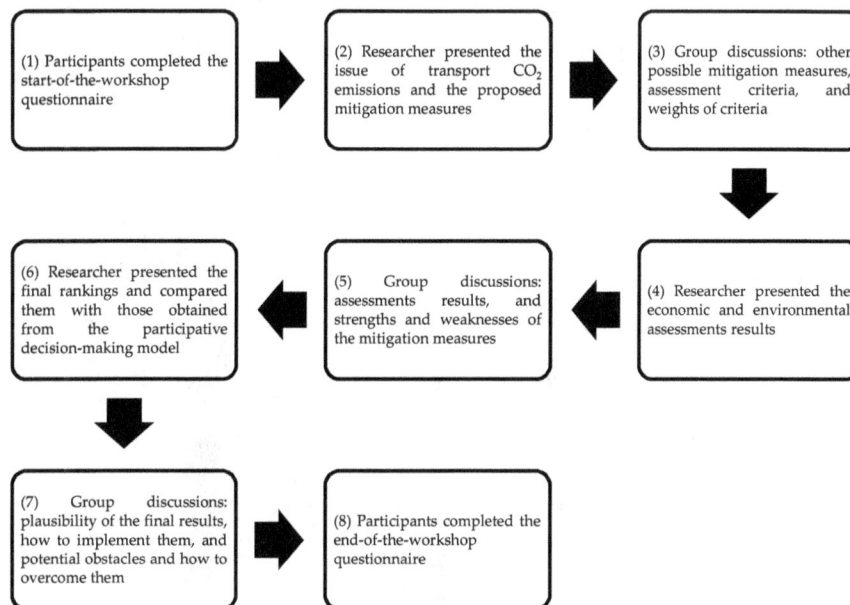

Figure 2. Flow of the workshops.

2.6. Decision-Making Models

A deliberative decision-making model was applied in this paper for assessing transport CO_2 mitigation measures for the road passenger transport sector in Bahrain. The results from this model were compared to those of a participative decision-making model developed in [34] (see Figure 3). These models, which were based on the Analytic Hierarchy Process (AHP), a Multi-Criteria Analysis (MCA) methodology, concurrently assessed different scenarios for the same mitigation measure [34]. The multiple models for Bahrain differed with respect to a single mitigation measure: annual vehicle registration fees based on CO_2 emissions, entailing three possible alternative scenarios [34]. The first scenario, designed by the author (the original registration fee (RF) scenario), assumed a maximum annual fee of USD 600, whereas the second and the third scenarios assumed lower maximum fees of USD 190 and 100 (RF 190 scenario and RF 100 scenario), respectively, as suggested by stakeholders (see Table 2).

Quantitative and qualitative data constituted inputs for both decision-making models. The results from economic and environmental assessments constituted quantitative inputs (as calculated by Alsabbagh et al. [35]), whereas the preferences of policymakers, experts and the general public constituted qualitative data collected via pair-wise comparisons.

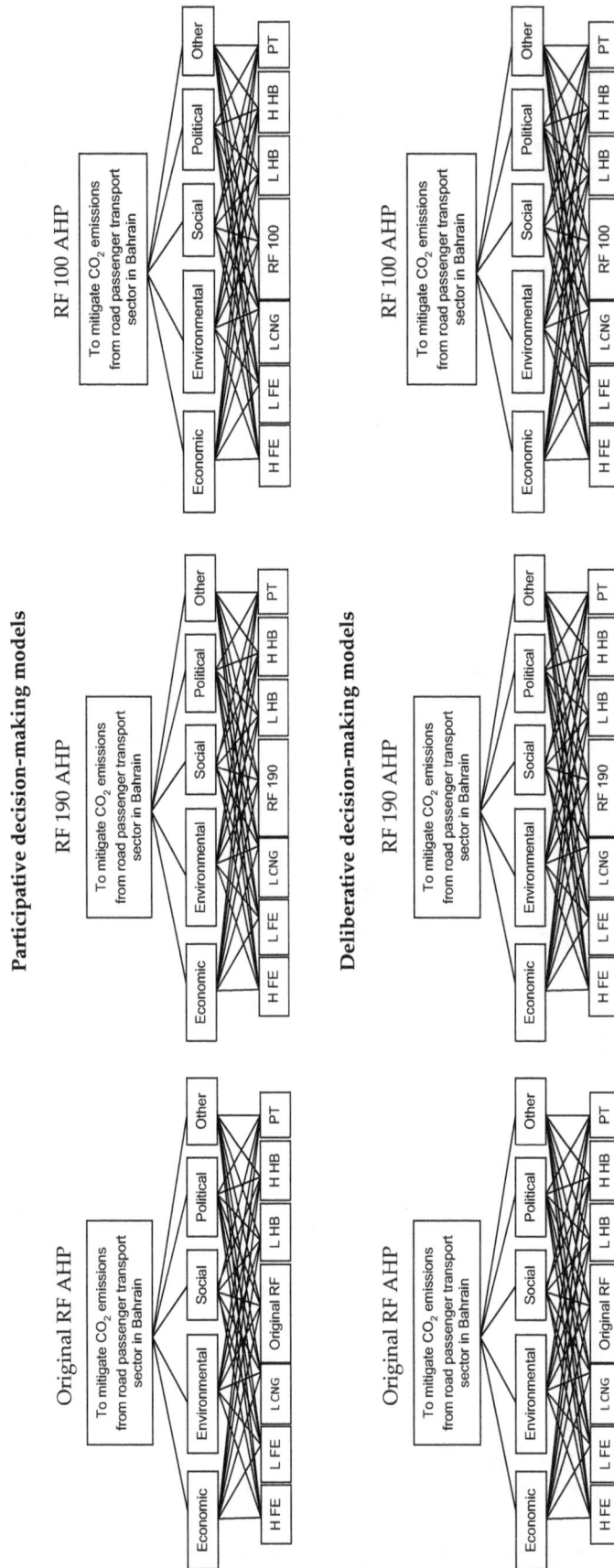

Figure 3. The design of the participative and deliberative decision-making models.

Three AHP models were designed for the participative decision-making models and another three AHP models were designed for the deliberative decision-making models. The participative and deliberative decision-making models differed in two aspects: who assigned criteria weights and how social preferences were elicited. In the participative model, introduced by Alsabbagh et al. [34], policymakers and experts assigned the criteria weights, and the public preferences were collected through semi structured interviews. However, in the deliberative model, workshop participants from the general public assigned criteria weights and provided social preferences of the transport CO_2 mitigation scenarios.

Notes: Original RF AHP means AHP model where the maximum annual vehicle registration fee is USD 600, RF 190 AHP means AHP model where the maximum annual vehicle registration fee is USD 190, RF 100 AHP means AHP model where the maximum annual vehicle registration fee is USD 100.

The two decision-making models differed in how the general public's preferences were elicited. Semi-structured interviews were used by Alsabbagh et al. [34] for the participative model. However, public preferences gathered during social learning-oriented workshops were used for the deliberative model. A comparison of the two decision-making models provided empirical evidence on the extent to which social learning influenced the decision-making process (Figure 3).

2.7. Data Analysis

Statistical analysis and non-parametric tests (e.g., the Wilcoxon non-parametric test and the Friedman test) were conducted using the IBM SPSS package (version 21). The Expert Choice software was used to calculate the final priorities and rankings obtained from the decision-making models. Scenario packages were developed using the scores obtained from the latter software. The plausibility of the scenario packages was assessed using a modified Delphi ranking type method entailing questionnaires completed by the workshop participants.

3. Results

The results are arranged according to the research objectives. First, the outcomes of social learning in the transport CO_2 emissions mitigation arena are presented, based on the four main dimensions described in Table 1. I then discuss the empirical evidence on the extent to which social learning contributes to decision making.

3.1. Outcomes of Social Learning in the Transport CO_2 Emissions Mitigation Arena

3.1.1. Cognitive Dimension

Cognitive changes were evident among participants during the workshops. For instance, their views shifted with the generation of new knowledge. Knowledge was acquired from four main channels: interactions with workshop materials and presentations, discussions with experts, discussions with participants and exchanges of views and preferences across groups. Heterogeneous grouping provided participants with opportunities to learn about the views and preferences of people from diverse cultures and countries. This knowledge acquisition, in particular, was critically important in enriching discussions. At the conclusion of the workshops, most participants stated that they had learnt about different dimensions related to the topic of the workshop.

Additionally, participants acquired knowledge about other participants' preferences. Through group discussions, participants learnt about and acknowledged each other's preferences, which provided a basis for social learning. An example of this learning occurred during one of the group sessions when participants gained the insight that Bahraini women could be interested in sports cars. Although this was a novel insight for male participants, they respected and accepted the concerned woman's choice. This evidenced a double-loop learning cycle [16].

Another evident outcome of social learning is related to the changes in views. The responses in the two questionnaires completed by participants at the beginning and end of the workshops were comprehensively analysed. The results, shown in Table 3, demonstrate that while almost all questionnaire responses had changed, only those related to knowledge of climate change and hybrid cars, support to changing registration fees and willingness to use public transport were of statistical significance at an alpha level of 0.1. By the end of the workshop, all participants were familiar with climate change, its causes and impacts. Support for changing registration fees, setting fuel economy standards and using public transport had all increased. However, opposition to raising fuel prices and to Compressed Natural Gas (CNG) car ownership had also increased. This implies that participants who initially did not have a clear view regarding these measures were convinced by the majority to oppose them, as observed during discussions. Another important observation was an increase in supporters of the RF scenario at the conclusion of workshops. An analysis of responses in the initial and concluding questionnaires revealed a 30% increase in the number of proponents of the policy.

Table 3. A comparison of participants' views obtained from questionnaires completed at the beginning and conclusion of workshops.

No.	Item	Response	Start	End	Significance of Change
1	Have you ever heard/known about causes and impacts of climate change?	Yes	27	42	0.03
		Heard about climate change but not mitigation	15	0	
		No	0	0	
2	In your view, who should be responsible for reducing the impacts of car-use on climate change?	Government	6	9	0.11
		Public	0	3	
		All are responsible	24	27	
		Government and manufacturers	12	3	
3	Do you support imposing a new registration fees system based on the car's CO_2 emissions?	Yes	30	39	0.08
		No	12	3	
4	Would you be prepared to pay extra on the annual registration fee to keep your current car?	Yes	18	27	0.08
		No	24	15	
5	Would you consider changing to smaller and more efficient car if the suggested registration fees system is implemented?	Yes	27	27	1.00
		No	12	12	
		Don't know	3	3	
6	Would you support the setting of controls over the efficiency of cars, in terms of fuel use, entering the country?	Yes	39	42	0.32
		Don't know	3	0	
		No	0	0	
7	In your view, will such a control make a difference with regard to saving environment and non-renewable resources?	Yes	42	42	1.00
		No	0	0	
8	Are you willing to use public transport if reliable and affordable services are offered?	Yes	27	36	0.08
		Don't know	12	3	
		NA	3	3	
		No	0	0	
9	Have you ever heard about hybrid cars?	Yes	33	42	0.08
		No	9	0	
10	Would you consider buying a hybrid car in the future?	Yes	27	33	0.71
		No	6	0	
		Don't know	6	6	
		Maybe	3	3	
11	Do you think that such hybrid car technology fits within the Bahraini context?	Yes	36	30	0.10
		No	3	0	
		Don't know	3	12	
12	Have you ever heard about natural gas cars?	Yes	33	39	0.32
		No	9	3	
13	Would you consider buying a natural gas car in the future?	Yes	6	6	0.56
		No	33	36	
		Don't know	3	0	

Table 3. *Cont.*

No.	Item	Response	Start	End	Significance of Change
14	Do you think that such technology fits within the Bahraini context?	Yes	3	3	0.10
		No	30	39	
		Don't know	6	0	
		Maybe	3	0	
15	Do you support raising the fuel price?	Yes	12	15	0.14
		No	21	27	
		Don't know	3	0	
		Re-direct subsidy	6	0	
16	Do you think that raising fuel price will help reducing CO_2 emissions and fuel consumption?	Yes	15	27	0.22
		No	21	12	
		Maybe	6	3	

Change in preferences was also observed in workshops and can be considered as an outcome that falls within the cognitive dimension. For the initial questionnaire, participants assigned the highest score to improving the public transport system (Table 4). Conversely, in both the initial and concluding questionnaires, the CNG cars option was ranked lowest by all participants. Further in-depth analysis showed that although the ranking order for the transport CO_2 mitigation scenarios was not an exact match for the multi-AHP models, variations in these rankings were not statistically significant at an alpha level of 0.1.

Table 4. Preferences of workshop participants (un-normalised weights).

Mitigation Scenario	Original RF AHP		RF 100 AHP		RF 190 AHP	
	Start	End	Start	End	Start	End
H FE	0.53	0.55	0.83	0.74	0.37	0.37
L FE	0.36	0.37	0.68	0.41	0.19	0.19
L CNG	0.16	0.15	0.22	0.12	0.08	0.08
RF	0.22	0.60	0.44	0.49	0.33	0.33
L HB	0.52	0.48	0.97	0.73	0.67	0.67
H HB	0.25	0.17	0.97	0.23	0.32	0.32
PT	1.00	1.00	1.00	1.00	1.00	1.00

Notes: Original RF AHP denotes an AHP model in which the maximum annual vehicle registration fee is USD 600, RF 190 AHP denotes an AHP model in which the maximum annual vehicle registration fee is USD 190, RF 100 AHP denotes an AHP model in which the maximum annual vehicle registration fee is USD 100, H FE denotes setting high fuel economy standards, L FE denotes setting low fuel economy standards, L CNG denotes low penetration of natural gas cars, RF denotes setting annual vehicle registration fees based on CO_2 emissions, L HB denotes low penetration of hybrid cars, H HB denotes high penetration of hybrid cars, PT denotes improving the public transport system.

Changes in views and preferences of the participants, as described above, resulted in behavioural change. After the workshops, I was contacted by one of the participants who told me that attending the workshop had influenced his decision regarding the purchase of a car. When he came to the workshop, he was planning to buy a used car. He was concerned about the price of the car and the mileage and had almost reached an agreement with the seller. However, after attending the workshop, he checked the car's fuel economy and its technical specifications and cancelled the purchase. He found out that his old car was actually more fuel efficient than the one he was going to buy and he understood that buying the used car would result in more fuel consumption and, consequently, higher carbon emissions. A participant also contacted me saying that she was going to buy a sport utility vehicle (SUV) before she attended the workshop and that she was wealthy enough to pay for it. However, after her participation in the workshop, discussions with the experts and with other participants, she had changed her mind and purchased a sedan car that was more fuel-efficient than the SUV. These two examples provide evidence of a triple-loop learning cycle [16].

3.1.2. Moral Dimension

Mutual understanding can be identified as an outcome of social learning that is related to the moral dimension. The small group settings during the workshops facilitated mutual understanding. Throughout the workshops, participants were encouraged to express their opinions, share their experiences and explain and justify their selections. They were also asked to provide clarifications, examples and reasons rather than just stating their views. This approach enhanced their learning and understanding of each other's views as evidenced in the literature [12]. During group discussions, they also learnt about each other's cultural values and beliefs. Regardless of whether or not they agreed with each other, they consistently respected each other's views and opinions. This demonstrated a single-loop learning cycle [16].

3.1.3. Relational Dimension

Participants in the workshops showed willingness to cooperate and act collectively to reach an applicable and feasible solution to mitigate transport CO_2 emissions in Bahrain. Participants took the initiative to propose other mitigation measures not mentioned in the questionnaires and attempted to assess their applicability to Bahrain. They also showed commitment to the issue in hand, attended the full workshop and participated in group discussions. Participants also developed trust in each other and in experts as well.

3.1.4. Agreement Dimension

Consensus about the final ranking of the mitigation scenarios and scenario packages was reached by the end of the workshops. In fact, this consensus was reflected in the individually-completed questionnaire at the end of the workshop. The agreement on the ranking of the scenario packages improved considerably in comparison to that completed at the start of the workshop. Further details are presented in Section 3.2.

Section 3.1 highlighted outcomes of social learning in the transport CO_2 emissions mitigation arena, which is explored for the first time in this research, thus achieving the first objective of the study. The following section addresses the second objective, focusing on the contribution of social learning to decision making.

3.2. Extent to Which Social Learning Influences Decision Making

Table 5 provides details of a comparison of rankings of transport CO_2 mitigation scenarios obtained from the multi-AHP models used in the deliberative and participative decision-making models [34]. Table 5 illustrates variations in the rankings; however, social learning does not appear to have had a statistically significant impact on the final rankings of the scenarios with an alpha level of 0.1. When integrated with the results of environmental and economic assessments, and political preferences, setting high fuel economy standards ranked highest for both models. This demonstrated that its performance was consistently the best, regardless of who assigned the criteria weights and how social preferences were elicited.

Further analysis showed no statistically significant (at an alpha level of 0.1) variations between the results of the decision-making models based on the conduct of initial and concluding workshop surveys. However, high penetration of hybrid cars was clearly ranked lower in the concluding workshop surveys compared with the initial surveys. This can be interpreted in terms of the influence of social learning and interactions among participants. Discussions within groups during the workshops emphasised the need to start with small trials before incentivising high penetration of hybrid cars. Additionally, there were no statistically significant differences between the results of the deliberative and participative models at 0.1 level. This implies that the participatory technique used to elicit social preferences (i.e., workshops), and consequently social learning, did not significantly impact on the results of the decision-making models. However, the priorities were not perfectly matched.

Table 5. A comparison of ranking orders of mitigation scenarios in the first and second decision-making models.

Mitigation Scenarios	Original RF AHP			RF 100 AHP			RF 190 AHP		
	1st Model	2nd Model Start	2nd Model End	1st Model	2nd Model Start	2nd Model End	1st Model	2nd Model Start	2nd Model End
H FE	1	1	1	1	1	1	1	1	1
L FE	2	5	4	3	5	4	2	5	5
L CNG	7	7	7	7	7	7	7	7	7
RF	6	6	5	6	6	6	5	6	6
L HB	5	4	3	5	3	3	4	4	3
H HB	4	3	6	4	2	5	3	3	4
PT	3	2	2	2	4	2	6	2	2

Notes: Original RF AHP denotes an AHP model in which the maximum annual vehicle registration fee is USD 600, RF 190 AHP denotes an AHP model in which the maximum annual vehicle registration fee is USD 190, RF 100 AHP denotes an AHP model in which the maximum annual vehicle registration fee is USD 100, 1st model refers to the participative model entailing collection of the general public's preferences through interviews, 2nd model start refers to the deliberative decision-making model entailing collection of the public's preferences through pre-workshop questionnaires, 2nd model end refers to the deliberative decision-making model entailing collection of the public's preferences through post-workshop questionnaires, H FE denotes setting high fuel economy standards, L FE denotes setting low fuel economy standards, L CNG denotes low penetration of natural gas cars, RF denotes setting annual vehicle registration fees based on CO_2 emissions, L HB denotes low penetration of hybrid cars, H HB denotes high penetration of hybrid cars, PT denotes improving the public transport system.

3.3. Plausibility of Mitigation Scenario Packages

Mitigation scenarios were combined into scenario packages, because no single scenario significantly outranked the others. When these were presented to workshops' participants, social learning apparently contributed to building consensus among participants in ranking scenario packages. The results of the modified Delphi ranking type revealed a considerable improvement in consensus from 0.536 in the initial workshop survey to 0.854 in the concluding survey. Interestingly, the consensus level among workshops participants was considerably higher than that of policymakers, experts and the general public when obtained through surveys conducted for the participative model (Table 6).

Table 6. Analysis of the plausibility results for participants in the first and second decision-making models.

Results		First Model		Second Model
		Policymakers and Experts	General Public	General Public
	SP1	5	4	5
	SP2	3	3	3
Overall ranking	SP3	1	1	1
	SP4	2	2	2
	SP5	4	5	4
W coefficient		0.566	0.674	0.854

Note: 1st model refers to the participative model entailing collection of the general public's preferences, 2nd model refers to the deliberative model entailing collection of the general public's preferences through post-workshop questionnaires, SP means scenario package.

4. Discussion

Although this study applied a case study methodology, the findings can be generalisable to other socio-economic contexts [39]. This is particularly appropriate in this case as the literature on climate change advocates the transference of mitigation actions and learning gained from small-scale studies [40]. Furthermore, policy transfer studies in the environmental arena are very limited [41], especially those related to transport measures between cities in developing countries [42]. Accordingly, this study makes the following contributions to knowledge.

4.1. Implications for Decision Making

The literature on participation recommends public involvement from the early stages of the decision-making process [7,43,44]. However, not all participation-oriented decision-making models are suitable in particular contexts, especially those characterised by top-down decision-making [45]. Nonetheless, empirical evidence presented here suggests that a deliberative approach can contribute to improved decision making related to the mitigation of transport CO_2 emissions in developing countries. Our results indicate that deliberation and social learning have the capacity to make statistically significant changes in how the public perceives mitigation measures. Social learning, as reported by one study [46], did not significantly alter participants' views on issues about which they were already knowledgeable. However, it did contribute to changing their views on issues about which they were misinformed. This study shows that social learning can contribute to improving public acceptance towards taxation policies which are less favoured by the general public [46]. Moreover, the study's findings suggest that while this may, to a certain extent, be true, the public may be willing to pay when their feedback is considered during the decision-making process. For instance, general public participants demonstrated their acceptance of extra vehicle registration fees after attending the workshops. Social learning contributed to a 30% increase in acceptance of this policy. This implies that it could be used as a tool for improving acceptance of a specific policy. Accordingly, improving acceptance of policies can be added to the list of social learning outcomes that are described in Table 1.

Additionally, social learning can contribute to building consensus. Although this has already been documented in the literature, as in the case of flood risk management [14], this study is the first to provide empirical data that supports this finding. Its findings show that social learning has significantly contributed to building consensus throughout the decision-making process. However, the empirical evidence suggests that it may have a limited influence in terms of changing the general public's ranking of preferences regarding transport CO_2 mitigation measures. Although it can contribute to increasing acceptance of specific policies, it does not appear to contribute to high prioritisation of those policies. Therefore, social learning may have limited role—at least in the short term—in producing a list of the general public's priorities that completely matches that of policymakers.

Another area of contribution of these results relates to obtaining the general public's preferences using two different participatory techniques: semi-structured interviews (in the participative model) and workshops (in the deliberative model) (Figure 3). The preferences of the general public in both models were not an exact match; however, differences in the ranking of the mitigation preferences can still be noticed. Accordingly, future research should be undertaken to investigate how other participatory techniques influence the decision-making process.

These findings are critical to the decision-making process. Although this study was conducted in a developing country, its methodology and findings may be utilised within both developing and developed countries. Social learning may be targeted during the decision-making process, as in this study, and empirically proved effective in increasing acceptance and consensus. Accordingly, a process of involving the public in decision making can be designed and tools and materials can be selected based on these findings.

4.2. Implications for Bahrain

The findings of this study are of special relevance to the government's policy goal of "preparing action plans to reduce carbon emissions from the Kingdom of Bahrain" (p. 48) [47]. In general, they provide the government with evidence on feasibility for formulating environmentally effective, economically feasible and socially accepted policies to promote reduction of transport CO_2 emissions.

Drawing on the findings of this study, the government's next step could be to either adopt a single mitigation measure or a policy package. It could introduce high fuel economy standards, which were ranked highest in the multi-AHP models for both decision-making models, and were prioritised by policymakers, experts and general public participants.

The mitigation policy package options are either to adopt the one that performs best against environmental, economic, social, political and other criteria (the fifth policy package in this case), or the one with the greatest social and political acceptance (the third policy package) (Table 6). Adopting the former would require the provision of financial incentives as this entails the widespread adoption of hybrid cars by the public. Additionally, it would require considerable efforts to ensure public acceptance of the RF 190 scenario as its acceptance level is low. As suggested here, this could be achieved through social learning. However, adopting the latter package would be easier as this has already been ranked highest by the various stakeholder groups. This indicates acceptability regarding implementation of the entailed policies.

However, important issues need to be addressed prior to implementing any of these measures. From an ethical perceptive aimed at ensuring that less well-endowed individuals are not affected, alternatives to private transport first need to be provided before introducing taxation policies. Social equity and inclusion are other considerations for ensuring a fair distribution of impacts (costs and benefits). This would also ensure that public participation is not hindered by transport constraints [48]. It is also important to note that taking small steps at a time can still contribute to CO_2 emission reductions [49]. In particular, the results of this study have indicated that public acceptance of taxation policies has improved, in addition to acquisition of knowledge and change in preferences.

The changes that were identified as outcomes of social learning can contribute to reducing CO_2 emissions from the transport sector in Bahrain and also in the other GCC countries. Being high-income, oil-exporting countries that rely almost completely on fossil fuels for their energy production, GCC countries show high values for both per capita energy consumption and CO_2 emissions. The results of this study can inform policymaking regarding how to change the preferences and the behaviour of the general public. This is of extreme importance to the GCC countries for two main reasons. First, the general public in the GCC countries are habituated to energy-intensive lifestyles, and acceptance of radical technological change is needed to achieve significant emission reductions [50,51]. Second, the GCC countries are obliged to reduce their CO_2 emissions and meet the legally binding climate agreement (i.e., the Paris Agreement), which entered into force in November 2016.

A future longitudinal study, utilising participatory techniques aimed at assessing the impacts of social learning on the decision-making process, would provide valuable insights. Another area for future research is an examination of the impacts of other participatory techniques on the final results of the decision-making model. Further, replicating these methods in a different context entailing bottom-up decision-making would be useful for identifying the extent to which results may change when these methods are applied in different contexts. Lastly, experiments could be conducted on whether social learning can contribute to a complete change in public preferences within a short timeframe.

Acknowledgments: The study was financially supported by both the Arabian Gulf University, Bahrain, and the L'Oréal-UNESCO for Women in Science Middle East fellowship.

Conflicts of Interest: The author declares no conflict of interest. The founding sponsors had no role in the design of the study; in the collection, analyses, or interpretation of data; in the writing of the manuscript, and in the decision to publish the results.

References

1. IPCC. *Climate Change 2014: Mitigation of Climate Change. Contribution of Working Group III to the Fifth Assessment Report of the Intergovernmental Panel on Climate Change*; Edenhofer, O., Pichs-Madruga, O., Sokona, R., Farahani, Y., Kadner, E., Seyboth, S., Adler, K., Baum, A., Brunner, I., Eickemeier, S., et al., Eds.; Cambridge University Press: Cambridge, UK, 2014.

2. EIA. *International Energy Outlook 2014: World Petroleum and Other Liquid Fuels*; EIA: Washington, DC, USA, 2014.

3. Vergragt, P.J.; Brown, H.S. Sustainable mobility: From technological innovation to societal learning. *J. Clean. Prod.* **2007**, *15*, 1104–1115. [CrossRef]

4.   IEA. *Key World Energy Statistics*; IEA: Paris, France, 2014.

5.   Cundill, G.; Rodela, R. A search for coherence: Social learning natural resource management. In *(Re) Views on Social Learning Literature: A monograph for Social Learning Researchers in Natural Resources Management and Environmental Education*; Lotz-Sisitka, H., Ed.; Rhdes University/EEASA/SADC REEP: Grahamstown, South Africa, 2012.

6.   Cundill, G.; Lotz-Sisitka, H.; Mukute, M.; Belay, M.; Shackleton, S.; Kulundu, I. A reflection on the use of case studies as a methodology for social learning research in sub Saharan Africa. *NJAS-Wagening. J. Life Sci.* **2014**, *69*, 39–47. [CrossRef]

7.   Reed, M.; Evely, A.; Cundill, G.; Fazey, I.; Glass, J.; Laing, A.; Newig, J.; Parrish, B.; Prell, C.; Raymond, C.; et al. What is social learning? *Ecol. Soc.* **2010**, *15*, 1–10.

8.   Schusler, T.; Decker, D.; Pfeffer, M. Social learning for collaborative natural resource management. *Soc. Nat. Res.* **2003**, *16*, 309–326. [CrossRef]

9.   Van Bommel, S.; Röling, N.; Aarts, N.; Turnhout, E. Social learning for solving complex problems: A promising solution or wishful thinking? A case study of multi-actor negotiation for the integrated management and sustainable use of the Drentsche Aa Area in the Netherlands. *Environ. Policy Gov.* **2009**, *19*, 400–412.

10. Khalil, K.; Ardoin, N.; Wojcik, D. Social learning within a community of practice: Investigating interactions about evaluation among zoo education professionals. *Eval. Progr. Plan.* **2017**, *61*, 45–54. [CrossRef] [PubMed]

11. Van der Wal, M.; Kraker, J.; Kroeze, C.; Kirschner, P.; Valkering, P. Can computer models be used for social learning? A serious game in water management. *Environ. Model. Softw.* **2016**, *75*, 119–132. [CrossRef]

12. Stagl, S. Multicriteria evaluation and public participation: The case of UK energy policy. *Land Use Policy* **2006**, *23*, 53–62. [CrossRef]

13. Harvey, B.; Ensor, J.; Carlile, L.; Garside, B.; Patterson, Z.; Naess, L. *Climate Change Communication and Social Learning—Review and Strategy Development for CCAFS*; CCAFS: Copenhagen, Denmark, 2012.

14. Benson, D.; Lorenzoni, I.; Cook, H. Evaluating social learning in England flood risk management: An 'individual-community interaction' perspective. *Environ. Sci. Policy* **2016**, *55*, 326–334. [CrossRef]

15. Siebenhüner, B.; Rodela, R.; Ecker, F. Social learning research in ecological economics: A survey. *Environ. Sci. Policy* **2016**, *55*, 116–126. [CrossRef]

16. Henly-Shepard, S.; Gray, S.A.; Cox, L.J. The use of participatory modeling to promote social learning and facilitate community disaster planning. *Environ. Sci. Policy* **2015**, *45*, 109–122. [CrossRef]

17. Broto, V.; Dewberry, E. Economic crisis and social learning for the provision of public services in two Spanish municipalities. *J. Clean. Prod.* **2016**, *112*, 3018–3027. [CrossRef]

18. Muro, M.; Jeffrey, P. A critical review of the theory and application of social learning in participatory natural resource management processes. *J. Environ. Plan. Manag.* **2008**, *51*, 325–344. [CrossRef]

19. Romina, R. Social learning, natural resource management, and participatory activities: A reflection on construct development and testing. *NJAS-Wagening. J. Life Sci.* **2014**, *69*, 15–22. [CrossRef]

20. Muro, M. *The Role of Social Learning in Participatory Planning & Management of Water Resources*; Cranfield University: Cranfield, UK, 2008.

21. Rodela, R.; Cundill, G.; Wals, A.E.J. An analysis of the methodological underpinnings of social learning research in natural resource management. *Ecol. Econ.* **2012**, *77*, 16–26. [CrossRef]

22. Parkins, J.R.; Mitchell, R.E. Public participation as public debate: A deliberative turn in natural resource management. *Soc. Nat. Res.* **2005**, *18*, 529–540. [CrossRef]

23. Keen, M.; Mahanty, S. Learning in sustainable natural resource management: Challenges and opportunities in the Pacific. *Soc. Nat. Res.* **2006**, *19*, 497–513. [CrossRef]

24. Cheng, A.S.; Mattor, K.M. Place-based planning as a platform for social learning: Insights from a national forest landscape assessment process in Western Colorado. *Soc. Nat. Res.* **2010**, *23*, 385–400.

25. De los Ríos, I.; Rivera, M.; García, C. Redefining rural prosperity through social learning in the cooperativesector: 25 years of experience from organic agriculture in Spain. *Land Use Policy* **2016**, *54*, 85–94. [CrossRef]

26. Barrantes, C.; Yagüe, J.L. Adults' education and agricultural innovation: A social learning approach. *Proced.—Soc. Behav. Sci.* **2015**, *191*, 163–168. [CrossRef]

27. IPCC. *Climate change 2007: Mitigation. Contribution of Working Group III to the Fourth Assessment Report of the Intergovernmental Panel on Climate Change*; Metz, B., Davidson, O., Bosch, P., Dave, R., Meyer, L., Eds.; Cambridge University Press: Cambridge, UK, 2007.

28. Nilsson, A.; Swartling, Å.G. Social Learning about Climate Adaptation: Global and Local Perspectives. 2009. Available online: http://www.sei-international.org/mediamanager/documents/Publications/ Policyinstitutions/social_learning_wp_091112.pdf (accessed on 12 January 2016).

29. Albert, C.; Zimmermann, T.; Knieling, J.; Von haaren, C. Social learning can benefit decision-making in landscape planning: Gartow case study on climate change adaptation, Elbe valley biosphere reserve. *Landsc. Urban Plan.* **2012**, *105*, 347–360. [CrossRef]

30. Salvini, G.; Van Paassenb, A.; Ligtenberga, A.; Carreroc, G.; Bregt, A. A role-playing game as a tool to facilitate social learning and collective action towards Climate Smart Agriculture: Lessons learned from Apuí, Brazil. *Environ. Sci. Policy* **2016**, *63*, 113–121. [CrossRef]

31. Goetzke, F. Social Interactions and Social Learning in Transportation Behaviour. 2013. Available online: http://www.dahrendorf-forum.eu/wp-content/uploads/2016/03/Social_Interactions_and_Social_ Learning_in_Transportation_Behaviour__GOETZKE.pdf (accessed on 12 January 2016).

32. Cundill, G.; Rodela, R. A review of assertions about the processes and outcomes of social learning in natural resource management. *J. Environ. Manag.* **2012**, *113*, 7–14. [CrossRef] [PubMed]

33. Schwilch, G.; Bachmann, F.; Valente, S.; Coelho, C.; Moreira, J.; Laouina, A.; Chaker, M.; Aderghal, M.; Santos, P.; Reed, M. A structured multi-stakeholder learning process for sustainable land management. *J. Environ. Manag.* **2012**, *107*, 52–63. [CrossRef] [PubMed]

34. Alsabbagh, M.; Siu, Y.L.; Guehnemann, A.; Barrett, J. Integrated approach to the assessment of CO_2 e-mitigation measures for the road passenger transport sector in Bahrain. *Renew. Sustain. Energy Rev.* **2016**. [CrossRef]

35. Alsabbagh, M.; Siu, Y.L.; Guehnemann, A.; Barrett, J. Mitigation of CO_2 emissions from the road passenger transport sector in Bahrain. *Mitig. Adapt. Strateg. Glob. Chang.* **2015**. [CrossRef]

36. Muro, M.; Jeffrey, P. Time to talk? How the structure of dialog processes shapes stakeholder learning in participatory water resources management? *Ecol. Soc.* **2012**, *17*. [CrossRef]

37. Rauschmayer, F.; Wittmer, H. Evaluating deliberative and analytical methods for the resolution of environmental conflicts. *Land Use Policy* **2006**, *23*, 108–122. [CrossRef]

38. Munda, G. Social multi-criteria evaluation: Methodological foundations and operational consequences. *Eur. J. Oper. Res.* **2004**, *158*, 662–677. [CrossRef]

39. Bassey, M. *Case Study Research in Educational Settings*; Open University Press: Buckingham, UK, 1999.

40. World Bank. *Cities and Climate Change: An Urgent Agenda*; World Bank: Washington, DC, USA, 2010.

41. Marsden, G.; Stead, D. Policy transfer and learning in the field of transport: A review of concepts and evidence. *Trans. Policy* **2011**, *18*, 492–500. [CrossRef]

42. Rahman, M.S. *Integrating BRT with Rickshaws in Developing Cities: A Case Study on Dhaka City, Bangladesh*; University of Leeds: Leeds, UK, 2013.

43. Rosenstrom, U.; Kyllonen, S. Impacts of a participatory approach to developing national level sustainable development indicators in Finland. *J. Environ. Manag.* **2007**, *84*, 282–298. [CrossRef] [PubMed]

44. Banister, D. The sustainable mobility paradigm. *Transp. Policy* **2008**, *15*, 73–80. [CrossRef]

45. Schwilch, G.; Bachmann, F.; Liniger, H.P. Appraising and selecting conservation measures to mitigate desertification and land degradation based on stakeholder participation and global best practices. *Land Degrad. Dev.* **2009**, *20*, 308–326. [CrossRef]

46. Xenias, D.; Whitmarsh, L. Dimensions and determinants of expert and public attitudes to sustainable transport policies and technologies. *Transp. Res. Part A: Policy Pract.* **2013**, *48*, 75–85. [CrossRef]

47. Bahrain, G.O. Parnamaj Amal Alhokoma 2015–2018. 2015. Available online: http://bna.bh/pdf/gov_ program_2015_2018.pdf (accessed on 12 January 2016).

48. Annema, J.A. Transport policy. In *The Transport System and Transport Policy: An Introduction*; Van Wee, B., Annema, J.A., Banister, D., Eds.; Edward Elgar: Cheltenham, UK, 2013.

49. Hickman, R.; Saxena, S.; Banister, D.; Ashiru, O. Examining transport futures with scenario analysis and MCA. *Transp. Res. Part A: Policy Pract.* **2012**, *46*, 560–575. [CrossRef]

50. Kohler, J.; Whitmarsh, L.; Nykvist, B.; Schilperoord, M.; Bergman, N.; Haxeltine, A. A transitions model for sustainable mobility. *Ecol. Econ.* **2009**, *68*, 2985–2995. [CrossRef]

51. Hickman, R.; Ashiru, O.; Banister, D. Transitions to low carbon transport futures: Strategic conversations from London and Delhi. *J. Transp. Geogr.* **2011**, *19*, 1553–1562. [CrossRef]

Long-Term Climate Trends and Extreme Events in Northern Fennoscandia (1914–2013)

Sonja Kivinen [1,2,*], **Sirpa Rasmus** [3], **Kirsti Jylhä** [4] and **Mikko Laapas** [4]

[1]   Department of Geographical and Historical Studies, University of Eastern Finland, P. O. Box 111, FI-80101 Joensuu, Finland

[2]   Department of Geography and Geology, University of Turku, FI-20014 Turku, Finland

[3]   Department of Biological and Environmental Science, University of Jyväskylä, P.O. Box 35, FI-40014 Jyväskylä, Finland; sirpa.rasmus@jyu.fi

[4]   Finnish Meteorological Institute, P. O. Box 503, FI-00101 Helsinki, Finland; kirsti.jylha@fmi.fi (K.J.); mikko.laapas@fmi.fi (M.L.)

*   Correspondence: sonja.kivinen@uef.fi

Academic Editor: Christina Anagnostopoulou

Abstract: We studied climate trends and the occurrence of rare and extreme temperature and precipitation events in northern Fennoscandia in 1914–2013. Weather data were derived from nine observation stations located in Finland, Norway, Sweden and Russia. The results showed that spring and autumn temperatures and to a lesser extent summer temperatures increased significantly in the study region, the observed changes being the greatest for daily minimum temperatures. The number of frost days declined both in spring and autumn. Rarely cold winter, spring, summer and autumn seasons had a low occurrence and rarely warm spring and autumn seasons a high occurrence during the last 20-year interval (1994–2013), compared to the other 20-year intervals. That period was also characterized by a low number of days with extremely low temperature in all seasons (4%–9% of all extremely cold days) and a high number of April and October days with extremely high temperature (36%–42% of all extremely warm days). A tendency of exceptionally high daily precipitation sums to grow even higher towards the end of the study period was also observed. To summarize, the results indicate a shortening of the cold season in northern Fennoscandia. Furthermore, the results suggest significant declines in extremely cold climate events in all seasons and increases in extremely warm climate events particularly in spring and autumn seasons.

Keywords: climate trends; climate warming; cold season; extreme events; northern Fennoscandia

1. Introduction

Observations from northern regions have shown a significant warming trend during the past decades [1,2]. Temperatures at northern high latitudes have increased more than twice as fast as the global average. This phenomenon, usually referred as arctic or polar amplification, is driven by complex climate system feedbacks, such as reductions of sea ice and snow cover and changes in atmospheric and ocean circulation [3–5]. Nevertheless, the magnitude and spatiotemporal pattern of warming and associated precipitation changes show a strong variation across circumpolar areas [6–8].

In addition to increasing temperatures, climate warming is expected to lead to increases in frequency, intensity and duration of several extreme weather and climate events [9–11]. Earlier observational and climate model studies focused mainly on changes in mean climate, and possible changes in weather and climate extremes started to be analyzed only since the early 1990s [12,13]. An extreme weather event is an event that is rare within its statistical reference distribution at a particular place, whereas an extreme climate event is an average of a number of weather events over a certain period of time. This means an average which is itself extreme [14]. Statistics of past climate and model simulations for the future both suggest, e.g. higher maximum and minimum temperatures, more hot summer days and heavier precipitation events or more severe drought, in different parts of the world [9,15,16]. Recent papers have discussed influences of changes in atmospheric circulation patterns and in thermodynamic aspects, such as sea level, sea surface temperature and atmospheric moisture content, on the occurrence and intensity of extreme events [17,18]. The issue of connections between Arctic amplification and mid-latitude weather extremes is also actively debated [19–22].

In northern Europe, the temperature rise is expected to keep on being much larger than the global mean [2]. The warming has a wide variety of impacts on northern ecosystems and significant consequences for agriculture, forestry, human health, and infrastructure [23–26]. It is assumed that especially more frequent and intense extreme weather events increase the vulnerability of northern environment and human activities to warming [24]. Direct impacts of temperature and precipitation changes are experienced in northern hydrological regimes, such as groundwater levels, permafrost, and snow and ice conditions [27–29]. Dependency of northern key species on snow and ice increase the northern terrestrial ecosystems vulnerability to warming [26,30].

Traditional livelihoods, like hunting, fishing, harvesting and herding economies, depend on healthy ecosystems with habitats suitable for plant and animal populations. Climate change together with other drivers decrease the availability and quality of these habitats [31,32]. Reindeer herding is one of the most important traditional livelihoods in northern Fennoscandia and northwestern Russian tundra and boreal environments. It is strongly dependent on climatic conditions, and particularly the winter season is a critical period for reindeer grazing [33–35]. Exceptional winters with difficult snow and ice conditions have long-term negative impacts on reindeer populations and consequently, economic and cultural viability of reindeer herding livelihoods [36–38]. In addition, modern livelihoods may be sensitive to changes. For example, tourism is an important part of the northern economies and is very vulnerable to changing winter conditions [39].

Understanding climate variability, trends and changes in frequency, intensity and duration of extreme events requires long-term climate observation data. Temperature records have generally a good coverage since the 1950s in northern regions, but the availability of instrumental measurements extending to the beginning of the 20th century is considerably limited [6,7,40]. In northernmost Europe, long-term climate trends (ca 100 years) calculated from weather station data have usually been studied within one country or a region with a relatively limited spatial extent ([40–46], but see, e.g. [47,48] for a wider Northern Russian area). Furthermore, long-term changes in extreme weather events in northern Fennoscandia are relatively little studied [49–51]. The study period of [49,50] ended in the year 1995, and most of the study sites were located to the south of the Arctic Circle. On the other hand, [51] focused on wintertime warm weather episodes alone.

In this work, we studied climate trends and extreme events in northern Fennoscandia over the 100-year period 1914–2013 using long-term weather data from nine weather stations located in Finland, Norway, Sweden and Russia. Our aims were to: (1) examine trends in monthly mean temperature variables, monthly precipitation sums and in the spring and autumn, number of frost days at different geographical locations, and (2) study temporal distributions of the occurrence of rare and extreme climate events, also discussing their links to atmospheric circulation and reductions in the Barents Sea ice. The following climate events were considered: rarely cold and warm seasons, extremely high and low daily temperatures, and extremely high daily precipitation sums. The results will be considered from the viewpoint of high-latitude species, ecosystem services and local livelihoods.

2. Materials and Methods

2.1. Study area

The study area covered the northern parts of Finland, Norway and Sweden and the Kola Peninsula in Russia (67.15–70.37° N, 18.83–40.68° E). Nine weather stations located in the region were included in the analyses: (1) Abisko (SWE), (2) Sodankylä (FIN), (3), Sihccajavri (NOR), (4) Tromsø (NOR), (5) Vardø (NOR), (6) Teriberka (RUS), (7) Kandalaksa (RUS), (8) Svyatoy Nos (RUS), and (9) Ostrov Sosnovez (RUS) (Figure 1, Table 1).

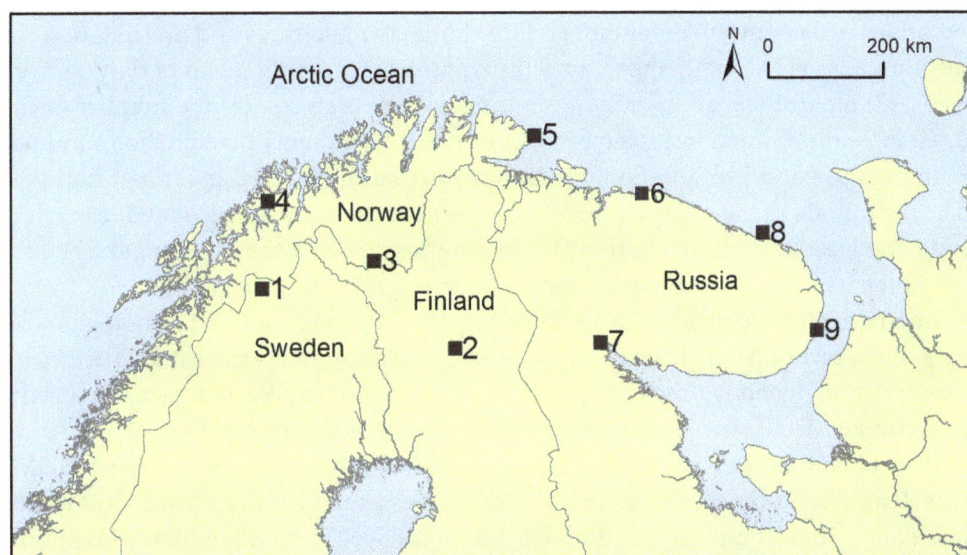

Figure 1. The location of weather stations in northern Fennoscandia. (1) Abisko, (2) Sodankylä, (3) Sihccajavri, (4) Tromsø, (5) Vardø, (6) Teriberka, (7) Kandalaksa, (8) Svyatoy Nos, (9) Ostrov Sosnovez.

Table 1. Weather stations included in the study and their general climate characteristics. Annual mean temperature (Tmean), January mean temperature (TJan), July mean temperature (TJul) and annual precipitation sum (Prec) were calculated for the period 1914–2013.

	Station	Country	Altitude (m.a.s.l.)	Tmean (°C)	TJan (°C)	TJul (°C)	Prec (mm)
1	Abisko	Sweden	388	−0.5	−10.7	11,7	307
2	Sodankylä	Finland	179	−0.5	−13.7	14.8	485
3	Sihccajavri	Norway	382	−2.5	−14.4	12.6	383
4	Tromsø	Norway	100	2.8	−3.6	12.0	1015
5	Vardø	Norway	14	1.6	−4.5	9.3	570
6	Teriberka	Russia	33	0.8	−7.9	11.4	470
7	Kandalaksa	Russia	26	0.4	−11.9	14.7	462
8	Svyatoy Nos	Russia	70	0.2	−7.3	9.2	412
9	Ostrov Sosnovez	Russia	15	−0.6	−9.2	8.6	345

2.2. Climate Data

Climate data for the period 1914–2013 were derived from the Finnish Meteorological Institute's open data service (Sodankylä) [52], the Eklima web portal of Norwegian Meteorological Institute (Sihccajavri, Tromsø, Vardø) [53] and the European Climate Assessment & Dataset project (Teriberka, Kandalaksa, Svyatoy Nos, Ostrov Sosnovez) [54]. The Abisko dataset was provided by the Abisko Scientific Research Station. The number of months excluded from the analyses of temperature and precipitation due to lack of data is given in Supplementary Material (Table S1).

In addition to temperature and precipitation data in northern Fennoscandia, we used seasonal time series of two indices of the North Atlantic Oscillation (NAO). The station-based index has been calculated as the difference of normalized sea level pressure between Lisbon, Portugal and Stykkisholmur/Reykjavik, Iceland, and the principal-component index is defined by the leading empirical orthogonal function of sea level pressure over the Atlantic sector

(20°–80° N, 90° W–40° E) [55]. Both indices are provided by the Climate Analysis Section, NCAR, Boulder, USA [56].

2.3. Statistical Analyses

Observations of daily mean, minimum and maximum temperatures and precipitation were used to calculate annually: (1) monthly mean air temperature, (2) monthly mean of daily maximum air temperatures, (3) monthly mean of daily minimum air temperatures, (4) the number of frost days (TMin < 0 °C) in April-May and in September–October, and (5) monthly precipitation sum. Long-term trends in temperature and precipitation conditions were studied using linear and non-parametric models. The magnitude of the trend with 95% confidence intervals was calculated for each calendar month using the least squares method and non-parametric Sen's trend estimate. The statistical significance of the trends was calculated using the Mann-Kendall trend test.

Most long-term climatic time series are influenced by inhomogeneities, i.e., non-climatic factors, like changes in observational instruments and practices, station relocations and changes in station environment. A homogenous time series is defined as one where variations are caused only by variations in climate. Time series influenced by inhomogeneities may lead to deceptive conclusions about the state of climate and possible trends included. In this study, we applied HOMER [57], software for detecting inhomogeneities in time series of essential climate variables at monthly and annual time scales. It is an outcome of the COST Action ES0601 HOME, which carried out a blind benchmarking experiment to compare and validate homogenization algorithms, resulting in a group of recommended methods [58]. HOMER integrates the best characteristics of some of these recommended methods into one interactive open access software tool. It is a semi-automatic multiple breakpoint method utilizing state-of-the-art homogenization algorithms that are able to work with inhomogenous reference series [57,59,60].

We limited detection of inhomogeneities with respect to monthly mean, minimum and maximum temperatures, since time series of precipitation are more difficult to homogenize due to lower cross-correlations [58], especially with a sparse station network like ours (Figure 1). Because of the lack of metadata and the sparse station network and in order to make sure that no possible false detections are given importance, we took a rather reserved approach to isolated inhomogeneities, i.e., to inconsistencies detected in one of the temperature series but not in two others. This demand of having inhomogeneities simultaneously in more than one of the temperature variables (difference of one year accepted) is one way to provide some additional evidence for detection based solely on statistics. We decided not to adjust any time series even if inhomogeneities were detected. Instead, when reporting the results, we place more emphasis on homogeneous stations and stations with only isolated inhomogeneities than on stations with coinciding inhomogeneities.

In the analysis of the time series, seasonal mean temperature values in (1) December-February, (2) March-May, (3) June-August and (4) September-November were first calculated for each year. Separately for each weather station, a rarely cold (or warm) season was then defined as a case belonging to the 10th (or 90th) percentile of the mean seasonal temperature values, respectively (10 rarely cold/warm seasons per station during the 100-year study period).

Second, extremely low and high daily temperatures and extremely high daily precipitation sum in (1) January, (2) April, (3) July and (4) October were studied. Extremely low daily temperatures were defined as belonging to the 1st percentile of minimum daily temperatures, and extremely high daily temperatures were defined as values belonging to the 99th percentile of maximum daily temperatures. Extremely low and high temperatures were defined separately for each station (30 or 31 extremely cold/warm days per station during the study period). Extremely high precipitation was defined as values belonging to the 99th percentile of daily precipitation sum. The Sihccajavri, Tromsø and Vardø weather stations were excluded from temperature analyses, because maximum temperature values were available only since 1925–1955.

Temporal distribution of rarely cold and warm seasons and extreme daily events was studied over the past 100 years by examining their total occurrence (all included weather stations) during five subsequent 20-year intervals, the first covering the years 1914–1933 and the last, years 1994–2013. A temporally even distribution of the occurrence would imply that each 20-year interval had 20% of the cases. Averages of the seasonal mean NAO indices were calculated for the same 20-year periods.

3. Results

3.1. Temperature and Precipitation Variability and Trends

The weather stations were located over a wide range of latitudinal, altitudinal and continental-oceanic values (Figure 1, Table 1). The 100-year mean annual temperature ranged from −0.6 to 2.8 °C across the stations. Tromsø (4) and Vardø (5) weather stations were characterized by oceanic climate with high annual mean temperatures and precipitation sums. Sodankylä (2), Sihccajavri (3) and Kandalaksa (7) weather stations were characterized by continental climate with a large difference in winter and summer temperatures. The Abisko weather station (1) was located at the highest altitude in the Scandinavian mountain chain. Teriberka (6), Svyatoy Nos (8) and Ostrov Sosnovez (9) were the most eastern weather stations located on the shore of the Arctic Sea.

In the context of homogeneity testing, three stations were found to be homogenous, i.e., no inhomogeneities were detected in any of the temperature time series (stations 1, 6 and 9). Only isolated inhomogeneities were found from stations 2 and 4. Inhomogeneities consistent between at least two of the three temperature variables were found from stations 3, 5, 7 and 8 mainly before the 1950s and 1960s. Inhomogeneities detected from the temperature series of stations 3 and 7 mean that the period 1914–1957 (1914–1961) was on average about 0.5 °C too warm compared to the latter part of the studied time period. At station 5, the period 1914–1954 was on average 0.3 °C too cold compared to the period 1955–2014. At station 8, the period 1914–1980 was on average 0.2 °C too cold compared to the period of 1981–2013.

Statistically significant warming trends across the period 1914–2013 were found in March–May in all studied locations and in September-October in most of the locations (Table 2). The monthly mean minimum temperatures showed the greatest increases, 0.1–0.5 °C decade^{-1}. Focusing on the five stations with no or only isolated inhomogeneities, we can see that the mean and maximum temperatures increased by 0.1–0.3 °C decade^{-1}. Monthly mean minimum temperatures, in particular, increased at some stations also in summer, whereas nearly no increasing trends were found during winter months. Annual mean minimum temperatures increased by 0.1–0.3 °C decade^{-1} at four of the five stations with no or only isolated inhomogeneities. Statistically significant trends in annual mean temperature and in annual mean maximum temperature were less commonly detected. It is worth noting that inhomogeneities detected in the time series of stations 5 and 8 potentially mean weaker warming trends than shown in Table 2 and vice versa for stations 3 and 7.

The number of frost days varied between 29 and 50 days in April–May and between 13 and 37 days in September-October (Figure 2). The number of frost days declined by 7–11 days century^{-1} in April–May and 0–12 days century^{-1} in September-October in five stations with no or only isolated inhomogeneities.

Statistically significant trends in monthly precipitation sums were less common compared to those in temperatures (Table 2). Monthly precipitation sums increased significantly particularly in October and December (1.1–3.7 mm decade^{-1}) and in July (3.3–4.0 mm decade^{-1}) at several eastern and coastal stations, but only at a few stations in the other months. Annual precipitation sums increased by 5–38 mm decade^{-1} at four stations (1, 3, 4, 7).

Table 2. Trends (p < 0.05) and the 95% confidence intervals over the period 1914–2013 for monthly and annual means of daily mean, minimum and maximum temperatures (°C/decade) and monthly and annual precipitation sums (mm/decade) using a least-squares method. Sen's trend estimate produced nearly identical results (not shown in the table). Homogeneous stations (1, 6, 9) or stations with only isolated inhomogeneities (2, 4) are marked in bold. Blanks indicate the absence of statistically significant trends.

Station	Jan	Feb	Mar	Apr	May	Jun	Jul	Aug	Sep	Oct	Nov	Dec	Annual
TMean (°C)													
1				**0.2 ± 0.1**	**0.2 ± 0.1**	**0.2 ± 0.1**			**0.1 ± 0.1**	**0.2 ± 0.1**			**0.1 ± 0.1**
2			**0.3 ± 0.2**	**0.1 ± 0.1**									
3				0.2 ± 0.2	0.2 ± 0.1		-0.1 ± 0.1						
4					**0.2 ± 0.1**								
5			0.2 ± 0.1	0.2 ± 0.1	0.2 ± 0.1	0.1 ± 0.1			0.1 ± 0.1	0.1 ± 0.1			0.1 ± 0.1
6			**0.3 ± 0.2**	**0.1 ± 0.1**	**0.1 ± 0.1**					**0.1 ± 0.1**			
7			0.2 ± 0.2		0.1 ± 0.1	0.1 ± 0.1							
8	0.2 ± 0.2		0.4 ± 0.2	0.2 ± 0.2	0.2 ± 0.1				0.1 ± 0.1	0.2 ± 0.1		0.2 ± 0.2	0.2 ± 0.1
9			**0.3 ± 0.2**		**0.1 ± 0.1**		**0.1 ± 0.1**			**0.2 ± 0.1**			
TMin (°C)													
1			**0.2 ± 0.2**	**0.3 ± 0.1**	**0.3 ± 0.1**	**0.2 ± 0.1**	**0.2 ± 0.1**	**0.2 ± 0.1**	**0.1 ± 0.1**	**0.2 ± 0.1**			**0.1 ± 0.1**
2			**0.4 ± 0.3**	**0.3 ± 0.2**	**0.2 ± 0.1**	**0.2 ± 0.1**	**0.2 ± 0.1**	**0.2 ± 0.1**	**0.2 ± 0.1**	**0.3 ± 0.2**			**0.2 ± 0.1**
3			0.2 ± 0.1	0.3 ± 0.2	0.2 ± 0.1								
4													
5	0.1 ± 0.1	0.1 ± 0.1	0.3 ± 0.1	0.2 ± 0.1	0.2 ± 0.1	0.1 ± 0.1	0.1 ± 0.1	0.1 ± 0.1	0.1 ± 0.1	0.2 ± 0.1			0.2 ± 0.1
6			**0.3 ± 0.2**	**0.2 ± 0.2**	**0.1 ± 0.1**	**0.1 ± 0.1**				**0.1 ± 0.1**			**0.1 ± 0.1**
7			0.3 ± 0.3					0.1 ± 0.1					
8	0.2 ± 0.2	0.3 ± 0.2	0.5 ± 0.2	0.3 ± 0.2	0.3 ± 0.1	0.2 ± 0.1	0.2 ± 0.1		0.2 ± 0.1	0.3 ± 0.1		0.2 ± 0.2	0.3 ± 0.1
9			**0.5 ± 0.2**	**0.2 ± 0.2**	**0.2 ± 0.1**	**0.1 ± 0.1**	**0.2 ± 0.1**		**0.1 ± 0.1**	**0.2 ± 0.2**			**0.2 ± 0.1**
TMax (°C)													
1			**0.2 ± 0.2**	**0.2 ± 0.1**	**0.3 ± 0.1**	**0.2 ± 0.1**				**0.2 ± 0.1**			**0.1 ± 0.1**
2	**n.d.**	**n.d.**		**0.2 ± 0.1**	**0.2 ± 0.2**					**0.1 ± 0.1**			
3	n.d.	n.d.	n.d.	n.d.	n.d.	n.d.	n.d.	n.d.	n.d.	n.d.	n.d.	n.d.	n.d.
4			**0.2 ± 0.1**	**0.2 ± 0.1**	**0.3 ± 0.1**			**n.d.**	**n.d.**	**n.d.**	**n.d.**		
5			0.2 ± 0.2	0.2 ± 0.1	0.2 ± 0.1				0.2 ± 0.1				
6			**0.2 ± 0.2**	**0.2 ± 0.2**	**0.2 ± 0.2**	**0.2 ± 0.1**							
7			0.3 ± 0.2	0.2 ± 0.2	0.3 ± 0.1								
8								-0.2 ± 0.2		0.2 ± 0.1		0.2 ± 0.2	
9			**0.3 ± 0.2**	**0.1 ± 0.1**	**0.1 ± 0.1**	**0.1 ± 0.1**	**0.1 ± 0.1**		**0.1 ± 0.1**	**0.2 ± 0.1**			
Prec (mm)													
1										**1 ± 1**			**5 ± 4**
2													
3													
4										**3 ± 1**		**3 ± 4**	**8 ± 7**
5				1 ± 1			4 ± 2		3 ± 2	2 ± 2			18 ± 11
6							**4 ± 2**		**3 ± 2**	**2 ± 2**		**1 ± 1**	
7	4 ± 1	3 ± 1	3 ± 1	1 ± 1			3 ± 2		2 ± 2	4 ± 2	4 ± 1	4 ± 1	
8					3 ± 2		3 ± 2					2 ± 1	38 ± 7
9	**2 ± 1**	**2 ± 1**	**1 ± 1**							**2 ± 2**	**2 ± 2**	**3 ± 1**	

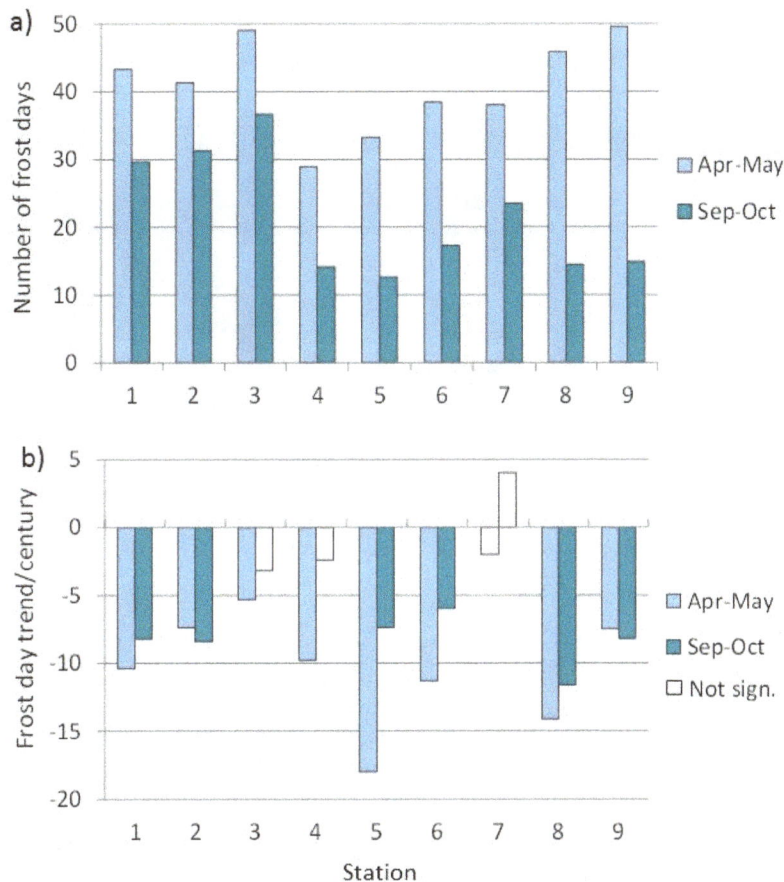

Figure 2. (a) The average number of frost days (TMin < 0 °C) and (**b**) a frost day trend (change in the number of frost days/100 years) in April-May and September-October in nine studied weather station. Stations 1, 6, and 9 are homogeneous, and only isolated inhomogeneities were detected in stations 2 and 4.

3.2. Rarely Cold and Warm Seasons

The rarely cold seasons (defined as cases belonging to the 10th percentile of the mean seasonal temperature values) did not occur temporally evenly among the five 20-year intervals of the study period. In particular, the last interval (1994–2013) had low proportions of rarely cold seasons (2%–9% of all rare events, depending on the season) (Figure 3). Rarely cold winter and summer seasons occurred in the study region, particularly between 1954 and 1993 (69%–71% of seasonal rare events), and rarely cold spring and autumn seasons occurred especially in 1913–1934 and 1954–1973 (67%–72% of seasonal rare events).

On average, 37% of rarely warm spring seasons and 48% of rarely warm autumn seasons (defined as cases belonging to the 90th percentile of the mean seasonal temperature values) occurred in 1994–2013 (Figure 3). On the contrary, rarely warm winter and summer seasons had no significant peaks during the last 20-year interval, but occurred more evenly during the 100-year study period. For example, 59% of rarely warm summer seasons occurred during the second and third 20-year periods (1934–1973).

A tentative comparison between the occurrence of rarely cold or warm seasons and the average NAO indices during the five 20-year intervals suggested that although the portion of cold winters had a tendency to decrease with increasing wintertime NAO, as could be expected, the period 1974–1993 did not fit into that general picture. This interval was characterized by a large portion of cold winters, a typical number of mild winters and a high NAO value. Rarely warm winters were most uncommon during the period of the most negative winter NAO (1954–1973), and the share of rarely cold springs clearly decreased with increasing principle-component spring NAO index. Otherwise, no clear linkages were found between the occurrence of extreme seasons and the oscillation pattern.

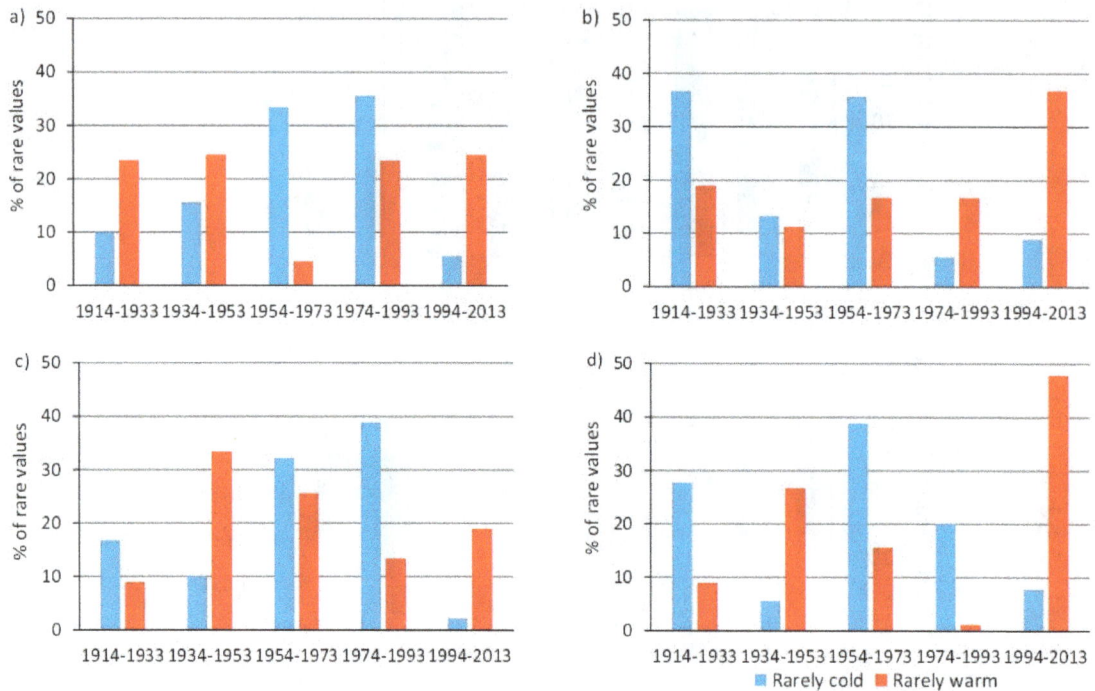

Figure 3. The total proportion of rarely cold and warm seasons (10th and 90th percentile of mean seasonal temperatures defined separately for each weather station) observed in 20-year study intervals in northern Fennoscandia. (**a**) December-February, (**b**) March–May, (**c**) June-August, and (**d**) September–November.

3.3. Extreme Daily Events

The number of days with extremely low minimum temperatures in April, July and October (defined as 1st percentile of minimum daily temperatures) was noticeably smaller during the last 20-year interval compared to other time periods (4%–9% of all extremely cold days) (Figure 4). Extremely low minimum temperatures in April occurred particularly during the first 20-year interval of the study period (39% of days with extremely low temperatures), whereas extremely low temperatures in July and October were more evenly distributed between the time period of 1914–1993. Extremely low minimum temperatures in January occurred particularly in 1973–1994.

On average, 36% and 42% of days with extremely high maximum temperature in April and October (defined as 99th percentile of minimum daily temperatures, respectively, occurred during the last 20-year interval of the study period (Figure 4). Extremely high daily temperatures in January were more evenly distributed over the 20-year study intervals. Extremely high daily temperatures in July occurred particularly in the time period of 1954–1973 (43% of all extreme values).

Extreme daily precipitation sums were relatively evenly distributed over the studied 20-year intervals (Figure 5). However, a slight tendency of higher extreme daily precipitation sums towards the end of the study period could be observed.

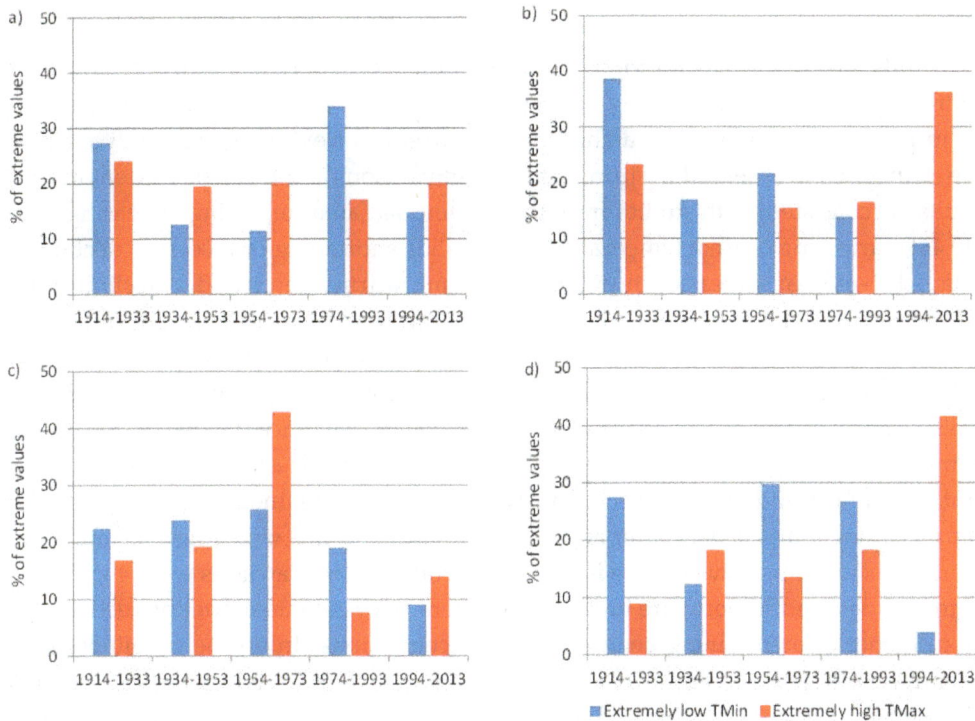

Figure 4. The total proportion of extremely low daily minimum temperatures and extremely high daily maximum temperatures (1st percentile of minimum temperatures and 99th percentile of maximum temperatures defined separately for each weather station) observed in 20-year study intervals in northern Fennoscandia. (**a**) January, (**b**) April, (**c**) July, and (**d**) October. (Stations 2, 4 and 5 excluded from maximum temperature calculations).

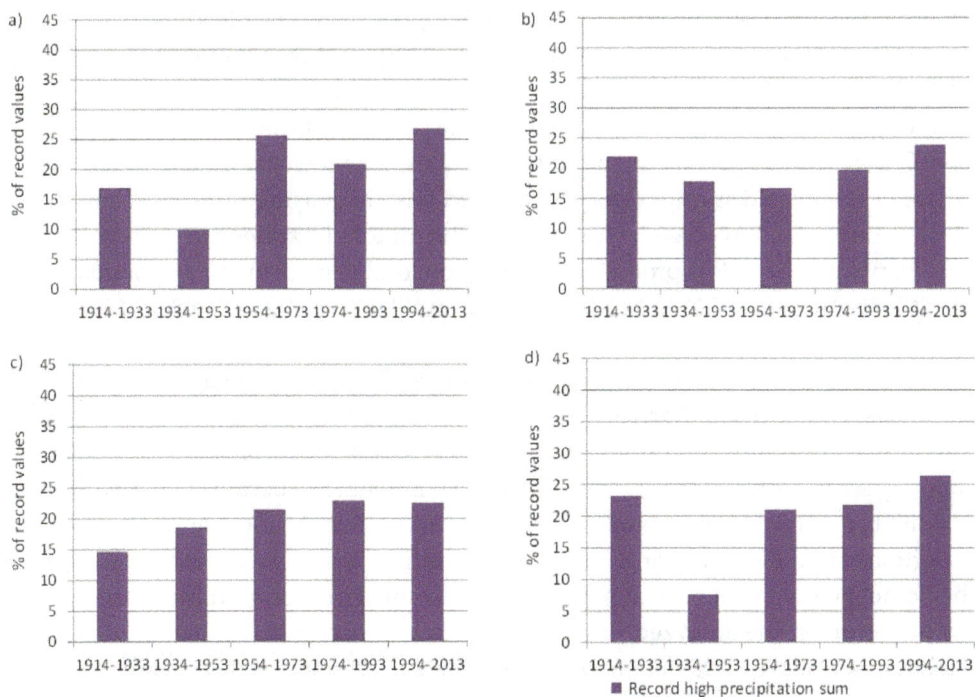

Figure 5. The total proportion of extremely high daily precipitation (99th percentile of daily precipitation sum defined separately for each weather station) observed in 20-year study intervals in northern Fennoscandia. (**a**) January, (**b**) April, (**c**) July, and (**d**) October.

4. Discussion

The results showed that northern Fennoscandia is experiencing increasing surface air temperatures, steeper trends in monthly mean minimum rather than maximum temperatures, slight increases in precipitation as well as a changing occurrence of extreme climate events. The most noticeable changes were observed in spring and autumn seasons and to a lesser extent in the summer. These results are consistent with studies from the circumpolar areas reporting increasing temperature trends [40,46], narrowing diurnal temperature ranges [50], shortening cold season, declining number of frost days [61–64], an increasing number of extreme warm events and a declining number of extreme cold events [11,48,51,65–69]. For example, the mean temperature in Finland has risen by 0.14 °C/decade during the years 1847–2013 with the highest increases in winter and spring months [45]. In Fennoscandia, a long-term decrease of diurnal temperature range has been observed. The warming trend has been most notable in March-May with a significant increase in mean maximum air temperature by 0.4 °C/decade during the period 1950–1995 [50]. The fact that hardly any statistically significant warming trends were detected in winter can be explained by the high variability of temperatures during that season. Fewer and less pronounced trends were likewise found in summer than in spring and autumn. This feature is common to other Nordic areas as well [40,50].

The trends detected hitherto are broadly consistent with climate model projections showing a warming influence of increasing greenhouse gas concentrations. In northern Fennoscandia under the high-emission RCP8.5 forcing scenario for the period 2040–2069, temperatures are projected to increase by 4–6 °C in winter and by 2–3 °C in summer, relative to 1981–2010, while the warming rate in spring and autumn is expected to fall between these values [70]. Simultaneously, the diurnal temperature range is projected to decrease by about 1 °C in winter, but less so in summer. The observed slight increase in extreme precipitation events is likewise consistent with projected increase in extreme precipitation on northern Fennoscandia [71,72].

Besides resulting from anthropogenic influences on the climate system, climate variations and trends may be due to natural internal processes within the climate system (internal variability), and natural external forcing (volcanic eruptions, solar activity). NAO is a leading pattern of weather and climate variability over the northern hemisphere [55], and it is known to have a significant influence on wintertime temperatures and precipitation in the Nordic region [73] and references therein. Although our simple comparison between the portions of rarely warm or cold seasons and seasonal NAO indices at 20-year averaging intervals does not allow any definite inferences, it seems as if the share of rarely cold springs clearly decreased with increasing principle-component spring NAO index. Stronger-than-average westerlies over the middle latitudes during a positive phase of NAO might, however, have a less pronounced impact on local climate in our study area than south of 65° N [74]. Of importance are also other atmospheric oscillation patterns, such as Siberian high [46] and Scandinavian blocking [75].

The reduced sea ice in the Barents and White Seas [75–77] and associated circulation changes have been regarded as likely principal drivers behind the observed warming in the Kola Peninsula [46]. During the most recent time span considered by us (1994–2013), the mean annual ice cover in Barents Sea was notably small [76], and substantial areas of the Barents Sea became ice-free year round [77]. Consistent with that, rarely warm spring and autumn seasons were frequent in northern Fennoscandia, while the portions of rarely cold spring, summer and autumn seasons were at their lowest (Figure 3).

In the homogeneity testing, we aimed to reveal the most pronounced inhomogeneities having the largest influence on the analyzed time series. This approach was driven by the sparse station network and the lack of available station history metadata. Although use was made of state-of-the-art homogenization algorithms in HOMER, the homogeneity testing performed in this study must be considered rather simplistic. However, we are confident that even this simplistic approach improves the robustness of analyzes and helps to assess uncertainties. First, at five stations out of nine, no or only isolated inhomogeneities were detected in the time series of monthly mean, minimum and maximum temperatures. Trend analyses for these stations can be regarded as rather reliable. Second, the inhomogeneities at the remaining four stations may be taken into account in a qualitative manner. One may argue that in the absence of no-climatic biases in the temperature data of stations 3 and 7, even greater warming trends had been analysed, whereas inhomogeneities detected in stations 5 and 8 potentially mean weaker warming trends.

A shortening of the cold season, declines in extremely cold climate events in all seasons and increases in extremely warm climate events, particularly in spring and autumn seasons, can have several consequences. In northern Fennoscandia, spring and autumn seasons are transitional periods of melting and formation of continuous seasonal snow, ice and ground frost. Snow covers the ground several months in the winter with the longest duration in the Scandinavian mountain range [78,79]. Increasing spring temperatures are likely to result in more efficient and earlier snow melt in the future. Furthermore, the observed increases in precipitation together with warmer air temperatures in the beginning of the cold season are likely to result in increases in rainfall and decreases in snowfall [80–82]. A shorter duration of snow cover and increasing air and soil temperatures alter the geophysical and hydrological characteristics of the high-latitude areas [3,78,83]. As the cryosphere, climate and terrestrial ecology are closely related, these changes have significant impacts on the structure and functioning of high-latitude ecosystems and human activities [79,84–86].

Decreases in the length of the cold season can result in major changes in vegetation and have a variety of impacts on distribution, abundance, migration and overwintering of high-latitude species [87–89]. Vegetation in northern Fennoscandia is characterized by coniferous forests, birch forests and open tundra. A warming climate is predicted to result in shrubification of tundra with important consequences for habitat availability and ecosystem functioning [90]. A shorter period of snow cover also significantly decreases surface albedo, which has a positive feedback on global climate change [3,91]. A majority of ecological studies have concentrated on studying impacts of warming during the spring and summer, whereas the autumn season has been largely neglected [92,93]. The significant temperature increases in the late autumn observed in this study highlights the need of more autumnal research in arctic and subarctic areas.

Hydrological changes in warming climate, such as increased river discharge and altered freeze-up and break-up timing for rivers, lakes and oceans, have important consequences in northern ecosystems and communities [94]. In the Barents region, for example, consequences of hydrological changes are assumed to include alterations in nutrient cycling, species distributions and animal fitness, which eventually are reflected in ecosystem services and local livelihoods [26]. Permafrost thawing has been reported in many parts of the circumpolar area as a consequence of warmer ground temperatures [95]. Permafrost and its dynamics are major drivers of hydrological and geomorphological processes [96]. The northern parts and mountain regions of Fennoscandia belong mainly to the zone of discontinuous permafrost. In these areas, thawing permafrost can cause various hazards, such as less stable ground and slope failures that can threaten local community infrastructure and become an increasing cost factor in the maintenance and construction of mountain infrastructure [97–99]. There is also evidence that palsas, peat covered permafrost mounds containing segregation ice and surrounded by wet mires, are shrinking in northern Fennoscandia [100].

Understanding the extreme events and their effects is one of the major challenges of current climate research [2]. For example, extreme weather events may affect the ecosystem functioning by causing or accelerating the shifts in species composition and distribution or by changing competitive interactions within the ecosystem. Sudden and extreme events can change the whole system characteristics through disturbance effects and can drive ecosystems beyond their stability and resilience [101–103]. It may be probable that reaching the tipping point of the near-zero °C in common cold season temperatures is critical, and northern ecosystems and societies are especially sensitive to this kind of regime shift [85].

Reindeer herding in northern Fennoscandia is a good example of the impacts of both gradual warming and extreme climate events on a socio-ecological system [104,105]. Reindeer access to lichen during the winter is strongly dependent on snow cover characteristics [106]. The formation of the snow cover during the autumn is considered crucial for all winter conditions. During unstable early winter with several 0-degree days, there is a high probability of icy snow cover or ground ice formation that prevents reindeer access to lichens [33–35,38]. Warmer and wetter winters have been reported to decrease reindeer condition and productivity and increase their mortality in Nordic countries and Russia. Extreme climate events, such as warm spells during the cold season and heavy rain-on-snow events can also result in formation of icy layers in the snowpack with negative impacts on reindeer populations [85]. A recent study [51] found that the rate of melt days in winter increased by 3–9 days decade^{-1} in Nordic arctic islands and mainland over the past 50 years. Furthermore, an increase in precipitation sum for winter melt days was observed [51]. Warm conditions in the late winter may lead to wet and slushy snow that hamper reindeer grazing and herding [35]. Increasing temperatures can also, e.g. lead to a declining extent of summer snowbeds and snow patches [82]. These isolated spots of snow in summer are important resources for reindeer through decreasing insect harassment and temperature stress and providing fresh high-quality fodder along the melting edges of snow [107].

5. Conclusions

Significant increasing trends in spring and autumn temperatures, increasing occurrence of extremely warm events and decreasing occurrence of extremely cold events suggest notable changes in northern Fennoscandian cold season characteristics. Based on climate model projections, these changes are not likely to decelerate in the future, rather vice versa, unless efficient mitigation actions will be realized. The shortening duration of the cold season together with higher minimum temperatures and more frequently occurring extremely warm climate events are likely to have major impacts on geophysical, hydrological and ecological as well as socio-economic systems of high latitude environments.

Acknowledgments: We thank Annika Kristoffersson for providing data from the Abisko weather station. Kai Rasmus and Eirik Forland are acknowledged for their help with Sodankylä and Russian data. We thank Minna Turunen with her help on socio-economic effects of warming. We also thank two anonymous reviewers for constructive comments on the manuscript. SK was funded by the NCoE Tundra (the Nordic Council of Ministers) and Maj and Tor Nessling Foundation. SR was funded by the Finnish Cultural Foundation. KJ was partially funded by the Academy of Finland (Decision No. 278067 for the PLUMES project).

Author Contributions: Sonja Kivinen and Sirpa Rasmus initiated the study and acquired climate data from different sources. Sonja Kivinen performed trend analyses and extreme event analyses, Mikko Laapas carried out homogeneity testing of temperature data, and Kirsti Jylhä conducted comparisons related to NAO. Kirsti Jylhä and Sirpa Rasmus provided regular feedback on the analyses and helped develop the methodology part of the study. All authors contributed to the discussion and interpretation of the results and writing of the manuscript.

Conflicts of Interest: The authors declare no conflict of interest.

References

1. ACIA. *Arctic Climate Impact Assessment*; Cambridge University Press: Cambridge, UK, 2005.
2. IPCC. *Climate Change 2013. The Physical Science Basis. Contribution of Working Group I to the Fifth Assessment Report of the Intergovernmental Panel on Climate Change*; Cambridge University Press: Cambridge, UK; New York, NY, USA, 2003.
3. Serreze, M.C.; Barry, R.G. Processes and impacts of Arctic amplification: A research synthesis. *Glob. Planet. Chang.* **2011**, *77*, 85–96. [CrossRef]
4. Pithan, F.; Mauritsen, T. Arctic amplification dominated by temperature feedbacks in contemporary climate models. *Nat. Geosci.* **2014**, *7*, 181–184. [CrossRef]
5. Vihma, T. Effects of Arctic sea ice decline on weather and climate: A review. *Surv. Geophys.* **2014**, *35*, 1175–1214. [CrossRef]
6. Przybylak, R. Temporal and spatial variation of surface air temperature over the period of instrumental observations in the Arctic. *Intern. J. Clim.* **2000**, *20*, 587–614. [CrossRef]
7. Overland, J.E.; Spillane, M.C.; Percival, D.B.; Wang, M.; Mofjeld, M. Seasonal and regional variation of Pan-Arctic surface air temperature over the instrumental record. *J. Clim.* **2004**, *17*, 3263–3282. [CrossRef]

8.   Ding, Q.; Walalce, J.M.; Battisti, D.S.; Steig, E.J.; Gallant, A.J.E.; Kim, H.-J.; Geng, L. Tropical forcing of the recent rapid Arctic warming in northeastern Canada and Greenland. *Nature* **2014**, *509*, 209–212. [CrossRef] [PubMed]

9.   Easterling, D.R.; Meehl, G.A.; Parmesan, C.; Changnon, S.A.; Karl, T.R.; Mearns, L.O. Climate extremes: Observations, modeling and impacts. *Science* **2000**, *289*, 2068. [CrossRef] [PubMed]

10. Dankers, R.; Hiederer, R. *Extreme Temperatures and Precipitation in Europe: Analysis of a High-Resolution Climate Change Scenario*; EUR 23291; Office for Official Publications of the European Communities: Luxembourg, 2008.

11. Rahmstorf, S.; Coumou, D. Increase of extreme weather events in a warming world. *Proc. Natl. Acad. Sci. USA* **2011**, *108*, 17905–17909. [CrossRef] [PubMed]

12. Meehl, G.A.; Zwier, F.; Evans, J.; Knutson, T.; Mearns, L.; Whetton, P. Trends in extreme weather and climate events: Issues related to modeling extremes in projections of future climate change. *B. Am. Meteorol. Soc.* **2000**, *81*, 427–436. [CrossRef]

13. Alexander, L.V.; Zhang, X.; Peterson, T.C.; Caesar, J.; Gleason, B.; Klein Tank, A.M.G.; Haylock, M.; Collins, D.; Trewin, B.; Rahimzadeh, F.; et al. Global observed changes in daily climate extremes of temperature and precipitation. *J. Geophys. Res.* **2006**, *111*, D05109.

14. IPCC. Annex III: Glossary (ed. Planton, S.). In *Climate Change 2013: The Physical Science Basis. Contribution of Working Group I to the fifth Assessment Report of the Intergovernmental Panel on Climate Change*; Stocker, T.F., Qin, D., Plattner, G.-K., Tignor, M., Allen, S.K., Boschung, J., Nauels, A., Xia, Y., Bex, V., Midgley, P.M., Eds.; Cambridge University Press: Cambridge, UK, 2013.

15. Wen, G.; Huang, G.; Hu, K.; Qu, X.; Tao, W.; Gong, H. Changes in the characteristics of precipitation over northern Eurasia. *Theor. Appl. Climatol.* **2015**, *119*, 653–665. [CrossRef]

16. Ye, H.; Fetzer, E.J.; Behrangi, A.; Wong, S.; Lambrigtsen, B.H.; Wang, C.Y.; Cohen, J.; Gamelin, B.L. Increasing daily precipitation intensity associated with warmer air temperatures over northern Eurasia. *J. Clim.* **2016**, *29*, 623–636. [CrossRef]

17. Horton, D.E.; Johnson, N.C.; Singh, D.; Swain, D.L.; Rajaratnam, B.; Diffenbaugh, N.S. Contribution of changes in atmospheric circulation patterns to extreme temperature trends. *Nature* **2015**, *522*, 465–469. [CrossRef] [PubMed]

18. Trenbeth, K.E.; Fasullo, J.T.; Shepherd, T.G. Attribution of climate extreme events. *Nat. Clim. Change* **2015**, *5*, 725–730. [CrossRef]

19. Francis, J.A.; Vavrus, S.J. Evidence linking Arctic Amplification to extreme weather in mid-latitudes. *Geophys. Res. Lett.* **2012**, *39*, L06801. [CrossRef]

20. Cohen, J.; Screen, J.A.; Furtado, J.C.; Barlow, M.; Whittleston, D.; Coumou, D.; Francis, J.; Dethloff, K.; Entekhabi, D.; Overland, J.; et al. Recent Arctic amplification and extreme mid-latitude weather. *Nat. Geosci.* **2014**, *7*, 627–637. [CrossRef]

21. Overland, J.E.; Francis, J.A.; Hall, R.; Hanna, E.; Kim, S.-J.; Vihma, T. The melting Arctic and midlatitude weather patterns: Are they connected? *J. Clim.* **2015**, *28*, 7917–7932. [CrossRef]

22. Overland, J.E. A difficult Arctic science issue: Midlatitude weather linkages. *Polar Sci.* **2016**, *10*, 210–216. [CrossRef]

23. Duarte, C.M.; Lenton, T.M.; Wadhams, P.; Wassmann, P. Abrupt climate change in the Arctic. *Nat. Clim. Change* **2012**, *2*, 60–62.

24. European Environment Agency EEA. *Climate Change, Impacts and Vulnerability in Europe*; Environment Agency EEA: København, Denmark, 2012.

25. Krug, J.; Eriksson, H.; Heidecke, C.; Kellomäki, S.; Köhl, M.; Lindner, M.; Saikkonen, K. Socio-economic Impacts—Forestry and Agriculture. In *Second Assessment of Climate Change for the Baltic Sea Basin*; Springer Regional Climate Studies; The BACC II Author Team, Ed.; Springer International Publishing: Cham, Switzerland, 2015; pp. 399–409.

26. AMAP. *Adaptation Actions for a Changing Arctic*; Arctic Monitoring and Assessment Programme (AMAP): Oslo, Norway, 2017. (in press)

27. AMAP. *Snow, Water, Ice and Permafrost in the Arctic (SWIPA): Climate Change and the Cryosphere*; Arctic Monitoring and Assessment Programme (AMAP): Oslo, Norway, 2011; p. 538.

28. Ye, H.; Cohen, J. A shorter snowfall season associated with higher air temperatures over northern Eurasia. *Environ. Res. Lett.* **2013**, *8*, 014052. [CrossRef]

29. The BACC II Author Team. *Second Assessment of Climate Change for the Baltic Sea Basin*; Springer Regional Climate Studies; Springer International Publishing: Cham, Switzerland, 2015.

30. Bjerke, J.W.; Karlsen, S.R.; Høgda, K.A.; Malnes, E.; Jepsen, J.U.; Lovibind, S.; Vikhamar-Schuler, D.; Tømmervik, H. Record-low primary productivity and high plant damage in the Nordic Arctic Region in 2012 caused by multiple weather events and pest outbreaks. *Environ. Res. Lett.* **2015**, *9*, 084006. [CrossRef]

31. Kivinen, S. Many a little makes a mickle: Cumulative land cover changes and traditional land use in the Kyrö reindeer herding district, northern Finland. *Appl. Geogr.* **2015**, *63*, 204–211. [CrossRef]

32. Turunen, M.; Reinert, E.; Kietäväinen, A. Indigenous and local economies. In *Adaptation Actions for a Changing Arctic*; Arctic Monitoring and Assessment Programme (AMAP): Oslo, Norway, 2017. (in press)

33. Kivinen, S.; Rasmus, S. Observed cold season changes in a Fennoscandian fell area over the past three decades. *Ambio* **2015**, *44*, 214–225. [CrossRef] [PubMed]

34. Rasmus, S.; Kivinen, S.; Bavay, M.; Heiskanen, J. Local and regional variability in snow conditions in northern Finland: A reindeer herding perspective. *Ambio* **2016**, *45*, 398–414. [CrossRef] [PubMed]

35. Turunen, M.; Rasmus, S.; Bavay, M.; Ruosteenoja, K.; Heiskanen, J. Coping with increasingly difficult weather and snow conditions: Reindeer herders' views on climate change impacts and coping strategies. *Clim. Risk Manag.* **2016**, *11*, 15–36. [CrossRef]

36. Lee, S.E.; Press, M.C.; Lee, J.A.; Ingold, T.; Kurttila, T. Regional effects of climate change on reindeer: A case study of the Muotkatunturi region in Finnish Lapland. *Polar Res.* **2000**, *19*, 99–105. [CrossRef]

37. Kumpula, J.; Colpaert, A. Effects of weather and snow conditions on reproduction and survival of semi-domesticated reindeer (R. t. tarandus). *Polar Res.* **2003**, *22*, 225–233. [CrossRef]

38. Helle, T.; Kojola, I. Demographics in an alpine reindeer herd: Effects of density and winter weather. *Ecography* **2008**, *31*, 221–230. [CrossRef]

39. Kietäväinen, A.; Tuulentie, S. Tourism strategies and climate change: Rhetoric at both strategic and grassroots levels about growth and sustainable development in Finland. *J. Sustain. Tour.* **2013**, *21*, 845–861. [CrossRef]

40. Tietäväinen, H.; Tuomenvirta, H.; Venäläinen, A. Annual and seasonal mean temperatures in Finland during the last 160 years based on gridded temperature data. *Int. J. Climatol.* **2010**, *30*, 2247–2256. [CrossRef]

41. Hanssen-Bauer, I.; Førland, E. Temperature and precipitation variations in Norway 1900–1994 and their links to atmospheric circulation. *Int. J. Climatol.* **2000**, *20*, 1693–1708. [CrossRef]

42. Førland, E.J.; Hanssen-Bauer, I. Past and future climate variations in the Norwegian Arctic: Overview and novel analyses. *Polar Res.* **2003**, *22*, 113–124. [CrossRef]

43. Vikhamar-Schuler, D.; Hanssen-Bauer, I.; Førland, E. *Long-Term Climate Trends of Finnmarksvidda, Northern-Norway*; Report no. 6/2010; Norwegian Meteorological Institute: Oslo, Norway, 2010.

44. Yang, Z.; Hanna, E.; Callaghan, T.V.; Jonasson, C. How can meteorological and microclimate simulations improve understanding of 1913–2010 climate change around Abisko, Swedish Lapland? *Meteorol. Appl.* **2011**, *19*, 454–463. [CrossRef]

45. Mikkonen, S.; Laine, M.; Mäkelä, H.; Gregow, H.; Tuomenvirta, H.; Lahtinen, M.; . Laaksonen, A. Trends in the average temperature in Finland, 1847–2013. *Stoch. Environ. Res. Risk Assess.* **2015**, *29*, 1521–1529. [CrossRef]

46. Marshall, G.J.; Vignols, R.M.; Rees, W.G. Climate Change in the Kola Peninsula, Arctic Russia, during the Last 50 Years from Meteorological Observations. *J. Clim.* **2016**, *29*, 6823–6840. [CrossRef]

47. Svyashchennikov, P.; Førland, E. *Long-Term Trends in Temperature, Precipitation and Snow Conditions in Northern Russia*; Report no. 9/2010; Norwegian Meteorological Institute: Oslo, Norway, 2010.

48. Bulygina, O.N.; Razuvaev, V.N.; Korshunova, N.N.; Groisman, P.Y. Climate variations and changes in extreme climate events in Russia. *Environ. Res. Lett.* **2007**, *2*, 044020. [CrossRef]

49. Heino, R.; Brázdil, R.; Førland, E.; Tuomenvirta, H.; Alexandersson, H.; Beniston, M.; Pfister, C.; Rebetez, M.; Rosenhagen, G.; Rösner, S.; et al. Progress in the study of climatic extremes in Northern and Central Europe. *Clim. Change* **1999**, *42*, 151–181. [CrossRef]

50. Tuomenvirta, H.; Alexandersson, H.; Drebs, A.; Frich, P.; Nordli, P.O. Trends in Nordic and Arctic temperature extremes and ranges. *J. Clim.* **2000**, *13*, 977–990. [CrossRef]

51. Vikhamar-Schuler, D.; Isaksen, K.; Haugen, J.E.; Tømmervik, H.; Luks, B.; Schuler, T.V.; Bjerke, J.W. Changes in winter warming events in the Nordic Arctic Region. *J. Clim.* **2016**, *29*, 6223–6244. [CrossRef]

52. The Finnish Meteorological Institute's open data. Available online: https://en.ilmatieteenlaitos.fi/open-data (accessed 10 May 2015).

53. The Eklima web portal of Norwegian Meteorological Institute. Available online: www.eklima.met.no (accessed 23 September 2015).

54. European Climate Assessment & Dataset. Available online: http://eca.knmi.nl (accessed 23 September 2015).

55. Hurrell, J.W.; Kushnir, Y.; Ottersen, G.; Visbeck, M. An Overview of the North Atlantic Oscillation. In *The North Atlantic Oscillation: Climate Significance and Environmental Impact*; Geophysical Monograph Series; American Geophysical Union: Washington, DC, USA, 2003; Volume 134, pp. 1–35.

56. NAO Index Data. Provided by the Climate Analysis Section, NCAR, Boulder, USA, Hurrell (2003). Updated regularly. Available online: https://climatedataguide.ucar.edu/climate-data/hurrell-north-atlantic-oscillation-nao-index-pc-based (accessed 1 February 2017).

57. Mestre, O.; Domonkos, P.; Picard, F. HOMER: A Homogenization Software—methods and Applications. *Időjárás* **2013**, *117*, 47–67.

58. Venema, V.K.C.; Mestre, O.; Aguilar, E.; Auer, I. Benchmarking homogenization algorithms for monthly data. *Clim. Past* **2012**, *8*, 89–115. [CrossRef]

59. Hawkins, D.M. Fitting Multiple Change-Point Models to Data. *Comput. Stat. Data Anal.* **2001**, *37*, 323–341. [CrossRef]

60. Engström, E.; Carlund, T.; Laapas, M.; Aalto, J.; Drebs, A.; Lundstad, E.; Gjelte, H.M.; Vint, K. NORDHOM—A Nordic Collaboration to Homogenize Long-Term Climate Data. In Proceedings of the EGU General Assembly 2015, Vienna, Austria, 12–17 April 2015.

61. Kaufman, D.S.; Schneider, D.P.; McKay, N.P.; Ammann, C.M.; Bradley, R.S.; Briffa, K.R.; Miller, G.H.; Otto-Bliesner, B.L.; Overpeck, J.T.; Vinther, B.M.; et al. Recent warming reverses long-term Arctic cooling. *Science* **2009**, *325*, 1236–1239. [CrossRef] [PubMed]

62. Vincent, L.A.; Zhang, X.; Brown, R.D.; Feng, Y.; Mekis, E.; Milewska, E.J.; Wan, H.; Wang, X.L. Observed trends in Canada's climate and influence of low-frequency variability modes. *J. Clim.* **2015**, *28*, 4545–4560. [CrossRef]

63. Anisimov, O.; Kokorev, V.; Zhil'tsova, Y. Temporal and spatial patterns of modern climatic warming: Case study of Northern Eurasia. *Clim. Change* **2013**, *118*, 871–883. [CrossRef]

64. Jones, P.D.; Lister, D.H.; Osborn, T.J.; Harpham, C.; Salmon, M.; Morice, C.P. Hemispheric and large-scale land-surface air temperature variations: An extensive revision and an update to 2010. *J. Geophys. Res. Atmos.* **2012**, *117*. [CrossRef]

65. Beniston, M.; Stephenson, D.B.; Christensen, O.; Ferro, C.A.; Frei, C.; Goyette, S.; Halsnaes, K.; Holt, T.; Jylhä, K.; Koffi, B.; et al. Future extreme events in European climate: An exploration of regional climate model projections. *Clim. Change* **2007**, *81*, 71–95. [CrossRef]

66. Frich, P.; Alexander, L.V.; Della-Marta, P.; Gleason, B.; Haylock, M.; Tank, A.K.; Peterson, T. Observed coherent changes in climatic extremes during the second half of the twentieth century. *Clim. Res.* **2002**, *19*, 193–212. [CrossRef]

67. Donat, M.G.; Alexander, L.V.; Yang, H.; Durre, I.; Vose, R.; Dunn, R.J.; Willett, K.M.; Aguilar, E.; Brunet, M.; Caesar, J.; et al. Updated analyses of temperature and precipitation extreme indices since the beginning of the twentieth century: The HadEX2 dataset. *J. Geophys. Res. Atmos.* **2013**, *118*, 2098–2118. [CrossRef]

68. Wang, X.L.; Feng, Y.; Vincent, L.A. Observed changes in one-in-20 year extremes of Canadian surface air temperatures. *Atmos.-Ocean* **2014**, *52*, 222–231. [CrossRef]

69. Beniston, M. Ratios of record high to record low temperatures in Europe exhibit sharp increases since 2000 despite a slowdown in the rise of mean temperatures. *Clim. Change* **2015**, *129*, 225–237. [CrossRef]

70. Ruosteenoja, K.; Jylhä, K.; Kämäräinen, M. Climate projections for Finland under the RCP forcing scenarios. *Geophysica* **2016**, *51*, 17–50.

71. Nikulin, G.; Kjellström, E.; Hansson, U.L.; Strandberg, G.; Ullerstig, A. Evaluation and future projections of temperature, precipitation and wind extremes over Europe in an ensemble of regional climate simulations. *Tellus A* **2011**, *63*, 41–55. [CrossRef]

72. Lehtonen, I.; Ruosteenoja, K.; Jylhä, K. Projected changes in European extreme precipitation indices on the basis of global and regional climate model ensembles. *Int. J. Climatol.* **2014**, *34*, 1208–1222. [CrossRef]

73. Seitola, T.; Järvinen, H. Decadal climate variability and potential predictability in the Nordic region. *Boreal Environ. Res.* **2014**, *19*, 387–407.

74. Blenckner, T.; Järvinen, M.; Weyhenmeyer, G.A. Atmospheric circulation and its impact on ice phenology in Scandinavia. *Boreal Environ. Res.* **2004**, *9*, 371–380.

75. Dobricic, S.; Vignati, E.; Russo, S. Large-scale atmospheric warming in winter and the Arctic sea ice retreat. *J. Clim.* **2016**, *29*, 2869–2888. [CrossRef]

76. Matishov, G.; Moiseev, D.; Lyubina, O.; Zhichkin, A.; Dzhenyuk, S.; Karamushko, O.; Frolova, E. Climate and cyclic hydrobiological changes of the Barents Sea from the twentieth to twenty-first centuries. *Polar Biol.* **2012**, *35*, 1773. [CrossRef]

77. Parkinson, C.L. Spatially mapped reductions in the length of the Arctic sea ice season. *Geophys. Res. Lett.* **2014**, *41*, 4316–4322. [CrossRef] [PubMed]

78. Brown, R.D.; Möte, P.W. The response of northern hemisphere snow cover to a changing climate. *J. Clim.* **2009**, *22*, 2124–2145. [CrossRef]

79. Callaghan, T.V.; Johansson, M.; Brown, R.D.; Groisman, P.Y.; Labba, N.; Radionov, V.; Barry, R.G.; Bulygina, O.N.; Essery, R.L.; Frolov, D.M.; et al. The changing face of Arctic snow cover: A synthesis of observed and projected changes. *Ambio* **2011**, *40*, 17–31. [CrossRef]

80. Räisänen, J. Warmer climate: Less or more snow? *Clim. Dynam.* **2008**, *30*, 307–319. [CrossRef]

81. Derksen, C.; Brown, R. Spring snow cover extent reductions in the 2008–2012 period exceeding climate model projections. *Geophys. Res. Lett.* **2012**, *39*, 19. [CrossRef]

82. Kivinen, S.; Kaarlejärvi, E.; Jylhä, K.; Räisänen, J. Spatiotemporal distribution of threatened high-latitude snowbed and snow patch habitats in warming climate. *Environ. Res. Lett.* **2012**, 034024. [CrossRef]

83. Choi, G.; Robinson, D.A.; Kang, S. Changing Northern Hemisphere Snow Seasons. *J. Clim.* **2010**, *23*, 5305–5310. [CrossRef]

84. Post, E.; Forchhammer, M.C.; Bret-Harte, M.S.; Callaghan, T.V.; Christensen, T.R.; Elberling, B.; Fox, A.D.; Gilg, O.; Hik, D.S.; Høye, T.T.; et al. Ecological dynamics across the Arctic associated with recent climate change. *Science* **2009**, *325*, 1355–1358. [CrossRef] [PubMed]

85. Hansen, B.; Isaksen, K.; Benestad, R.; Kohler, J.; Pedersen, Å.; Loe, L.E.; Coulson, S.; Larsen, J.; Varpe, Ø. Warmer and wetter winters: Characteristics and implications of an extreme weather event in the High Arctic. *Environ. Res. Lett.* **2014**, *9*, 114021. [CrossRef]

86. Euskirchen, E.; Turetsky, M.; O'Donnell, J.; Daanen, R.P. Snow, Permafrost, Ice Cover, and Climate Change. *Glob. Environ. Chang.* **2014**, 199–204.

87. Pauli, J.N.; Zuckerberg, B.; Whiteman, J.P.; Porter, W. The subnivium: A deteriorating seasonal refugium. *Front. Ecol. Environ.* **2013**, *11*, 260–267.

88. Pearson, R.G.; Phillips, S.J.; Loranty, M.M.; Beck, P.S.; Damoulas, T.; Knight, S.J.; Goetz, S.J. Shifts in Arctic vegetation and associated feedbacks under climate change. *Nature Clim. Change* **2013**, *3*, 673–677.

89. Ruosteenoja, K.; Räisänen, J.; Venäläinen, A.; Kämäräinen, M. Projections for the duration and degree days of the thermal growing season in Europe derived from CMIP5 model output. *Int. J. Climatol.* **2016**, *36*, 3039–3055. [CrossRef]

90. Myers-Smith, I.H.; Forbes, B.C.; Wilmking, M.; Hallinger, M.; Lantz, T.; Blok, D.; Tape, K.D.; Macias-Fauria, M.; Sass-Klaassen, U.; Esther, L.; et al. Shrub expansion in tundra ecosystems: Dynamics, impacts and research priorities. *Environ. Res. Lett.* **2011**, *6*, 045509. [CrossRef]

91. Sturm, M.; Holmgren, J.; McFadden, J.P.; Liston, G.E.; Chapin, F.S., III; Racine, C.H. Snow-shrub interactions in Arctic tundra: A hypothesis with climatic implications. *J. Clim.* **2001**, *14*, 336–344. [CrossRef]

92. Gallinat, A.S.; Primack, R.B.; Wagner, D.L. Autumn, the neglected season in climate change research. *Trends Ecol. Evol.* **2015**, *30*, 169–176. [CrossRef] [PubMed]

93. Liu, Q.; Fu, Y.H.; Zhu, Z.; Liu, Y.; Liu, Z.; Huang, M.; Janssens, I.A.; Piao, S. Delayed autumn phenology in the Northern Hemisphere is related to change in both climate and spring phenology. *Glob. Change Biol.* **2016**, *22*, 3702–3711. [CrossRef] [PubMed]

94. Olsen, M.S.; Callaghan, T.V.; Reist, J.D.; Reiersen, L.O.; Dahl-Jensen, D.; Granskog, M.A.; Goodison, B.; Hovelsrud, G.K.; Johansson, M.; Kallenborn, R.; et al. The changing Arctic cryosphere and likely consequences: An overview. *Ambio* **2011**, *40*, 111–118. [CrossRef]

95. Grosse, G.; Goetz, S.; McGuire, A.D.; Romanovsky, V.E.; Schuur, E.A. Changing permafrost in a warming world and feedbacks to the Earth system. *Environ. Res. Lett.* **2016**, *11*, 040201. [CrossRef]

96. Anisimov, O.A.; Nelson, F.E. Permafrost zonation and climate change in the northern hemisphere: Results from transient general circulation models. *Clim. Chang.* **1997**, *35*, 241–258. [CrossRef]

97. Nelson, F.E.; Anisimov, O.A.; Shiklomanov, N.I. Climate change and hazard zonation in the circum-Arctic permafrost regions. *Nat. Hazards* **2002**, *26*, 203–225. [CrossRef]

98. Isaksen, K.; Sollid, J.L.; Holmlund, P.; Harris, C. Recent warming of mountain permafrost in Svalbard and Scandinavia. *J. Geophys. Res. Earth Surface* **2007**, *112*. [CrossRef]

99. Harris, C.; Arenson, L.U.; Christiansen, H.H.; Etzelmüller, B.; Frauenfelder, R.; Gruber, S.; Isaksen, K. Permafrost and climate in Europe: Monitoring and modelling thermal, geomorphological and geotechnical responses. *Earth-Sci. Reviews* **2009**, *92*, 117–171. [CrossRef]

100. Luoto, M.; Heikkilä, R.K.; Carter, T.R. Loss of palsa mires in Europe and biological consequences. *Environ. Conserv.* **2004**, *31*, 30–37. [CrossRef]

101. Jentsch, A.; Kreyling, J.; Beierkuhnlein, C. A new generation of climate change experiments: Events, not trends. *Front. Ecol. Environ.* **2007**, *5*, 315–324. [CrossRef]

102. Bokhorst, S.; Phoenix, G.K.; Berg, M.P.; Callaghan, T.V.; Kirby-Lambert, C.; Bjerke, J.W. Climatic and biotic extreme events moderate long-term responses of above- and belowground sub-Arctic heathland communities to climate change. *Glob. Change Biol.* **2015**, *21*, 4063–4075. [CrossRef] [PubMed]

103. Vasseur, D.A.; DeLong, J.P.; Gilbert, B.; Greig, H.S.; Harley, C.D.; McCann, K.S.; Savage, V.; Tunney, T.D.; O'Connor, M.I. Increased temperature variation poses a greater risk to species than climate warming. *Proc. R. Soc. Lond. B: Biol. Sci.* **2014**, *281*, 20132612. [CrossRef] [PubMed]

104. Riseth, J.Å.; Tømmervik, H.; Helander-Renvall, E.; Labba, N.; Johansson, C.; Malnes, E.; Bjerke, J.W.; Jonsson, C.; Pohjola, V.; Sarri, L.E.; et al. Sámi traditional ecological knowledge as a guide to science: Snow, ice and reindeer pasture facing climate change. *Polar Rec.* **2011**, *47*, 202–217. [CrossRef]

105. Vuojala-Magga, T.; Turunen, M.; Ryyppö, T.; Tennberg, M. Resonance strategies of Sami reindeer herding during climatically extreme years in northernmost Finland in 1970–2007. *Arctic* **2011**, *64*, 227–241.

106. Horstkotte, T.; Roturier, S. Does forest stand structure impact the dynamics of snow on winter grazing grounds of reindeer (Rangifer t. tarandus)? *For. Ecol. Manage* **2013**, *291*, 162–171. [CrossRef]

107. Van Oort, B.; Rautio, P.; Denisov, D. Terrestrial and freshwater ecosystems. In *Adaptation Actions for a Changing Arctic*; Arctic Monitoring and Assessment Programme (AMAP): Oslo, Norway, 2017. (in press)

Future Climate of Colombo Downscaled with SDSM-Neural Network

Singay Dorji [1,*], **Srikantha Herath [2,3]** and **Binaya Kumar Mishra [1,4]**

[1]   United Nations University, Institute for the Advanced Study of Sustainability (UNU-IAS),
      5 Chome-53-70 Jingumae, Shibuya, Tokyo 150-8925, Japan; mishra@unu.edu
[2]   Ministry of Megapolis and Western Development, Battaramulla 10120, Sri Lanka;
      srikantha.herath@gmail.com
[3]   UNU-IAS, IR3S-University of Tokyo, and University of Peradeniya, Tokyo 150-8925, Japan
[4]   UNU-IAS, Tokyo 150-8925, Japan
[*]   Correspondence: singaydor@yahoo.co.in

Academic Editor: Yang Zhang

Abstract: The Global Climate Model (GCM) run at a coarse spatial resolution cannot be directly used for climate impact studies. Downscaling is required to extract the sub-grid and local scale information. This paper investigates if the artificial neural network (ANN) is better than the widely-used regression-based statistical downscaling model (SDSM) for downscaling climate for a site in Colombo, Sri Lanka. Based on seasonal and annual model biases and the root mean squared error (RMSE), the ANN performed better than the SDSM for precipitation. This paper proposes a novel methodology for improving climate predictions by combining SDSM with neural networks. This method will allow a user to apply SDSM with a neural network model for higher skills in downscaling. The study uses the Canadian Earth System Model (CanESM2) of the IPCC Fifth Assessment Report, reanalysis from the National Center for Environmental Prediction (NCEP), and the Asian Precipitation Highly Resolved Observational Data Integration towards Evaluation of Water Resources (APHRODITE) project data as the observation. SDSM and the focused time-delayed neural network (TDNN) models are used for the downscaling. The projected annual increase for Representative Concentration Pathway (RCP) is 8.5; the average temperature is 2.83 °C (SDSM) and 3.03 °C (TDNN), and rainfall is 33% (SDSM) and 63% (TDNN) for 2080's.

Keywords: downscaling; climate change; SDSM; neural networks; GCM; Sri Lanka

1. Introduction

The Fifth Assessment Report (AR5) of the Intergovernmental Panel on Climate Change (IPCC) concluded that the warming in the climate is 'unequivocal.' IPCC AR5 projects an increase of global mean surface temperature for 2081–2100 to 0.3–1.7 °C (RCP2.6) and 2.6–4.8 °C (RCP8.5) [1]. The year 2016 was the third consecutive hottest year on record according to the National Oceanic and Atmospheric Administration (NOAA) and National Aeronautics and Space Administration (NASA). Globally averaged temperatures in 2016 were 0.99 °C warmer than the mid-20th-century average [2]. The Paris Summit participants (Conference of Parties, COP21), in 2015, agreed to limit the rise in global temperature below 2 °C above the pre-industrial level till 2100. The global average temperature is already halfway to the target by 2016. Climate change is projected to increase the temperature and intensify the global water cycle, increasing both extreme events and non-rainy days, causing multiple stresses of floods and droughts. It is difficult to predict the future climate due to uncertainties from climate models and various other sources. Prediction of future climate research is important for impact studies and adaptation.

The Global Climate Model (GCM) are the models used for climate predictions and used to study climate variability and change. GCM are numerical coupled models that can simulate global climate features at the continental scale, such as atmospheric circulation cells, intertropical convergence zones, jet streams, and also simulate reasonably well the oceanic circulation like the conveyor belt and thermohaline circulation [3]. The model calculates the interactions based on the predefined physical laws within and across different grids (based on resolution) to represent the climate behavior in time. Higher resolution climate models enable potentially better representation of local features. The outputs from the GCM are at a coarse spatial resolution, typically 100's of km, while the resolution required for impact assessments are like the temperature and rainfall for a point location or a catchment. Downscaling is needed to bridge this difference and obtain the sub-grid scale information. Broadly downscaling can be divided into two main types, dynamic and statistical downscaling. In dynamic downscaling, a regional climate model (RCM) is nested within the GCM and run with boundary conditions from the GCM. Statistical methods relate the large-scale predictors with the local climate variables through some transfer function. Comprehensive comparisons of dynamic and statistical methods are available in [4–6]. Studies have shown the performance of statistical downscaling to be competitive compared to dynamical methods for climate change studies [6]. A significant advantage of the statistical methods is that they are computationally inexpensive.

The neural network (NN) has found a wide range of applications in climate science. The algorithms are inspired by the neuron structure and the way the human brain process information and learns from the past. There are many types of neural networks used to solve classification, regression and clustering problems. NN has been utilized for diverse applications like precipitation prediction [7–9], water resources studies [10], meteorology and oceanography [11], weather forecasting [12,13], climate variability [7,9] and other climate-related studies. Neural networks have been found useful to extract the non-linear relationships in climate variables [14–17]. NN has good nonlinear mapping, noise tolerance and predictive knowledge [9], believed to be more powerful than regression-based methods [16], and does not require a priori knowledge of the catchment [18]. Temporal Neural networks have been shown to be better than regression-based downscaling in climate variability and extremes [7]. NN can potentially be used to identify hidden relationships with the extraction of the time information.

The objective of this paper is to investigate if neural networks are better than SDSM for determining the relationships between GCM predictors and the local climate variables of temperature and rainfall. The paper will identify the optimal neural network that can be trained easily and applied for the given data and study site. The paper will propose a combined downscaling methodology of SDSM with the regression outputs from a neural network, allowing a user to apply SDSM for weather generator and other downscaling analysis. The following are used interchangeably: artificial neural network (ANN) and neural network (NN); precipitation and rainfall. The paper is organized as follows: Section 2 provides the overview on downscaling, study site and data used, Section 3 presents the results of SDSM, NN, and the combined methodology, and Section 4 summarizes the study with a discussion.

2. Materials and Methods

2.1. Downscaling Overview

Downscaling model relates the large-scale predictors with the local climate variables. The GCM predictors are then applied to this model to find the local scale predictands. Statistical downscaling is based on the view that regional climate occurs as an interplay of atmospheric, or oceanic circulation and regional topography, land-sea distribution and land-use, and it is conditioned by the climate on larger scales [16]. Conceptually, it can be written as

$$R = F(L),$$

where R is the predictand (regional climate variables such as temperature or rainfall), L is the predictor (large scale climate variables), and F is the transfer function deterministic or stochastic that is conditioned by L.

Several statistical downscaling methods with varying complexity have been proposed and used. Downscaling concepts, methodology and limitations are available in the literature [3,16,19,20]. Broadly they are sub-divided into weather typing, regression and generator methods. Some examples are ANN, self-organizing maps, regression, canonical correlation and principal component analysis, etc. Some limitations are stationarity assumptions; the relationship will remain valid outside the calibration period and requirement of large observation data. Statistical methods are computationally inexpensive and can be quickly used for impact assessments to provide onsite local information, for example, daily rainfall at a station to drive a hydrology model.

One of the popular statistical downscaling methods is the statistical downscaling model (SDSM). It is a combination of Multiple Linear Regression and the Stochastic Weather Generator [21]. SDSM is a widely-used downscaling model, and is relatively simple to apply. Extensive literature is available on the application of SDSM for climate-related studies and downscaling [22–25]. SDSM performs seven functions of quality control and transformations, screening, model calibration, weather generator, data analysis, graphical analysis and scenario generation [21]. Two optimization methods of Ordinary Least Squares and Dual Simplex is available along with various other features like bias correction and variance inflation. Predictors are selected through screening, using explained variance and partial correlations. Calibration builds the model using the selected predictors. Validation is performed for the new data subset and checked with statistical tests like t-tests, variance and mean. Validation is done by comparing with the observation and future scenarios built with scenario generator. Many statistical tests and analysis like frequency and time series analysis done within the Graphical User Interface (GUI). SDSM is used in this study first to assess the skills of the neural network model, and then is combined with the neural network.

Neural networks are defined by the interconnections, learning process and the activation functions. NN are a regression-based statistical method that learns from data to make predictions and solve complex problems. Multilayer perceptron (MLP) trained with backpropagation is probably the most commonly used topology. Feedforward networks and training with backpropagation algorithm is especially popular for hydrology-related studies [10]. Short-term memory of the network is the past information available as data, and long-term memory is the information contained in the weights. MLP have long-term memory while dynamic systems have short-term memory structures or recurrent connections. Fully-recurrent networks have memory inside the topology but are complex. Time lagged feed forward network (TLFN) are a special type of dynamic network with a short-term memory [26]. It is the most common temporal network consisting of multiple layers of processing elements (PE) with feed forward connection. The focused TLFN have memory only at the input layer, thus can still be adapted with the static backpropagation [26]. Time delayed neural network (TDNN) is a type of the TLFN. The TDNN is an MLP with the input PE replaced with a tap delay. The focused TDNN network with one hidden layer is shown in Figure 1 [26]. One advantage with the TDNN is that it can quickly be trained with the static backpropagation algorithm. TDNN has been used in non-linear system identification, time series prediction, and temporal pattern recognition [26]. Focused TDNN is selected as the best performing network.

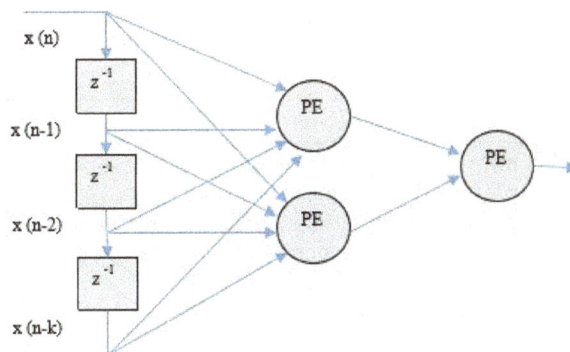

Figure 1. Focused time-delayed neural network (TDNN) with one hidden layer and tap delay line of k + 1 taps.

For scenarios, the IPCC AR5 uses representative concentration pathways (RCP), which replaced the Special Report on Emissions (SRES) of AR4. Radiative forcing, expressed as Watts/m^2 is the energy balance (the difference between the positive forcing due to the greenhouse gasses and the negative forcing due to aerosols) that stays in the atmosphere. There are four RCP's developed by different modeling groups, RCP's 2.6, 4.5, 6 and 8.5. RCP 8.5 is a high emissions scenario with heavy use of fossil fuel comparable to the SRES scenario A1F1. RCP 4.5 and 6 are intermediate emission scenarios similar to SRES B1 and B2 respectively. The study uses RCP 4.5 and 8.5 scenarios.

2.2. Study Area and Data

The study area used is Colombo, Sri Lanka. The observation data is from the Asian Precipitation Highly Resolved Observational Data Integration towards Evaluation of Water Resources (APHRODITE) project. The APHRODITE data is daily gridded precipitation dataset, analyzed from rain gauge observation data across Asia covering more than 57 years [27]. Data of 0.25° × 0.25° resolution is used for the grid point at lat/lon 6.875 × 79.875 centered at Colombo. IPCC recommends that the climate baseline period should be representative of the recent climate and should be of sufficient duration to include a range of climate variations and anomalies. The baseline period used in the study is from 1961 to 1990 (30 years). A 30-year normal period is a popular climatological baseline period, defined by the World Meteorological Organization (WMO). 1961–1990 is the current WMO normal period which serves as a standard reference for climate and impact studies. Models are validated for the period 1991–2005.

Models are built with the National Center for Environmental Prediction (NCEP) reanalysis [28]. Both the NCEP and the GCM have 26 large-scale predictors, shown in Table 1. The selection of the GCM for Colombo is based on an ongoing research at the University of Tokyo. From downscaling the CMIP5 GCM's, three models performed well for Sri Lanka in reproducing the seasonality. The three models were the Canadian Earth System Model (CanESM2) of the Canadian Centre for Climate Modelling and Analysis (CCCma), the Centro Euro-Mediterraneo sui Cambiamenti Climatici (CMCC.CM), Italy, and Institut Pierre Simon Laplace (IPSLCM5A-LR), France. This study used the CanESM2. CanESM2 is the second generation of the Earth System Model, which is the fourth generation coupled global climate model of the Coupled Model Intercomparison Project, Phase 5 (CMIP5) [29]. Daily predictor values are available for grid box (128 × 64), covering the whole globe along uniform longitude resolution of 2.8125° and nearly uniform latitude resolution of roughly 2.8125°. The GCM resolution is interpolated to the NCEP resolution of 2.5° × 2.5°, and data is normalized to 1961–1990 mean and standard deviation. Data is available from http://www.cccsn.ec.gc.ca/?page=pred-canesm2. The longitudinal and latitudinal index of the grid corresponds approximately to the centers of the grid boxes. Data used in the study corresponds to the cell no BOX 030X_35Y. Data is available for both temperature and precipitation, from 1961 to 2005 historical, 2006 to 2100 for three scenarios of RCP 2.6, 4.5 and 8.5 and the NCEP/NCAR predictors for 1961 to 2005.

Table 1. Predictor variables of the Global Climate Model (GCM) and the National Center for Environmental Prediction (NCEP).

SN	Name	Description
1	mslp	mean sea level pressure
2	p1_f	surface air flow strength
3	p1_u	surface zonal velocity component
4	p1_v	surface meridional velocity component
5	p1_z	surface vorticity
6	p1_th	surface wind direction
7	p1_zh	surface divergence
8	p5_f	500 hPa air flow strength
9	p5_u	500 hPa zonal velocity component
10	p5_v	500 hPa meridional velocity component
11	p5_z	500 hPa vorticity
12	p5_th	500 hPa wind direction
13	p5_zh	500 hPa divergence
14	p8_f	850 hPa air flow strength
15	p8_u	850 hPa zonal velocity component
16	p8_v	850 hPa meridional velocity component
17	p8_z	850 hPa vorticity
18	p8_th	850 hPa wind direction
19	p8_zh	850 hPa divergence
20	p500	500 hPa geopotential height
21	p850	850 hPa geopotential height
22	prcp	surface precipitation
23	s500	specific humidity at 500 hPa height
24	s850	specific humidity at 850 hPa height
25	shum	surface specific humidity
26	temp	surface mean temperature

p1 indicates near surface, p5, and p8 for 500 and 850 hPa heights.

3. Results

3.1. SDSM Downscaling

SDSM 4.2, an open source software, is used for the study. The user guide SDSM 4.2—A decision support tool for the assessment of regional climate change impacts [21] can be used for first-time users of the software. The model is calibrated with NCEP reanalysis from 1961 to 1990 and validated from 1991 to 2005. The selection of relevant predictors is an important task for the calibration of the model, for both the SDSM and neural network, and has large impacts on the result. Predictors chosen should not only have a strong correlation, but have a physically sensible meaning for the predictand being downscaled [30]. Studies have suggested that mid-tropospheric geopotential heights and humidity were the two most relevant predictors for daily precipitation [31], using MSLP for downscaling rainfall [32], and humidity is required to capture the climate change effects on the water-holding capacity of the atmosphere [20]. Screening, partial correlation and scatterplots in SDSM were used for the selection of the predictors. The predictors chosen are four for average temperature; p500, p8_v, shum, and temp; and six for precipitation; mslp, p1_f, p8_v, s500, shum, and temp. For the average temperature, monthly model was used. The coefficient of regression r^2 was 0.213, the sum of error SE 0.818. For the rainfall, a seasonal model was used with autoregression. The r^2 was 0.139 for conditional statistics, with SE of 0.510. Validation of the model results for average temperature and rainfall are presented in the final result section.

3.2. TDNN Downscaling

The neural network is developed using NeuroDimension's Neuro-Solutions [26]. The average temperature and the precipitation are modeled separately. First, the best network is selected with all the 26 NCEP predictors as inputs to the neural network for the outputs of temperature or precipitation. The best performing network is chosen after testing several networks with variations in memory type, activation functions, and backpropagation algorithm. Other network topologies tested were: Linear Regression, Multilayer Perceptron (MLP), and Time-lag Recurrent Network (TLRN). Performance is compared with the root mean squared error (RMSE), mean squared error (MSE), the coefficient of regression (r), and the hit score. TDNN network was selected as the best performing network. TDNN is a type of TLFN which is an MLP with memory components to store past values of the data, and allow the network to learn relationships over time. Different memory types for experimentation include GammaAxon, LaguarreAxon, ContextAxon, etc.

The type of memory is TDNN Axon, hyperbolic tangent tanhAxon transfer function is used in the hidden layer, bias axon at the output layer of the neural network, and trained with backpropagation RProp. The size of the memory layer (the tap delay) depends on the input, and the task and has to be determined on a case-by-case basis. Taps 5 and tap delay 1, and 10 PE's in the hidden layer, are used for both temperature and precipitation. Sensitivity analysis is a measure of relative importance of predictors, and it is a measure of standard deviation of the output divided by the standard deviation of the input [7]. The network is then retrained with the selected predictor variables. The predictors chosen are five for average temperature; p1_v, p8_v, p850, shum, and temp and eight for precipitation; mslp, p1_f, p1_u, p1_v, p1_zh, p8_v, s850, and shum. Data is tagged as: 1961–1985 for training, 1986–1990 for cross-validation and 1991–2005 for testing. The sensitivity analysis and scatter plot for the average temperature model is shown in Figure 2:

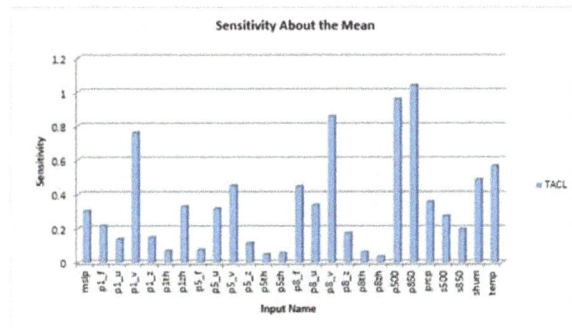

(a) Sensitivity analysis

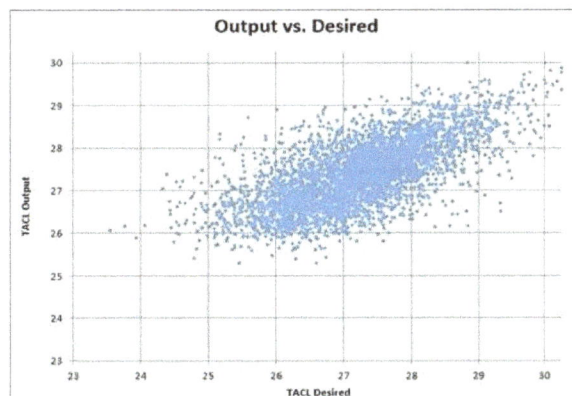

(b) Scatter plot.

Figure 2. Average temperature Colombo (TACL).

3.3. Final Results

The SDSM calibrated model is a parameter (.PAR) file which is used in the weather generator and scenario generator to create the output files as .OUT file and a .SIM file. The .OUT file contains the ensemble of outputs from the weather generator or the scenario generator. To apply the SDSM-NN combined methodology, the regression output of the SDSM is replaced by the output of the TDNN. A .OUT file, and .SIM file with the same name as the single column outputs from the TDNN is created. Scenario generation and the other statistical analysis is done in SDSM using the new .OUT files. With this method, SDSM can be used for scenario generation and to perform various downscaling and statistical analysis with a neural network model.

Seasonal/annual model biases and RMSE are used to compare the performance of SDSM and TDNN. The seasonal and annual mean model biases are given in Table 2 and RMSE in Table 3. For average temperature, the biases and RMSE were lower for the NN for the spring and annual mean. The SDSM errors were lower for winter, summer and autumn seasons. TDNN was better for winter and annual, whereas SDSM was better for spring, summer and autumn seasons' average temperature. The biases and RMSE were lower for the NN for all the seasons and the annual mean for the NN. TDNN performed better than SDSM for all the seasons and the annual mean. Figures 3 and 4 shows the monthly, seasonal and annual biases. For monthly temperature, NN bias was lower for the months of April, May and December, and for other months, SDSM bias was lower (Figure 3b). For monthly precipitation, SDSM bias is lower for the months of February, March, August, October and November, and in other months, NN bias was lower (Figure 4b). Overall, positive biases were observed for temperature and negative biases for precipitation by both the models.

Table 2. Seasonal and annual model biases for statistical downscaling model (SDSM) and TDNN for validation period.

Variable	Model	Winter	Spring	Summer	Autumn	Annual
Av. temp (°C)	SDSM	0.04	0.21	0.2	0.16	0.16
	TDNN	−0.09	0.11	0.3	0.27	0.15
Rainfall (mm)	SDSM	−0.5	−1.09	−0.68	−1.31	−0.89
	TDNN	−0.17	−0.09	0.08	−1.07	−0.31

Table 3. Seasonal and annual RMSE for SDSM and TDNN.

Variable	Model	Winter	Spring	Summer	Autumn	Annual
Av. temp (°C)	SDSM	0.321	0.495	0.312	0.383	0.223
	TDNN	0.34	0.467	0.398	0.418	0.22
Rainfall (mm)	SDSM	1.735	3.465	2.304	3.275	1.193
	TDNN	1.49	3.027	2.225	3.266	0.722

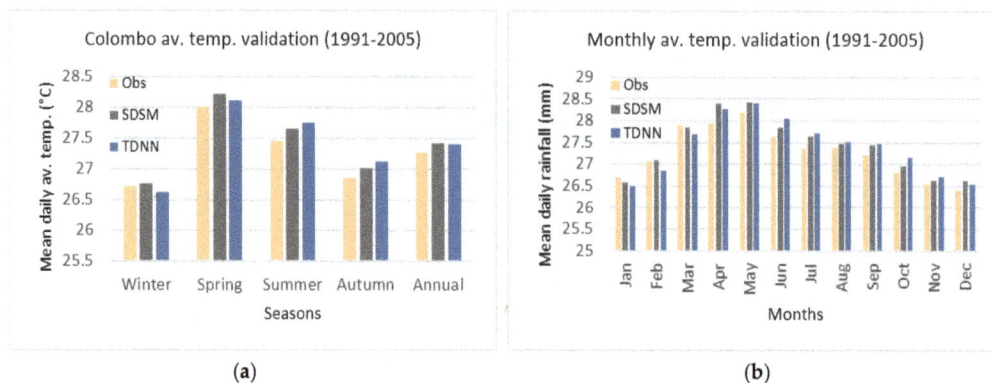

Figure 3. Average temperature validation (**a**) seasonal; (**b**) monthly.

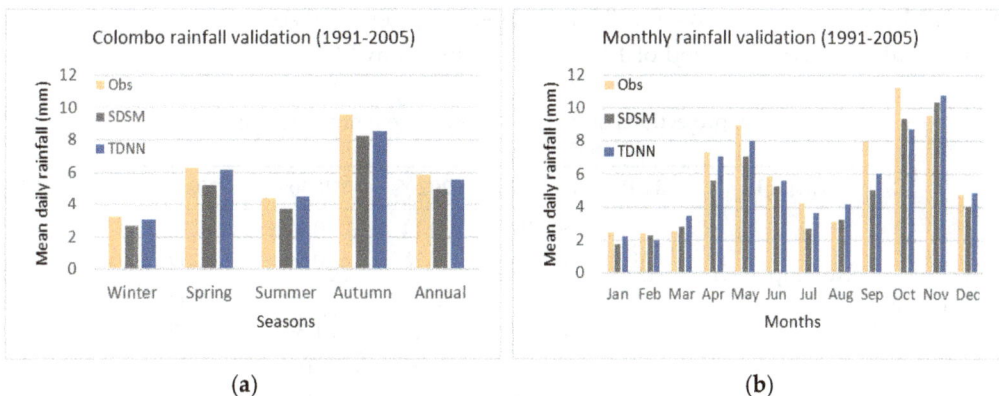

(a)   (b)

Figure 4. Rainfall validation (**a**) seasonal; (**b**) monthly.

Seasonal and annual statistical distribution is shown with the box plot, for temperature (Figure 5) and precipitation (Figure 6). The solid bar shows the monthly median value; boxes are an interquartile range (25th–75th percentile); the whiskers have 95% of the values and the circle showing outliers. Data distribution, skew and percentiles can be observed from the plot. Marginal differences between the two models are observed in the distribution. As expected SDSM compares well with the observed for temperature except for spring, where the NN median and distribution is closer to the observed. The changes of the median temperature are well reproduced by SDSM for three seasons. A closer agreement between NN and observations is found for the median precipitation for spring and summer. Both models do not fully capture the range of the precipitation events.

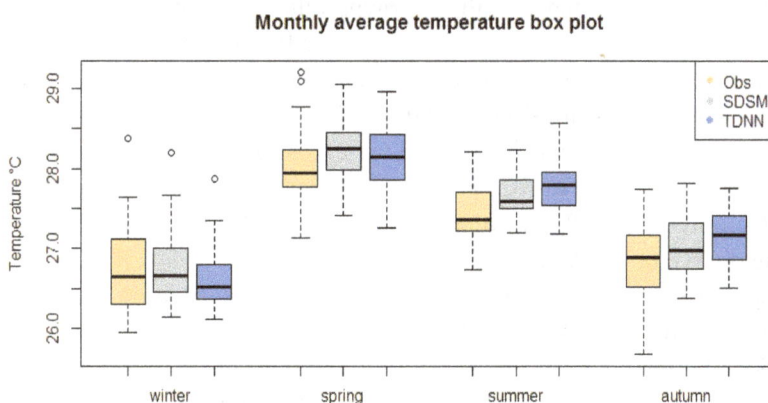

Figure 5. Average temperature box plot.

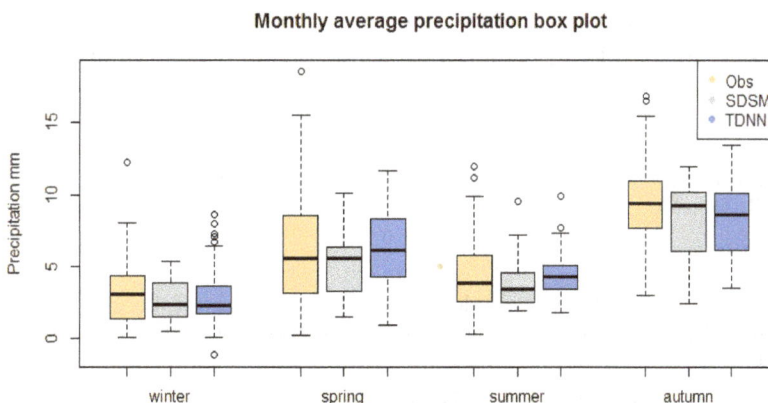

Figure 6. Average precipitation box plot.

The GCM future projections for 20's (2011–2040), 50's (2041–2070) and 80's (2071–2099), as compared with the current period of 1961–1990, is shown in Table 4.

Table 4. Future projections of increase in average temperature and rainfall.

Variable	RCP	Models	2020's	2050's	2080's
Av. temp. (°C)	4.5	SDSM	0.14	0.91	1.24
		TDNN	0.15	0.99	1.39
	8.5	SDSM	0.14	1.39	2.83
		TDNN	0.14	1.54	3.03
Rainfall (%)	4.5	SDSM	0.7	10	13
		TDNN	0.3	20	29
	8.5	SDSM	5	14	33
		TDNN	4	11	63

4. Discussion

The objective of this paper was to investigate if neural networks are better than the regression based SDSM for downscaling of temperature and rainfall in Colombo. Seasonal and annual model biases and the RMSE were used to assess the performance. With lower biases and RMSE for the winter, summer and autumn seasons, the SDSM was marginally better than the NN for downscaling average temperature. With lower biases and RMSE for all seasons and the annual mean, the NN performed better than SDSM for downscaling rainfall. The paper used a combined methodology, using SDSM with the regression outputs from the neural network. With this method, SDSM can still be used along with a neural network model for higher skills in downscaling for climate impact studies.

A limitation of the study is the use of one GCM for the downscaling, and the results will be largely dependent on the climate change signals from one GCM. Other limitations that are common to statistical methods are the assumptions of stationarity. Some of the further research directions to overcome these limitations are: to use this model for 2 or 3 selected GCMs, and to obtain ensemble average and uncertainty analysis. For RCP 8.5, the SDSM projections of an annual increase in average temperature for Colombo was 2.83 °C and 3.03 °C for TDNN. The annual increase in rainfall is projected at 33% and 63% for SDSM and TDNN.

Climate change is likely to cause more extreme weather events, flooding and droughts. Results of this study are indicating an increase in rainfall and flooding events in Colombo under an increased emissions scenario. The results from this study will augment other investigation and research for improving the prediction of rainfall. IPCC AR5 reports gaps in understanding the climate impacts on precipitation at the catchment scales [33]. Further work to downscale variability and extreme indices are important for impact studies. Within the stated limitations, the results provide daily values of temperature and rainfall for applications, like driving a hydrology model for the Colombo area. It also provides a scientific guideline for impact assessment studies, framing policies and long-term adaptation planning.

Acknowledgments: This study is a part of Ph.D. in Sustainability Science of the United Nations University, Institute for the Advanced Study of Sustainability (UNU-IAS), Tokyo, Japan and the scholarship provided by the jfScholarship for UNU-IAS. Model data from the Canadian Climate Data and observation data from APHRODITE is kindly acknowledged. We would also like to thank the reviewers for their thoughtful suggestions and comments that led to substantial improvement of the paper.

Author Contributions: Srikantha Herath provided guidance in all stages of the work, and arranged from the United Nations University, Japan, equipment and funds for the purchase of the software. Binaya Kumar Mishra assisted in data analysis and contributed to writing the article.

Conflicts of Interest: The authors declare no conflict of interest.

References

1. IPCC Summary for Policymakers. *Climate Change 2013: The Physical Science Basis. Contribution of Working Group I to the Fifth Assessment Report of the Intergovernmental Panel on Climate Change*; Stocker, T.F., Qin, D., Plattner, G.-K., Tignor, M., Allen, S.K., Boschung, J., Nauels, A., Xia, Y., Bex, V., Midgley, P.M., Eds.; Cambridge University Press: Cambridge, UK; New York, NY, USA, 2013; pp. 1–33.

2. Northon, K. NASA, NOAA Data Show 2016 Warmest Year on Record Globally. Available online: https://www.nasa.gov/press-release/nasa-noaa-data-show-2016-warmest-year-on-record-globally (accessed on 5 March 2017).

3. Zorita, E.; von Storch, H. The Analog Method as a Simple Statistical Downscaling Technique: Comparison with More Complicated Methods. *J. Clim.* **1999**, *12*, 2474–2489. [CrossRef]

4. Maraun, D.; Wetterhall, F.; Chandler, R.E.; Kendon, E.J.; Widmann, M.; Brienen, S.; Rust, H.W.; Sauter, T.; Themeßl, M.; Venema, V.K.C.; et al. Precipitation downscaling under climate change: Recent developments to bridge the gap between dynamical models and the end user. *Rev. Geophys.* **2010**, *48*, 1–38. [CrossRef]

5. Mearns, L.O.; Bogardi, I.; Giorgi, F.; Matyasovszky, I.; Palecki, M. Comparison of climate change scenarios generated from regional climate model experiments and statistical downscaling. *J. Geophys. Res.* **1999**, *104*, 6603. [CrossRef]

6. Murphy, J. An evaluation of statistical and dynamical techniques for downscaling local climate. *J. Clim.* **1999**, *12*, 2256–2284. [CrossRef]

7. Dibike, Y.B.; Coulibaly, P. Temporal neural networks for downscaling climate variability and extremes. *Neural Netw.* **2006**, *3*, 1636–1641. [CrossRef] [PubMed]

8. Silverman, D.; Dracup, J.A. Artificial Neural Networks and Long-Range Precipitation Prediction in California. *J. Appl. Meteorol.* **2000**, *39*, 57–66. [CrossRef]

9. Mendes, D.; Marengo, J.A. Temporal downscaling: A comparison between artificial neural network and autocorrelation techniques over the Amazon Basin in present and future climate change scenarios. *Theor. Appl. Climatol.* **2010**, *100*, 413–421. [CrossRef]

10. Maier, H.; Dandy, G. Neural Networks for the production and forecasting of water resource Environmental modelling and software variables: A review and modelling issues and application. *Environ. Model. Softw.* **2000**, *15*, 101–124. [CrossRef]

11. Hsieh, W.W.; Tang, B. Applying neural network models to prediction and data ananysis in meteorology and oceanography. *Bull. Am. Meteorol. Soc.* **1998**, *79*, 1855–1870. [CrossRef]

12. Abhishek, K.; Singh, M.P.; Ghosh, S.; Anand, A. Weather Forecasting Model using Artificial Neural Network. *Procedia Technol.* **2012**, *4*, 311–318. [CrossRef]

13. Shrivastava, G.; Karmakar, S.; Kumar Kowar, M.; Guhathakurta, P. Application of Artificial Neural Networks in Weather Forecasting: A Comprehensive Literature Review. *Int. J. Comput. Appl.* **2012**, *51*, 17–29. [CrossRef]

14. Cannon, A.J.; McKendry, I.G. A graphical sensitivity analysis for statistical climate models: Application to Indian monsoon rainfall prediction by artificial neural networks and multiple linear regression models. *Int. J. Climatol.* **2002**, *22*, 1687–1708. [CrossRef]

15. Liu, Z.; Peng, C.; Xiang, W.; Tian, D.; Deng, X.; Zhao, M. Application of artificial neural networks in global climate change and ecological research: An overview. *Chin. Sci. Bull.* **2010**, *55*, 3853–3863. [CrossRef]

16. Von Storch, H.; Hewitson, B.; Mearns, L. Review of Empirical Downscaling Techniques. In *Regional Climate Development under Global Warming*; General Technical Report; RegClim: Torbjørnrud, Norway, 2000; pp. 29–46.

17. Walter, A.; Schonwiese, C.D. Nonlinear statistical attribution and detection of anthropogenic climate change using a simulated annealing algorithm. *Theor. Appl. Climatol.* **2003**, *76*, 1–12. [CrossRef]

18. Nayak, D.; Mahapatra, A.; Mishra, P. A survey on rainfall prediction using artificial neural network. *Int. J. Comput. Appl.* **2013**, *72*, 32–40.

19. Hewitson, B.C.; Crane, R.G. Climate downscaling: Techniques and application. *Clim. Res.* **1996**, *7*, 85–95. [CrossRef]

20. Wilby, R.L.; Wigley, T.M.L. Downscaling general circulation model output: A review of methods and limitations. *Prog. Phys. Geogr.* **1997**, *21*, 530–548. [CrossRef]

21. Wilby, R.L.; Dawson, C.W. SDSM 4.2—A Decision Support Tool for the Assessment of Regional Climate Change Impacts User Manual. Available online: http://co-public.lboro.ac.uk/cocwd/SDSM/SDSMManual.pdf (accessed on 1 February 2017).

22. Tryhorn, L.; Degaetano, A. A comparison of techniques for downscaling extreme precipitation over the Northeastern United States. *Int. J. Climatol.* **2011**, *31*, 1975–1989. [CrossRef]

23. Tang, J.; Niu, X.; Wang, S.; Gao, H.; Wang, X.; Wu, J. Statistical downscaling and dynamical downscaling of regional climate in China: Present climate evaluations and future climate projections. *J. Geophys. Res. Atmos.* **2016**, *121*, 2110–2129. [CrossRef]

24. Mahmood, R.; Babel, M.S. Future changes in extreme temperature events using the statistical downscaling model (SDSM) in the trans-boundary region of the Jhelum river basin. *Weather Clim. Extrem.* **2014**, *5*, 56–66. [CrossRef]

25. Hassan, Z.; Shamsudin, S.; Harun, S. Application of SDSM and LARS-WG for simulating and downscaling of rainfall and temperature. *Theor. Appl. Climatol.* **2014**, *116*, 243–257. [CrossRef]

26. Principe, J.C.; Euliano, N.R.; Lefebvre, W.C. *Neural and Adaptive Systems: Fundamentals through Simulations*; John Wiley & Sons, Inc.: New York, NY, USA, 2000.

27. Yatagai, A.; Kamiguchi, K.; Arakawa, O.; Hamada, A.; Yasutomi, N.; Kitoh, A. Aphrodite constructing a long-term daily gridded precipitation dataset for Asia based on a dense network of rain gauges. *Bull. Am. Meteorol. Soc.* **2012**, *93*, 1401–1415. [CrossRef]

28. Kalnay, E.; Kanamitsu, M.; Kistler, R.; Collins, W.; Deaven, D.; Gandin, L.; Iredell, M.; Saha, S.; White, G.; Woollen, J.; et al. The NCEP/NCAR 40 year reanalysis project. *Bull. Am. Meteorol. Soc.* **1996**, *77*, 437–471. [CrossRef]

29. Taylor, K.E.; Stouffer, R.J.; Meehl, G.A. An overview of CMIP5 and the experiment design. *Bull. Am. Meteorol. Soc.* **2012**, *93*, 485–498. [CrossRef]

30. Wilby, R.L.; Charles, S.P.; Zorita, E.; Timbal, B.; Whetton, P.; Mearns, L.O. Guidelines for Use of Climate Scenarios Developed from Statistical Downscaling Methods. *Analysis* **2004**, *27*, 1–27.

31. Cavazos, T.; Hewitson, B.C. Performance of NCEP–NCAR reanalysis variables in statistical downscaling of daily precipitation. *Clim. Res.* **2005**, *28*, 95–107.

32. Timbal, B. Southwest Australia past and future rainfall trends. *Clim. Res.* **2004**, *26*, 233–249. [CrossRef]

33. Boucher, O.; Randall, D.; Artaxo, P.; Bretherton, C.; Feingold, G.; Forster, P.; Kerminen, V.-M.; Kondo, Y.; Liao, H.; Lohmann, U.; et al. Clouds and aerosols. In *Climate Change 2013: The Physical Science Basis*; Contribution of Working Group I to the Fifth Assessment Report of the Intergovernmental Panel on Climate Change; Cambridge University Press: Cambridge, UK; New York, NY, USA, 2013; pp. 571–658.

4

Tailoring Climate Parameters to Information Needs for Local Adaptation to Climate Change

Julia Hackenbruch [1,*]**, Tina Kunz-Plapp** [2]**, Sebastian Müller** [1] **and Janus Willem Schipper** [1]

[1] South German Climate Office, Institute of Meteorology and Climate Research,
Karlsruhe Institute of Technology, Hermann-von-Helmholtz-Platz 1,
76344 Eggenstein-Leopoldshafen, Germany; sebastian.mueller10@student.kit.edu (S.M.);
janus.schipper@kit.edu (J.W.S.)

[2] Institute of Meteorology and Climate Research, Karlsruhe Institute of Technology,
Hermann-von-Helmholtz-Platz 1, 76344 Eggenstein-Leopoldshafen, Germany; tina.kunz-plapp@kit.edu

* Correspondence: julia.hackenbruch@kit.edu

Academic Editor: Yang Zhang

Abstract: Municipalities are important actors in the field of local climate change adaptation. Stakeholders need scientifically sound information tailored to their needs to make local assessment of climate change effects. To provide tailored data to support municipal decision-making, climate scientists must know the state of municipal climate change adaptation, and the climate parameters relevant to decisions about such adaptation. The results of an empirical study in municipalities in the state of Baden-Wuerttemberg in Southwestern Germany showed that adaptation is a relatively new topic, but one of increasing importance. Therefore, past weather events that caused problems in a municipality can be a starting point in adaptation considerations. Deduction of tailored climate parameters has shown that, for decisions on the implementation of specific adaptation measures, it also is necessary to have information on specific parameters not yet evaluated in climate model simulations. We recommend intensifying the professional exchange between climate scientists and stakeholders in collaborative projects with the dual goals of making practical adaptation experience and knowledge accessible to climate science, and providing municipalities with tailored information about climate change and its effects.

Keywords: adaptation; climate change; municipalities; climate information; extreme weather events; climate parameters; user needs; Germany

1. Introduction

Climate change adaptation on the local level is of increasing relevance because of the regional variation in the effects of global climate change [1–4]. Local governments are increasingly being delegated responsibility to adapt to climate change because they are key state actors in the implementation of overarching national adaptation strategies, and have to develop, plan, and implement adaptation measures [5]. However, this responsibility is often difficult to implement given that the climate change information and science available are not necessarily easily translated into practical everyday decision-making. By comparison to climate mitigation plans, plans and strategies for climate change adaptation remain uncommon in European municipalities [6]. In the multi-scale, multi-sectoral, and multi-level challenge of adapting to climate change, many often intertwined political, economic, institutional, and technical or scientific obstacles involving different political levels have been identified and discussed that prevent or delay the development of local adaptation plans and strategies and the implementation of measures [7–15]. While many of these obstacles are not climate-specific and may emerge in other fields as well, Biesbroek et al. [11] identify three barriers related specifically and directly to adaptation to climate change: the long time scale of climate change, the reliance on scientific knowledge of climate change, and its inherent uncertainties and ambiguities. From a municipality's perspective, the fact that climate change is an ongoing process implies that adaptation strategies involve developments that reach far into the future, which may lead municipalities to the erroneous conclusion that they can postpone action [16], especially because decisions on the local level often focus on shorter time periods, in many cases, election periods [8,17]. Informational and cognitive barriers also can prevent decision makers from beginning or advancing the adaptation process [18]. A study on flood risk management in two Swedish municipalities [19] indicated that, although the stakeholders are aware of climate change, uncertainties remain about "what to adapt to" because of a lack of knowledge about local effects of climate change. This can result in uncertainty about the elements of an adaptation strategy, and, consequently, retard efforts in the task of municipal adaptation. A survey of municipal stakeholders in the German Baltic sea region also found problems associated with lack of knowledge, as well as uncertainty about regional effects of climate change [20]. In the same context, Lehmann et al. reported confusion about the terms "weather" and "climate", as well as "adaptation", and "mitigation", together with "actor-specific characteristics, as, for instance, limited individual (processing) capabilities" and a "lack of high-resolution data for the local level" ([8], p. 85). Finally, perceptions of climate change on the part of the public [21,22], local authorities [20,23,24], and local officials in specific fields of action [12,19] influence decisions about local adaptation to climate change.

With respect to increasing knowledge, scientifically sound, but tailored, climate information and data are essential for planning adaptation measures in municipalities. Many decision makers expect to obtain precise and small-scale climate projections [20,25]. The European Environment Agency (EEA) [4] underscored the " ... clear requirement for the information relating to adaptation to be tailored to the local level. This also includes the access to downscaled climate change scenarios and their impacts" ([4], p. 64). To develop this tailored information, both the political priority of the topic, as well as communication between knowledge providers and decision makers are crucial [19,24,26].

The resolution of regional climate models has advanced during the last years. Continuously increasing computing capacities allow modeling finer horizontal resolutions up to between 1 and 10 km [27–29]. However, urban planning impact models operate on the scale of urban districts or even below, and require climatological input data at spatial and temporal resolutions not yet available in current standard climate simulations. Nevertheless, the scale gap between regional climate models and local impact models has closed considerably [27]. Furthermore, simulation of longer time periods, and generation of ensembles of regional climate model projections based on different global climate model driving data or different emission scenarios are now possible [28–33]. All of these advances enhance the robustness of climate model simulations, allow the estimation of the range of possible future regional climates, and enable regional assessments of climate change. If local decision makers can access the data in an appropriate way, such data can contribute to an enlarged database for planning and implementation of adaptation strategies and measures. In addition, if these data are considered in different fields of action of climate change adaptation, they can support municipal decision-making.

To date, the parameters evaluated in climate model simulations and incorporated in reports and booklets have been defined based largely on meteorological expertise. These climate parameters include, for example, annual mean temperatures and precipitation, and deduced parameters such as summer days (days with a daily maximum temperature of at least 25 °C) or ice days (days with a daily maximum temperature below 0 °C). These climate parameters are useful in monitoring climate change scientifically. However, climate information inquiries made to the South German Climate Office at the Karlsruhe Institute of Technology—which offers regional climate services that can be described as "science–society interaction revolving around local and regional adaptation and mitigation" [34]—revealed that these climate parameters often do not offer enough information to be useful in specific decisions. Lemos et al. [26] identified three interconnected factors that affect individuals' use of scientific information based on a simple market model, and outline interaction, value-adding, retailing, wholesaling, and customization, as strategies to improve the usability of climate information. The three factors are defined as "the level and quality of interaction between producers and users of climate information; the fit, how users perceive that climate information meets their needs; and the interplay, how new knowledge interacts with other types of knowledge decision makers currently use" ([35], p. 402).

To address the "climate information usability gap" between "what scientists understand as useful information and what users recognize as usable in their decision-making" ([26], p. 789), we present the results of an empirical study in local municipalities in Baden-Wuerttemberg, Germany. We approach municipal adaptation from the perspective of climate science and with reference to the tools of a climate service. Drawing on the conceptual model of transforming climate information from useful to usable, we tested the way in which a regional climate service, as an "interacting actor", could contribute to the usability of climate information for local municipalities. We intend to "add value" to already existing climate model simulations by tailoring, or "customizing" ([26], p. 792), climate parameters from simulations to their needs so that municipalities can use them for decision-making in planning and implementing climate change adaptation.

In our empirical study, which was based on a questionnaire and additional interviews with experts, we investigated the municipalities' needs for climate information to identify the way in which climate science can provide tailored information and data support. To determine the untapped potential in exchanging knowledge and data between climate scientists and local authorities, the second objective of the study was to learn about the information sources in Baden-Wuerttemberg used today.

In Section 2, we introduce the regional context as well as study concept and data collection. We show the results in Section 3, addressing the state of climate change adaptation, the data and information used today, the climate parameters relevant to adaptation and the sensitivity to changes. The results are discussed in Section 4, followed by the main conclusions (Section 5).

2. Materials and Methods

To deduce tailored climate parameters that municipalities need for decision-making in planning and implementation of climate change adaptation, we conducted a survey based on a standardized questionnaire and additional interviews with experts. A prerequisite in identifying and defining tailored climate parameters important for adaptation is that a municipality already considers climate change and its effects. Because we cannot expect this to be true for all municipalities [6], and as various barriers can impede implementation [11], we assessed the state of climate adaptation and implementation of measures first. Consistent with the concept of the Intergovernmental Panel on Climate Change (IPCC), we referred to adaptation as "the process of adjustment to actual or expected climate and its effects" ([36], p. 40), as the study focused explicitly on supporting adaptation to "climate impacts as the major source of vulnerability" ([37], p. 49), and approached adaptation from the perspective of climate sciences. Thus, adaptation is not limited to a future (changed) climate, but also can refer to the variability and extremes in the current climate [16]. We used two approaches to deduce tailored climate parameters for municipalities' climate adaptation: first, direct deduction, which is possible if a municipality can relate the effects of climate change directly to climate variables. The second approach, indirect deduction, considers past weather events that caused problems in a municipality in recent years. Thereafter, a sensitivity assessment with respect to the effect of this change is required to use tailored climate parameters as a basis for the development of adaptation activities in municipalities. Referring to the concept of the IPCC, we defined sensitivity in our study as " ... the degree to which a system or species is affected, either adversely or beneficially, by climate variability or change" ([2], p. 1772). Therefore, we explored which change in a certain climate parameter would require action in a municipality or make adaptation measures urgently necessary or even impossible.

Thus, we investigated the following main research issues: the current state of climate change adaptation in municipalities in Baden-Wuerttemberg, the climate information used today, the climate parameters relevant and desired for adaptation, and the sensitivity to future changes in climate parameters. Thus, the local focus—in our case on municipalities in Southwestern Germany—was essential to investigate adaptation either to the current or future climate because the climate differs considerably between municipalities and urban agglomerations in the different climate regions of Baden-Wuerttemberg.

2.1. Investigation Area

The largest urban agglomerations in the state of Baden-Wuerttemberg in Southwestern Germany are Stuttgart (approximately 620,000 inhabitants) and the surrounding metropolitan area, Karlsruhe in the upper Rhine valley, and Mannheim in the metropolitan Rhine-Neckar region (both of the latter two have approximately 300,000 inhabitants). Approximately 250 cities have more than 10,000, approximately 50 more than 30,000, and nine cities more than 100,000 inhabitants [38].

Baden-Wuerttemberg (Figure 1a,b) is characterized by different landscapes, including parts of the upper and middle Rhine valley (approximately 100 m a.s.l.) and mid-range mountains, such as the Black Forest (Feldberg summit 1493 m a.s.l.) and the Swabian Jura (around 500 m a.s.l.). This orographically complex terrain causes large differences in the regional and local climate. The annual average temperature ranges from 4 °C (Black Forest) to 11 °C (Rhine Valley), and the annual precipitation from less than 600 mm (Rhine Valley) to more than 1500 mm (Black Forest) [39]. Climate change is evident in Baden-Wuerttemberg with respect to daily minimum and maximum temperatures, including heat, as well as heavy precipitation events [40–42]. Baden-Wuerttemberg is highly exposed to extreme meteorological events, including hot and dry summers [43,44], hailstorms [45], and heavy precipitation events [46], and several municipalities in Baden-Wuerttemberg have experienced negative effects of such events during years past. For the future, climate model simulations project a significant warming on seasonal and annual scales, an increase in the frequency, intensity, and duration of heat waves [47,48], and in mean winter precipitation, and a slight decrease in summertime mean precipitation; extreme precipitation events are expected to become more frequent [32].

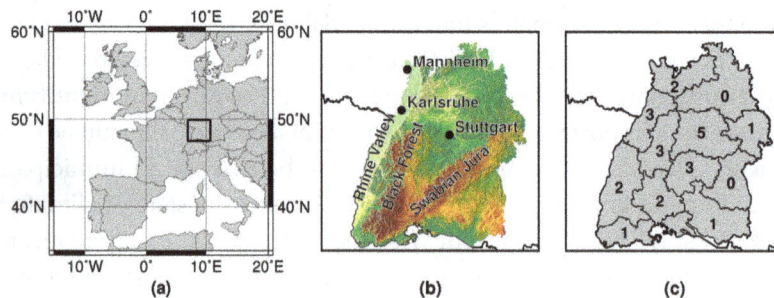

Figure 1. Investigation area. (**a**) location of Baden-Wuerttemberg in Europe (**b**) orography, landscapes and location of the three largest cities (**c**) administrative regions; numbers denote the number of questionnaire responses.

This implies a challenge for municipalities to adapt to long-term changes, as well as to more frequent or more intense extreme local events. The overarching national strategies and frameworks are the German Adaptation Strategy [49], the Action Plan for Adaptation [50], a monitoring report on the German Adaptation Strategy [51], and three planning laws [52]. Several months before the study was conducted, the Federal State adopted a Strategy on Adaptation to Climate Change in Baden-Wuerttemberg (SACC-BW) [53]. This strategy addresses nine fields of action: forestry, agriculture, soil, nature protection, water economy, tourism, health, urban and land use planning, and the economic and energy sectors. In addition, the "climate guide rails" ("Klimaleitplanken") developed in this context describe the future development of approximately 28 climate parameters for these fields of action [54] that are defined predominantly from a meteorological perspective. Under the research program KLIMOPASS [55], the Federal State supports research projects of short duration, such as the study presented here, and development of activities on adaptation and the implementation of SACC-BW in the nine fields of action. In combination with the different regional climates, these recent activities in climate change adaptation make Baden-Wuerttemberg suitable for investigating the way in which existing climate information that climate science believes is useful can be transformed into information that local municipalities can use.

2.2. Study Design

As described above, to investigate our research questions, we conducted a standardized survey and additional semi-structured interviews with experts.

After pre-tests, the final survey questionnaire comprised 26 closed and open-ended questions in four parts. The first contained questions about the state of adaptation in a municipality, the importance of the topic, and present activities in, and barriers to climate adaptation. Focusing on tailored climate parameters, the second part referred to weather events or climate variables that had caused problems in the past or are important to climate change adaptation in the nine fields of action identified in the SACC-BW [53]. Because the study was orientated to these nine fields of action, the climate variables in the final questionnaire included one parameter, air quality, which currently is not a direct output of climate simulations, but is part of the health action field in the SACC-BW. Thus, because compliance with limit values is a municipal task, the study took into account the air quality parameter. Furthermore, we asked for information not yet available on future developments of climate parameters that might be helpful to decision makers. The third part comprised an assessment of sensitivity to climate change to determine to what degree a single climate parameter, i.e., a climate stimulus, must change to require adaptation measures. The fourth and last part covered the location of the municipality and the information sources of the respondents.

We emailed the questionnaire to all of the approximately 180 members of the Association of Cities in Baden-Wuerttemberg ("Städtetag Baden-Württemberg") in spring 2015, together with a cover letter and a leaflet with information about the project, including the idea of providing tailored climate parameters and their sensitivity to support the development and implementation of adaptation on the municipal level. We asked the municipalities to return the completed questionnaires by email or ordinary mail. These were anonymized for further processing, mainly with univariate statistical analysis and analysis of the answers to the open-ended questions.

In addition, we conducted ten semi-structured interviews with experts [56,57] in four municipalities in the spring of 2015 to gain in-depth insights into the process of municipal climate change adaptation. The interviewees were either experts employed explicitly to address the topic of "climate" in (larger) municipalities or persons responsible for all issues of environmental protection (in smaller municipalities) and experts from several offices (for example, forest management, urban green space planning, and winter services) selected according to the nine fields of action of the SACC-BW. The interview protocols were analyzed and structured by the research team, in particular for tailoring usable climate parameters and their sensitivities or sensitivity thresholds.

2.3. Sample and Significance of the Study

In total, 23 questionnaires were included in the analyses. A response rate of nearly 13% is in the same range or even slightly better than that in similar studies, e.g., among political stakeholders in the German Baltic Sea region [20]. Of the 23 questionnaires, eight were completed by smaller municipalities with 10,000 to 30,000 inhabitants, and fifteen by larger municipalities with more than 30,000 inhabitants. This is a response rate of 30% for the larger municipalities, and of 4% for the smaller ones. Despite the relatively small absolute number of responses, the sample was suitable to answer the research questions proposed because the municipalities that participated are located in different regions in Baden-Wuerttemberg, as well as at different heights above sea level. The questionnaire responses cover ten of the twelve administrative regions in Baden-Wuerttemberg (for the regions and the number of participants per regions, see Figure 1c). One response from the Swiss Basel City Canton, which is located just at the border of Baden-Wuerttemberg, has been attributed to its neighboring region. Five municipalities lie at an altitude below 250 m a.s.l., twelve municipalities between 250 and 500 m a.s.l., five municipalities between 500 and 750 m a.s.l. and one municipality above 750 m a.s.l. Hence, they represent different regional climates. Because of the large variation in the numbers of inhabitants, they have different characteristics with respect to personal resources and administrative structures. Thus, generalizations proved to be difficult.

The interviews with experts provided insights into climate adaptation in practice, and its options and barriers in different municipalities. Direct dialogue during these interviews also helped identify weather and climate effects on specific fields of action that were important in particular in deducing climate parameters and their sensitivity. From the perspective of climate science, it also helped illustrate the wide range of applications of regional climate models.

3. Results

In the following, the results of the survey will be described with respect to the state of adaptation, the information used today, the climate parameters relevant to adaptation, and the sensitivity of municipalities. Results from the interviews supported the latter two.

3.1. State of Climate Change Adaptation in Municipalities in Baden-Wuerttemberg

The perceived importance of climate change adaptation varied in the municipalities in Baden-Wuerttemberg. While 10 of 23 municipalities that responded to the questionnaire consider the topic to be "important" or "very important" in their municipality, six rated it as "not important," "hardly important," or "of little importance". In general, climate change adaptation seemed to be more important in larger than in smaller municipalities (Figure 2).

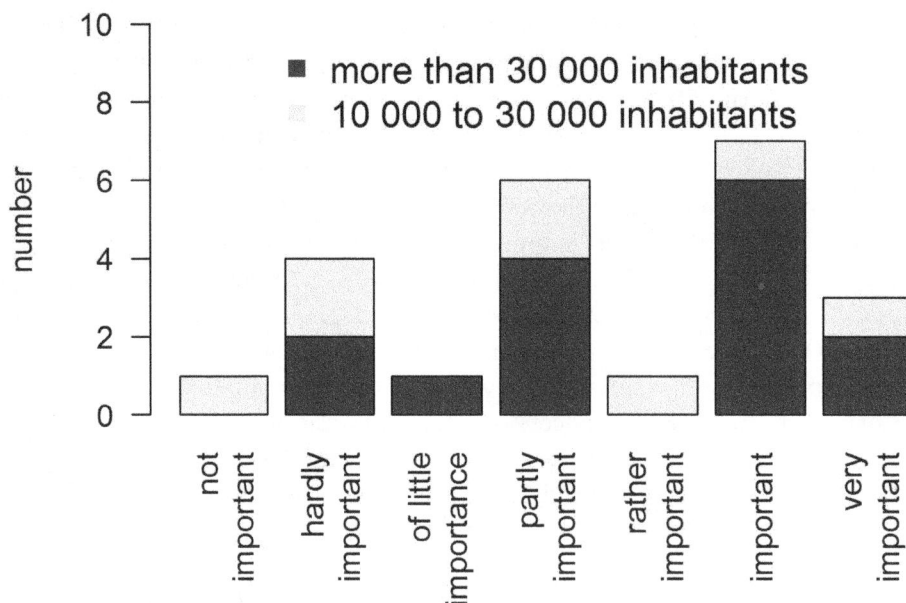

Figure 2. Answers to the question, "How important is the topic 'climate change adaptation' in your municipality?" (*n* = 23).

Two thirds of the municipalities indicated that they carry out diverse activities in the field of climate change adaptation. These comprise, for example, performing or ordering urban climate analyses and obtaining expert opinions, participating in research projects or working groups on climate and climate change, as well as enhancing public outreach by web presences to increase public acceptance of the topic. Climate adaptation plans exist in almost one third of the municipalities (some in preparation). In some cases, however, these are regional adaptation plans that extend beyond the fields of activity of a single municipality. Concrete implementation of measures consists, for example, of selecting tree species and plants for green areas, drawing up flood protection maps, and upgrading the sewage system. Several municipalities in the early stages of climate adaptation mentioned preliminary investigations or planned adaptation measures. Larger municipalities appeared to carry out both activities more frequently and to engage in more activities per municipality.

Although many municipalities would like to consider climate change in decision-making routines more effectively (14 municipalities chose "stronger" or "much stronger" in answering this question), they mentioned various barriers to doing so (Figure 3). One barrier reported by nearly half of the municipalities is that the topic is relatively new for local administrations. There also are major barriers of an internal nature, including personnel resources, the internal administrative structure, and lack of acceptance on the part of the administration itself. We also concluded from the answers to the open-ended questions that, although municipalities often regard climate adaptation as an important side issue, it covers several fields of action that require additional coordination that often is unavailable. In addition, some municipalities reported a "lack of acceptance in implementation of adaptation measures". With this answer option, we intended to address the possible barrier of a perceived lack of acceptance when measures are implemented, e.g., acceptance by parties involved or affected by measures concerning urban planning or construction plans ("lack of acceptance in implementation" in Figure 3). With respect to data accessibility, a quarter of the municipalities complained that they lacked availability to data and/or, did not know where to obtain them.

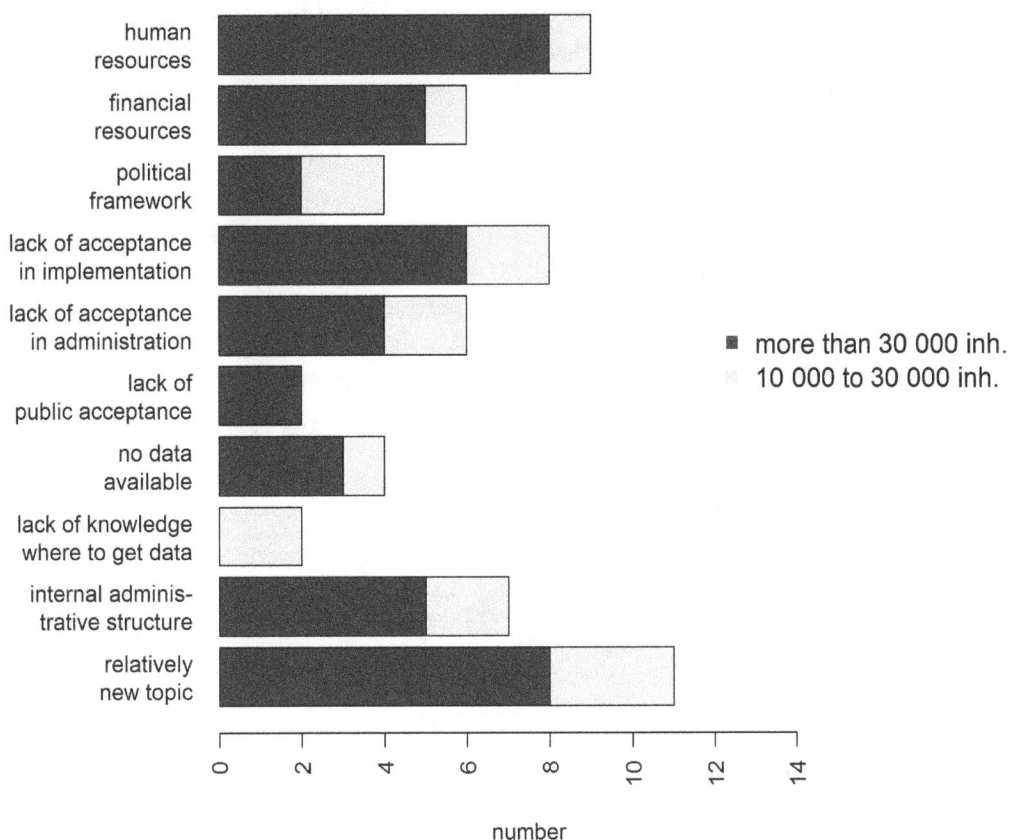

Figure 3. Answers to the question, "If you would like to consider climate aspects in a stronger or much stronger way, which barriers prevent you from doing so?" (number of responses, multiple answers possible, $n = 23$).

3.2. Data and Information Used Today

The municipalities in our sample stated that they use climate data derived primarily from measurement networks of higher authorities, including the State Office for the Environment, Measurements, and Nature Conservation of the Federal State of Baden-Wuerttemberg (LUBW, named by 13 municipalities), and the German Weather Service (DWD, mentioned by 10). Seven municipalities perform their own permanent or temporary meteorological measurements, including measurements of snowfall and ice on streets, and those related to the sewage system, technical operational services, or construction planning. In addition, nine municipalities perform or order analyses of urban climate, air quality, or areas of cold air production. Other climate data sources used include regional climate atlases and analyses, maps of climate functions, or other master plans. Nearly all of these data used refer to the current climate. Respondents mentioned data that refer to the future climate or effects of climate change only occasionally, for example, in one case, a master plan for climate change adaptation.

Different administrative as well as scientific institutions offer information accessible freely, as well as data on climate change and its effects, in most cases on websites or in booklets or leaflets; most municipalities in our survey know and use the information offered by higher authorities on the federal (DWD, Federal Environmental Agency (UBA)) or state level (LUBW). However, according to their responses, they seldom or rarely use climate information offered by scientific institutions and climate services, including the IPCC, universities in Baden-Wuerttemberg, and national and regional climate services. On the one hand, in specific cases, the respondents were looking for short-term weather and air quality forecasts, for example, weather warnings, and ozone and flood forecasts when addressing information sources. On the other hand, they expressed the need for projections of future climate change and its regional effects, for example, future frequency of heat events, development of ambrosia, as well as for best-practice examples in which municipalities comparable to their own are implementing adaptation measures and activities.

3.3. Climate Parameters Relevant to Climate Change Adaptation

Past weather events mentioned in the responses to the open-ended question in the survey about events "having caused problems in the municipality" were heat waves, hot periods, heavy precipitation events, flooding, storms, and dry periods (Table 1). Flooding was included in the list of weather events, because, although it is a hydro-meteorological event, it can cause severe problems. The respondents rated the occurrence of hot periods, as well as of heavy precipitation and precipitation in the summer most frequently as the most relevant to considerations about climate change adaptation. In addition, they listed air quality as one of the most important factors in considerations of adaptation (Figure 4).

Table 1. Grouped answers to the open-ended question, "Which weather events or changes in frequency caused problems at your municipality in the past?".

Number of Answers	Weather Events or Changes in Frequency	Single Events Indicated
11	Heat waves/hot periods/hot days/tropical nights	2003, 2005
9	Heavy precipitation events	Winter 2011, 2013
7	Floods *	June 2013, 1990s
7	Storm (winter gales)	Lothar (December 26, 1999), Wiebke (1 March 1990), Winter of years 2011, 2013
6	Dry periods	2003
4	Hail	June 2013, May 2015
3	Snow	-
2	Stationary temperature inversions	-
2	Thunderstorms	-
2	Cold periods	-

* Hydro-meteorological event, included because of the high number of times mentioned.

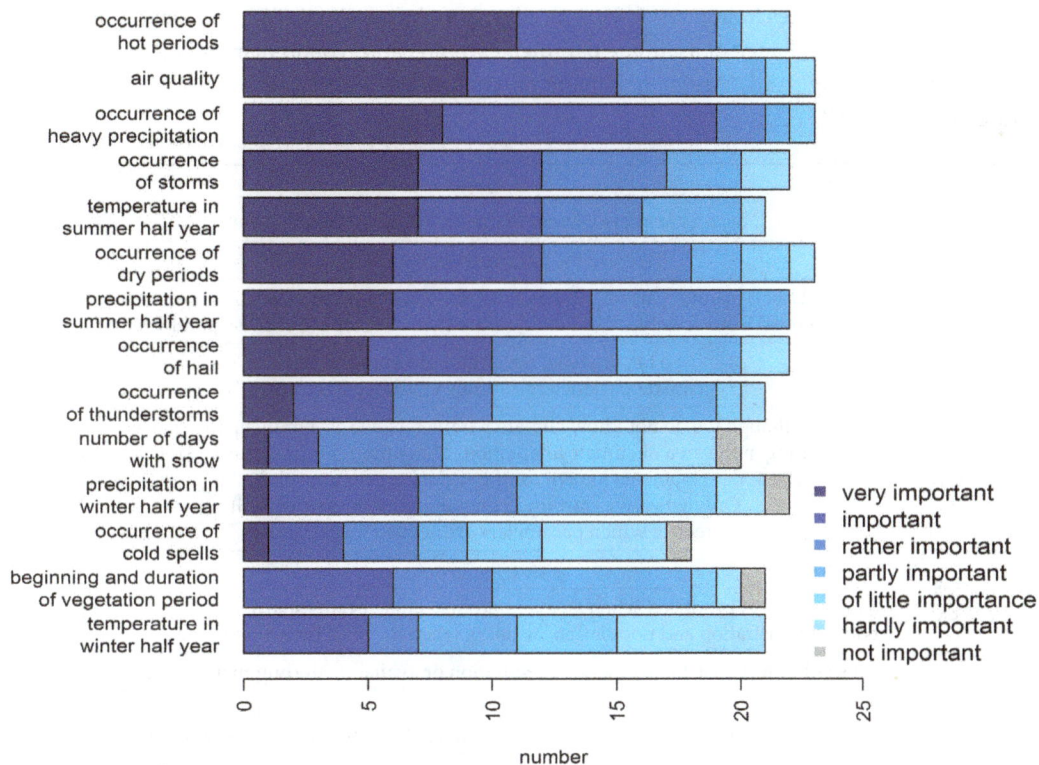

Figure 4. Answers to the question, "How important are the following events and parameters to your considerations relating to adaptation?" (*n* = 23).

The problematic past weather events mentioned corresponded largely to a municipality's climate adaptation considerations. Many municipalities consider additional weather events or changes in frequency to be important regardless of whether they produced negative effects in the past years. In some cases, however, municipalities deemed weather and climate factors irrelevant, despite problems in the past. In summary, slow-onset events that last a certain time and can cause health impairments, such as heat waves and poor air quality, are highly relevant to climate adaptation considerations. In addition, and as the expert interviews also confirmed, short-term events are important, such as heavy precipitation or storms, which can cause flash floods in settlements, and damage harvests, houses, or infrastructure (Table 2, statement S5). Generally, weather conditions during the summer half year seemed to be more important for adaptation considerations than did those during the winter half year. The regional climate in Southwestern Germany, which has relatively high temperatures in summer and mild winters—particularly in the Rhine valley—most likely is one reason for this. In addition, the responses most likely reflect the remembered experiences of the last several years, e.g., several hot summers.

Despite this, many of the survey respondents had difficulty naming specific climate parameters considered in climate adaptation today. Only some municipalities were able to provide climate parameters they use for current climate adaptation decisions. A quarter of the municipalities answered explicitly that they do not use parameters that can be described numerically, and half of the municipalities did not answer the respective question at all. The climate parameters given by municipalities included:

- Heat warnings by the German Weather Service (including the two warning levels "strong" and "extreme heat stress") in the health sector
- Snowfall and number of frost days for planning winter services or closing of (sports) halls due to snow loads on the roof
- Short-term weather forecasts for heat and precipitation to adjust the irrigation of green areas.

Table 2. Illustrative quotes from interviews with experts to the questions "Which climate parameters do you use?", "Which change of a climate parameter would cause you to plan or implement (more) adaptation measures?", and "Are there climate parameters for which you would like to know the future development?".

	Quotes
	Climate parameters relevant to climate change adaptation
S1	"We use the parameters given by science."
S2	"Urban planning among others requires 'work' with the climate parameters obtained from scientific discussion rather than developing of these parameters."
S3	"So far, we have never thought about individual climate parameters–but this would be interesting."
S4	"Even though we do not know climate parameters, we as a municipality have to do something in the area of climate adaptation. Many thoughts of a municipality may be based on basic developments known for the region, e.g., as a result of the 'guard rails' [54]. Only when we start implementing these measures, will we realize which parameters are needed for prognoses."
	Sensitivity
S5	"Sensitivity is produced when an event occurs: single events sharpen the awareness of both administration and population, as this is when the need for action becomes obvious."
S6	"It's either response to an emergency situation or available funding that trigger actions."
S7	"Generally, the climate adaptation process in cities is not based on fixed parameters, limits, or thresholds, but on general developments to be expected or already observed."
S8	"So far, measures considered or partly implemented have been lacking 'trigger criteria' (when do we have to start implementing measures, which priority do they have?)."

The questionnaire revealed that weather, climate, and its associated parameters play a role in practical applications in different fields of action. Examples include planning winter services, energy costs for heating and cooling, sewage management, selection of tree species for urban forests and the choice of plants for urban green areas. Climate parameters are also relevant in different planning contexts, for example, in drawing up the development plan and when planning green roofs, facades, or rainwater retention basins.

During the interviews, many experts were able to describe current weather effects precisely. It was easier for them to describe effects, either of high-impact weather events or of changes in frequencies of events in the past than to identify directly the climate parameters that could be relevant to the planning and implementation of adaptation. However, some experts were able to specify climate parameters for which they would like to have future projections. If they consider climate issues in their municipality, many rely on personal experiences or general knowledge about climate change in the form of rising temperatures or changes in precipitation (Table 2, statements S4 and S7). For this reason, it was difficult initially to identify their specific data needs. Only a few answers in the questionnaire indicate direct deduction of climate parameters. It was primarily the direct dialogue in the expert interviews that allowed us to deduce tailored climate parameters relevant to practical application in different fields of action directly, although some interviewees considered deducing climate parameters as a new task or even a task for scientists, not them (Table 2, statements S1–S3).

Among the parameters deduced were "standard" climate parameters that have been analyzed already on a regional scale. These include, for example, the number of tropical nights, hot days, hot periods, and the beginning and end of heating periods or vegetation periods. Beyond that, experts also mentioned new parameters that have not yet or rarely been evaluated from current climate models. These include, for example, the number of days per year in which winter services are necessary in the form of salting roads and clearing snow, the number of consecutive dry days between May and September, and the periods with weather conditions favorable for the spread of agricultural parasites, which temperature and humidity influence directly. Table 3 shows a selection of eight of approximately 25 such tailored climate parameters originating from both modes of data collection. Not only the mean values of the 30-year climatological period play a role in these tailored climate parameters, but also the range between extreme values and the inter-annual variability. The possible effect attributable to climate change (Table 3, right column) refers in some cases to adaptation practices and in some cases to climate change impacts. This originates from the explanations of the respondents to the question which climate parameter they desire. As mainly the interviews allowed for defining desired climate parameters, the experts often additionally explained in the dialogue why they desire a certain parameter.

Table 3. Selection of climate parameters desired for local adaptation and their possible effects on municipalities attributable to climate change.

	Desired Climate Parameters	**Possible Effect Attributable to Climate Change**
(1)	Hot days combined with high solar radiation	Effect on people's well-being; health risks attributable to heat and solar radiation
(2)	Number of hot and dry summers and years between them	Selection of tree species for streets Irrigation necessary for urban green areas Drying out of water bodies
(3)	Dry years	Selection of forest tree species
(4)	Number of consecutive dry days between May and September	Selection of plant species and irrigation of urban green areas
(5)	Days with precipitation in March and April	Change in best time to plant new trees
(6)	Storms and wind gusts	Infrastructure damage, damage to roofs, uprooted trees, high insurance payments
(7)	Cold temperatures with simultaneous precipitation (snow or rain)	Change in number of days when winter services necessary
(8)	Favorable conditions for cherry fruit fly (Rhagoletis cerasi)	Spread of agricultural fruit parasites

With respect to the way in which to provide tailored climate parameters, most survey respondents would prefer to obtain tailored information on specific climate parameters via a website. In addition, they requested datasets that geographical information systems can process. During the interviews, the experts expressed the expectation that this information should also include datasets that are well prepared, reliable, and evaluated for direct application. They also indicated that they need condensed data applicable readily to decision-making and available in a clearly processed form. Furthermore, they suggested that an event should be organized to provide information about the availability and quality of climate data.

3.4. Sensitivity

In the questionnaire responses, several municipalities indicated certain sensitivities to flooding, heat waves, or flash floods attributable to heavy precipitation events, but it was largely difficult or even impossible for the experts to describe their sensitivity in the sense of changes in parameters that would require adaptation actions. At present, they use specific numerical parameters randomly, for which, in addition to the past statistics, they also can estimate a range of possible future developments via regional climate projections. They indicated that an exact definition of sensitivity would require complex statistical evaluations, while decisions for specific adaptation actions rely currently on "subjective" event thresholds. Furthermore, answers to an open-ended question showed clearly that it can be the administrative and legislative framework that define the need for action, as illustrated by the following quote: "Municipalities cannot act alone in areas where measures of increased financial volume have to be implemented, such as flood protection. Municipalities depend on the prompt updating of regulations with respect to the assessment basis (e.g., frequencies of floods, rainfall series, etc.), even more so when measures have to be implemented in accordance with European Union (EU) directives (Water Framework Directive, Flood Risk Management Directive)".

In addition, the in-depth interviews revealed that the process of climate adaptation in municipalities largely is not associated with fixed climate parameters or thresholds (Tables 2, statement S7). Just as often, experiences of past events alone have formed the basis for considerations of climate change adaptation. The experts stated that single extreme events (Table 2, statements S5 and S6), as well as external influences such as project funding, rather than climate impacts, can trigger actions and be the starting point for adaptation (Table 2, statement S6). For the future, often only general projected climatic developments, such as higher temperatures, more hot days, or fewer frost days, are considered. Even if some stakeholders wish to have information on tailored climate parameters to plan and implement concrete adaptation measures, past and future development of these tailored and newly defined climate parameters has not yet been made available. Hence, we lack a basis for the experts' sensitivity assessment (Table 2, statement S8).

4. Discussion

4.1. State of Adaptation in Municipalities Varies

The observation that the state of climate change adaptation differs considerably among the municipalities that participated in our study is consistent with the results of other studies. Our study showed that approximately two thirds of the municipalities engage in activities related to climate change adaptation and one third has an adaptation plan. The latter is consistent with other studies in Germany that found adaptation plans in one third of the cities [6], and showed that one third of the regional planning authorities considered climate adaptation in their regional plans [23]. In combination with the survey result that climate change adaptation is considered a "relatively new topic" in many municipalities in Baden-Wuerttemberg, our study contributes further evidence to Lorenz et al.'s conclusion that, "Despite progress on adaptation at [the] national level [in Germany], adaptation at the local level still seems to be in the early stages" ([58], p. 7). This confirms the "gap between the perceived urgency of proactive adaptation to climate change by scientists and the perceptions of planners" observed in Dutch urban areas ([12], p. 777).

As shown elsewhere (e.g., [8]), lack of personnel resources is a relevant barrier in municipal adaptation. While some larger cities have several persons or offices that handle climate change adaptation, have established working groups on climate change impacts, or have employed persons explicitly responsible for the topic, in smaller communities, a single person is often responsible for all environmental issues. This likely explains in part the varied importance municipalities assign to adaptation. According to Lehmann et al. ([8], p. 87), "the lack of personnel, finances and time" is associated closely with information deficits. Hence, these aspects inhibit adaptation in two ways: the lack of resources prevents municipalities from adapting to climate change directly, and from informing themselves in detail about climate change and its effects. This lack of information then becomes a direct obstacle to adaptation [8]. We also found large differences in knowledge among respondents. Nevertheless, it is possible that they consider climate issues unconsciously or under a different heading, e.g., flood protection, because "climatic factors are often an integral part of environmental and comfort aspects of planning, and therefore, not necessarily perceived to be important in themselves (... [and urban planners] often included some climatic considerations in their work without being aware of it" ([59], p. 37). This unconscious use of knowledge must be transferred to explicit considerations in decision-making to institutionalize climate change adaptation in a formal administrative framework. Climate adaptation should become an independent aspect that is explicitly addressed in, e.g., planning processes instead of being just part of other environmental issues. In addition, the systematic implementation of a concept and its mainstreaming in municipalities, for example, ecosystem-based adaptation [60,61], could achieve progress, because "neither adaptation nor ecosystem-based approaches are labelled or systematized in any way" ([60], p. 76) so far.

4.2. Adaptation Often Relies on Previous Experiences

The decisions about climate adaptation measures today often rely on previous experiences, and in many cases, adaptations are considered initially after a single event, when awareness of the need for action increases after an extreme weather event causes damage in a municipality. Such "event-driven risk management" ([19], p. 461) also has been reported in former studies [19,62]. In our study, one example of a single event that triggered action is the summer of 2003. Temperatures were exceptionally high over a protracted period during that summer in Central Europe, and caused high mortality rates [43,63]. Consequently, the city of Karlsruhe in Baden-Wuerttemberg established a working group on climate change and later published a report on climate change adaptation [64] as well as a climate change adaptation strategy [65].

The climate data municipalities use consist primarily of measurements and urban climate analyses, which in most cases refer exclusively to the current climate. The fact that municipalities use analyses of the current climate and documents such as climate function maps or planning recommendation maps more often than climate projections also has been described in previous studies [23,58]. However, municipalities use (online) information sources frequently to obtain information about future climate change, its effects, and concrete measures for adaptation. Several municipalities use information on classical "standard" climate parameters, such as the number of hot days and tropical nights, and the number of days with heat warnings, or number of frost days. At the same time, however, other respondents also wished to have these parameters, even though future projections of these and certain other parameters desired are already available in several climate atlases or similar publications, although they are often provided on a larger horizontal scale (25 to 50 km, e.g., [54,66]). Obviously, to date, municipalities are largely unaware of these information sources.

4.3. Tailored Climate Parameters Can Narrow the Usability Gap

In the "seamless process" ([16], p. 28) of building climate change adaptation on adaptation to the current climate, a combination of data on current and future climates is desirable. This study showed that the capacities of regional climate modeling in providing data that can be used as a basis for adaptation planning are widely unknown, even though ensembles of climate model simulations of high horizontal resolution (less than 10 km) provide a tool to assess the range of possible future developments and the associated uncertainties for tailored climate parameters. Earlier studies revealed that climate projections have rarely been used so far in urban planning in Germany because of inappropriate data. In detail, respondents consider scientific robustness of the data, their obvious relevance to planning actions [23] and spatial resolutions appropriate for the local level and as input data for impact models are lacking [8].

It was clear that experts in administrative bodies expected to receive, preferably, a definite and fixed reference as a basis for decision-making. A discursive policy analysis in Germany that focused on the relation between administrative actors and uncertain scientific knowledge also found this to be the case [67]. In our study, their request for data that were processed, aggregated, evaluated professionally, and ready to be integrated into decisions reflected this. Municipalities expressed their wish for GIS (geographical information system) data repeatedly—in addition to information on a website and in a booklet—for information on the future development of tailored climate parameters. This underscores the importance of direct transferability of climate information into planning processes to improve the usability of climate information.

Although lack of data was not the barrier mentioned most frequently in the responses, transforming useful into usable knowledge [26] by evaluating already available climate model simulations in a user-oriented way promises to increase the value of municipalities' current use of data. In this context, the perceived novelty of the topic offers an opportunity to support municipal adaptation by using tailored climate information at an early stage, as well as providing a database to municipalities that are just beginning to think about adaptation.

Coupling the profound experiences of stakeholders with the tools of climate science by translating weather events or weather-related influences into tailored climate parameters has the significant potential to enlarge the dataset(s) municipalities use in different fields of action. This approach can help narrow the climate information usability gap [26] if climate simulation data thus far unused widely become usable for municipalities. In this way, interaction—one of the three factors that Lemos et al. [26] addressed in their conceptual model—between climate scientists from a regional climate service and stakeholders during the expert interviews was critical. Furthermore, our study enhanced our knowledge about the large range of possible applications and uses of climate model data in municipalities. In doing so, we addressed one of the direct, climate-specific barriers Biesbroek et al. ([11], p. 1124) identified, the " . . . reliance on scientific models to identify, understand, and communicate the problem and propose solutions".

The climate parameters that were deduced from experts' experiences were related either to one or more climate variables (e.g., temperature, precipitation). In addition to frequencies of occurrence, durations of certain weather situations are important. These parameters affect people (for example, high temperatures) and influence capital investments or municipal budgets (for example, winter services). Furthermore, the occurrence of events that climate parameters describe can cause property damage (heavy precipitation events, storms) or health risks (heat waves, high pollen levels). For example, knowledge about the direction and intensity of climate change can support decisions in urban planning, although further investigations might be necessary before taking concrete measures. The evaluation of tailored climate parameters for past periods also could underpin municipalities' previous experiences of climate change (for example "we have observed an increasing number of dry periods"). However, some climate parameters desired describe effects of weather events. Such parameters, e.g., flash floods in cities due to heavy precipitation or river flooding, cannot be assessed directly with regional climate models, but require coupling to impact models. Air quality assessments are available already from weather forecasting model data [68,69], but are not yet available widely from climate models for long periods.

4.4. The Concept of Sensitivity Is Difficult to Apply in Municipalities

For the respondents in the municipalities, it was largely unclear which change in a climate parameter requires action or causes damage in their municipalities, for example, what temperature increase will have negative effects on human's health. Although several municipalities exhibited certain sensitivities to extreme events, such as flooding, flash floods, or heat waves, it was difficult for them to assess thresholds for the need for action. The study showed that, from a climate science perspective, asking for thresholds that would imply a need for action is a concept difficult to apply in municipalities.

In addition, the range of possible future developments in climate parameters provided by an climate model ensemble approach does not provide a single number for planning, for example, a projected temperature increase of x,y K. Hence, these uncertainties in climate model projections [70] may be a barrier to making concrete decisions about climate change adaptation [25] and increase the difficulty in defining a threshold for action. Therefore, communication of the uncertainties in climate projections is an important aspect because there is a fundamental discrepancy between scientists' goals to increase knowledge constantly and administrative actors' demand and need for reliable, self-contained, and unambiguous data that they can use in decision-making [67].

4.5. Interaction along Climate Information in Context of Climate Change Adaptation Decisions in Local Municipalities

However, an improved database on regional climate change will not be sufficient to initiate planning and implementation of adaptation measures in municipalities. As indicated in previous studies, the political and organizational framework is even more important [24], as is guidance from higher authorities, mandates, and regulations [8]. In addition, local climate change decision-making is " … necessarily part of a larger local sustainability challenge" ([13], p. 173). Nevertheless, " … close collaboration of climate and impact scientists, sectoral practitioners, decision makers and other stakeholders" ([71], p. 273) can increase awareness of the topic of climate change adaptation and the availability of tailored information for decision-making. At this interface of science and society, more projects should be initiated to make practical expert knowledge accessible and detect information needs because personal contact between local authorities and scientists can increase access to information [8]. Research or pilot projects in municipalities contribute to a large extent to the progress in adaptation [60]. The three recommendations in considering urban climate in planning issues also can be applied to adaptation: "Improve awareness, improve communication and argumentation, develop tools and courses suitable for urban planners" ([59], p. 42). Numerous measures for training and capacity building already exist in the form of guidelines, handbooks, and websites. For example, the European Commission initiated Climate-ADAPT, the goal of which is to help users access and share data and information about climate change [72]. The EU-project, RAMSES, as well as the German federal initiatives "deeds bank" ("Tatenbank"), and "city climate guide" ("Klimalotse"), focus on urban climate adaptation measures [73–75]. For example, the latter is an online tool that supports medium-sized and smaller municipalities, in particular, in their decision-making processes with a database that contains approximately 140 suggestions for adaptation measures.

4.6. Limitations of the Study

As mentioned earlier, the relatively small sample size did not allow us to generalize our results to other regions of Germany. While we cannot say that the quite low response rate was attributable to the study design, work overload in the municipalities, or simply their non-interest in participating in a scientific study, the extremely different response rates of smaller (4%) and larger (30%) municipalities may be an indication. Combined with the finding that there are more activities in larger municipalities , the difference in participation rate suggests a self-selection effect in the sample, in that only those municipalities that are already aware of the topic of climate change adaptation participated in the study. At first glance, climate adaptation is neither a common nor a high-priority topic in the daily work of most municipalities in Baden-Wuerttemberg. However, several measures planned or implemented may be related to the adaptation context, even though they are not covered under the heading of adaptation, but are aspects of flood protection or nature and environment protection. As Overbeck et al. ([23] p. 198) indicated, "The fields of work in regional planning in Germany often are not yet considered to be related to climate change and it is not clear to regional planners what can be regarded as adaptation strategy or measure".

With respect to the two approaches taken to deduce tailored climate parameters for adaptation, the questionnaire responses, as well as the expert interviews, showed that indirect deduction of tailored climate parameters from extreme hydro-meteorological events experienced in the past was easier than direct deduction by naming parameters desired. As many of the respondents never had been asked what climate information they desired, the expert interviews proved to be more suitable in identifying users' needs because the more interactive format allowed further explanations and clarifications and thus enhanced the fit to previous knowledge and experiences. From the perspective of a regional climate service, this interactive approach to identifying municipalities' needs for tailored climate parameters proved to be suitable as a case study on the local level. However, such a deliberative and interactive process also requires human and financial resources [13,26], as do the further steps necessary to provide the data desired and its eventual coupling to impact models. This is especially the case if the parameters are tailored to very specific applications relevant in only few municipalities. Nevertheless, the "iterativity between scientists and users of knowledge is critical to the successful production of usable science" ([76], p. 687).

Future studies should test participative methods, for example workshops or focus group interviews, in order to encourage more interaction among the stakeholders. They also should focus on one or a few fields of action in the context of adaptation rather than taking such a broad approach, as this could yield more in-depth results. In addition, the legal and regulatory planning system, in which climate science and local stakeholders interact, should be considered [58] when identifying suitable formats for iterative, dialogue-based approaches.

5. Conclusions

Our survey of municipalities in Baden-Wuerttemberg in Southwestern Germany showed that climate change adaptation is of increasing importance, but that the progress in adaptation on the part of the municipalities differed strongly.

The novelty of the topic and the fact that municipalities, especially smaller ones, do not know where to obtain information and data, offer an opportunity to co-develop tailored information to information needs for local adaptation to climate change. Municipalities have considerable practical experience in, and knowledge of adaptation, as well as a need for information, which can be identified in close interaction and dialogue with climate scientists. At present, regional climate services help institutionalize this dialogue, meet information needs, and provide tailored data with a focus on the specific impacts of climate change on the regional scale [34].

Regional climate models offer reliable data on a regional scale that allow climate information to be tailored to municipalities' needs and thus transform useful climate data to usable information by adding value to the data and customizing the information. The deductive approach described is designed to translate local municipal experiences into tailored climate parameters described by thresholds, durations, or combinations of meteorological variables. Then, these tailored climate parameters can be evaluated with already existing regional climate projections. Therefore, we recommend intensifying the communication of scientific results and developing those results continuously in a user-oriented way. Regional cooperation of municipalities, scientific institutions, and companies should be developed further and collaborative projects that focus on practical application can be an incentive for municipalities to initiate adaptation activities.

Acknowledgments: The project was funded by the Baden-Wuerttemberg Ministry of the Environment, Climate Protection and the Energy Sector under the program KLIMOPASS (project number 347083) and by the Stiftung Umwelt und Schadenvorsorge (Environment and Damage Precaution Foundation). We thank all participants who contributed with their professional expertise to the study. Furthermore, we thank the two anonymous reviewers for valuable comments that helped improve the manuscript.

Author Contributions: Janus Willem Schipper conceived the study; Julia Hackenbruch, Janus Willem Schipper, and Tina Kunz-Plapp designed the survey; Julia Hackenbruch and Janus Willem Schipper performed the expert interviews; Sebastian Mueller and Julia Hackenbruch analyzed the data; and Julia Hackenbruch and Tina Kunz-Plapp wrote the paper with contributions from Janus Willem Schipper and Sebastian Mueller.

Conflicts of Interest: The authors declare no conflict of interest.

References

1.   IPCC. *Climate Change 2013: The Physical Science Basis. Contribution of Working Group I to the Fifth Assessment Report of the Intergovernmental Panel on Climate Change*; Stocker, T.F., Qin, D., Plattner, G.K., Tignor, M., Allen, S.K., Boschung, J., Nauels, A., Xia, Y., Bex, V., Midgley, P.M., Eds.; Cambridge University Press: Cambridge, UK; New York, NY, USA, 2013.

2.   IPCC. *Climate Change 2014: Impacts, Adaptation, and Vulnerability. Part A: Global and Sectoral Aspects. Contribution of Working Group II to the Fifth Assessment Report of the Intergovernmental Panel on Climate Change*; Field, C.B., Barros, V.R., Dokken, D.J., Mach, K.J., Mastrandrea, M.D., Bilir, T.E., Chatterjee, M., Ebi, K.L., Estrada, Y.O., Genova, R.C., et al., Eds.; Cambridge University Press: Cambridge, UK; New York, NY, USA, 2014.

3.   Revi, A.; Satterthwaite, D.E.; Aragón-Durand, F.; Corfee-Morlot, J.; Kiunsi, R.B.R.; Pelling, M.; Roberts, D.C.;
     Solecki, W. Urban areas. In *Climate Change 2014: Impacts, Adaptation, and Vulnerability. Part A: Global and
     Sectoral Aspects. Contribution of Working Group II to the Fifth Assessment Report of the Intergovernmental Panel on
     Climate Change*; Field, C.B., Barros, V.R., Dokken, D.J., Mach, K.J., Mastrandrea, M.D., Bilir, T.E., Chatterjee, M.,
     Ebi, K.L., Estrada, Y.O., Genova, R.C., et al., Eds.; Cambridge University Press: Cambridge, UK; New York,
     NY, USA, 2014; pp. 535–612.
4.   European Environment Agency. *Urban Adaptation to Climate Change in Europe*; EEA Report No 2/2012;
     European Environment Agency: Copenhagen, Denmark, 2012; Volume 2, pp. 62–94.
5.   Wilbanks, T.J.; Romero Lankao, P.; Bao, M.; Berkhout, F.; Cairncross, S.; Ceron, J.-P.; Kapshe, M.;
     Muir-Wood, R.; Zapata-Marti, R. Industry, settlement and society. In *Climate Change 2007: Impacts, Adaptation
     and Vulnerability. Contribution of Working Group II to the Fourth Assessment Report of the Intergovernmental
     Panel on Climate Change*; Parry, M.L., Canziani, O.F., Palutikof, J.P., van der Linden, P.J., Hanson, C.E., Eds.;
     Cambridge University Press: Cambridge, UK, 2007; pp. 357–390.
6.   Reckien, D.; Flacke, J.; Dawson, R.J.; Heidrich, O.; Olazabal, M.; Foley, A.; Hamann, J.J.P.; Orru, H.; Salvia, M.;
     De Gregorio Hurtado, S.; et al. Climate change response in Europe: What's the reality? Analysis of adaptation
     and mitigation plans from 200 urban areas in 11 countries. *Clim. Chang.* **2014**, *122*, 331–340. [CrossRef]
7.   Dewulf, A.; Meijerink, S.; Runhaar, H. Editorial: The governance of adaptation to climate change as a
     multi-level, multi-sector and multi-actor challenge: A European comparative perspective Art. *J. Water
     Clim. Chang.* **2015**, *6*, 1–8. [CrossRef]
8.   Lehmann, P.; Brenck, M.; Gebhardt, O.; Schaller, S.; Süßbauer, E. Barriers and opportunities for urban
     adaptation planning: Analytical framework and evidence from cities in Latin America and Germany.
     Mitig. Adapt. Strateg. Glob. Chang. **2015**, *20*, 75–97. [CrossRef]
9.   Klein, R.J.T.; Midgley, G.F.; Preston, B.L.; Alam, M.; Berkhout, F.G.H.; Dow, K.; Shaw, M.R. Adaptation
     opportunities, constraints and limits. In *Climate Change 2014: Impacts, Adaptation, and Vulnerability.
     Part A: Global and Sectoral Aspects.Contribution of the Working Group II to the Fifth Assessment Report
     of the Intergovernmental Panel on Climate Change*; Field, C.B., Barros, V.R., Dokken, D.J., Mach, K.J.,
     Mastrandrea, M.D., Bilir, T.E., Chatterjee, M., Ebi, K.L., Estrada, Y.O., Genova, R.C., et al., Eds.;
     Cambridge University Press: Cambridge, UK; New York, NY, USA, 2014; pp. 899–943.
10. Reisinger, A.; Kitching, R.L.; Chiew, F.; Hughes, L.; Newton, P.C.D.; Schuster, S.S.; Tait, A.; Whetton, P.
     Australasia. In *Climate Change 2014: Impacts, Adaptation, and Vulnerability. Part B: Regional Aspects. Contribution
     of Working Group II to the Fifth Assessment Report of the Intergovernmental Panel on Climate Change*; Barros, V.R.,
     Field, C.B., Dokken, D.J., Mastrandrea, M.D., Mach, K.J., Bilir, T.E., Chatterjee, M., Ebi, K.L., Estrada, Y.O.,
     Genova, R.C., et al., Eds.; Cambridge University Press: Cambridge, UK; New York, NY, USA, 2014;
     pp. 1371–1438.
11. Biesbroek, G.R.; Klostermann, J.E.M.; Termeer, C.J.A.M.; Kabat, P. On the nature of barriers to climate change
     adaptation. *Reg. Environ. Chang.* **2013**, *13*, 1119–1129. [CrossRef]
12. Runhaar, H.; Mees, H.; Wardekker, A.; van der Sluijs, J.; Driessen, P.P.J. Adaptation to climate change-related
     risks in Dutch urban areas: Stimuli and barriers. *Reg. Environ. Chang.* **2012**, *12*, 777–790. [CrossRef]
13. Corfee-Morlot, J.; Cochran, I.; Hallegatte, S.; Teasdale, P.-J. Mulitlevel risk governance and urban adaptation
     policy. *Clim. Chang.* **2011**, *104*, 169–197. [CrossRef]
14. Measham, T.G.; Preston, B.L.; Smith, T.F.; Brooke, C.; Gorddard, R.; Withycombe, G.; Morrison, C. Adapting to
     climate change through local municipal planning: Barriers and challenges. *Mitig. Adapt. Strateg. Glob. Chang.*
     2011, *16*, 889–909. [CrossRef]
15. Moser, S.C.; Ekstrom, J.A. A framework to diagnose barriers to climate change adaptation. *Proc. Natl. Acad.
     Sci. USA* **2010**, *107*, 22026–22031. [CrossRef] [PubMed]
16. Burton, I. Climate change and the adaptation deficit. In *Climate Change: Building the Adaptive Capacity*;
     Fenech, A., MacIver, D., Auld, H., Rong, R.B., Yin, Y., Eds.; Ministry of Public Works and Government
     Services: Ottawa, ON, Canada, 2004.
17. Laukkonen, J.; Blanco, P.K.; Lenhart, J.; Keiner, M.; Cavric, B.; Kinuthia-Njenga, C. Combining climate change
     adaptation and mitigation measures at the local level. *Habitat Int.* **2009**, *33*, 287–292. [CrossRef]

18. Adger, W.N.; Agrawala, S.; Mirza, M.M.Q.; Conde, C.; O'Brien, K.; Pulhin, J.; Pulwarty, R.; Smit, B.; Takahashi, K. Assessment of adaptation practices, options, constraints and capacity. In *Climate Change 2007: Impacts, Adaptation and Vulnerability. Contribution of Working Group II to the Fourth Assessment Report of the Intergovernmental Panel on Climate Change*; Parry, M.L., Canziani, O.F., Palutikof, J.P., van der Linden, P.J., Hanson, C.E., Eds.; Cambridge University Press: Cambridge, UK; New York, NY, USA, 2007; pp. 717–743.

19. Storbjörk, S. Governing climate adaptation in the local arena: Challenges of risk management and planning in Sweden. *Local Environ.* **2007**, *12*, 457–469. [CrossRef]

20. Martinez, G.; Bray, D. *Befragung politischer Entscheidungsträger zur Wahrnehmung des Klimawandels und zur Anpassung an den Klimawandel an der deutschen Ostseeküste*; RADOST-Berichtsreihe 4; Ecologic Institut: Berlin, Germany, 2011.

21. Hornsey, M.J.; Harris, E.A.; Brain, P.G.; Fielding, K.S. Meta-analyses of the determinants and outcomes of belief in climate change. *Nat. Clim. Chang.* **2016**. [CrossRef]

22. Ratter, B.M.W.; Phillipp, K.H.I.; von Storch, H. Between hype and decline: Recent trends in public perception of climate change. *Environ. Sci. Policy* **2012**, *18*, 3–7. [CrossRef]

23. Overbeck, G.; Sommerfeldt, P.; Köhler, S.; Birkmann, J. Klimawandel und Regionalplanung. *Raumforsch. Raumordn.* **2009**, *67*, 193–203. [CrossRef]

24. Demeritt, D.; Langdon, D. The UK Climate Change Programme and communication with local authorities. *Glob. Environ. Chang.* **2004**, *14*, 325–336. [CrossRef]

25. Kropp, J.P.; Daschkeit, A. Planungshandeln im Lichte des Klimawandels. *Inf. Raumentwickl.* **2008**, *6*, 353–361.

26. Lemos, M.C.; Kirchhoff, C.J.; Ramprasad, V. Narrowing the climate information usability gap. *Nat. Clim. Chang.* **2012**, *2*, 789–794. [CrossRef]

27. Hackenbruch, J.; Schädler, G.; Schipper, J.W. Added value of high resolution regional climate simulations for regional impact studies. *Meteorol. Z.* **2016**, *25*, 291–304.

28. Sedlmeier, K. Near Future Changes of Compound Extreme Events from an Ensemble of Regional Climate Simulations. Ph.D.Thesis, Karlsruher Institut für Technologie, Karlsruher, Germany, 2015.

29. Prein, A.F.; Langhans, W.; Fosser, G.; Ferrone, A.; Ban, N.; Goergen, K.; Keller, M.; Tölle, M.; Gutjahr, O.; Feser, F.; et al. A review on regional convection-permitting climate modeling: Demonstrations, prospects, and challenges. *Rev. Geophys.* **2015**, *53*, 323–361. [CrossRef] [PubMed]

30. Junk, J.; Matzarakis, A.; Ferrone, A.; Krein, A. Evidence of past and future changes in health-related meteorological variables across Luxembourg. *Air Qual. Atmoms. Health* **2014**, *7*, 71–81. [CrossRef]

31. Berg, P.; Wagner, S.; Kunstmann, H.; Schädler, G. High resolution regional climate model simulations for Germany: Part 1—Validation. *Clim. Dyn.* **2013**, *40*, 401–414. [CrossRef]

32. Feldmann, H.; Schädler, G.; Panitz, H.J.; Kottmeier, C. Near future changes of extreme precipitation over complex terrain in Central Europe derived from high resolution RCM ensemble simulations. *Int. J. Climatol.* **2013**, *33*, 1964–1977. [CrossRef]

33. Déqué, M.; Rowell, D.; Lüthi, D.; Giorgi, F.; Christensen, J.; Rockel, B.; Jacob, D.; Kjellström, E.; de Castro, M.; van den Hurk, B. An intercomparison of regional climate simulations for Europe: Assessing uncertainties in model projections. *Clim. Chang.* **2007**, *81*, 53–70. [CrossRef]

34. Von Storch, H.; Meinke, I.; Stehr, N.; Ratter, B.; Krauss, W.; Pielke, R.A., Jr.; Grundmann, R.; Reckermann, M.; Weisse, R. Regional climate services illustrated with experiences from Northern Europe. *Z. Umweltpolit. Umweltr.* **2011**, *34*, 1–15.

35. Kirchhoff, C.J.; Lemos, M.C.; Dessai, S. Actionable knowledge for environmental decision making: broadening the usability of climate science. *Annu. Rev. Environ. Resour.* **2013**, *38*, 393–414. [CrossRef]

36. Field, C.B.; Barros, V.R.; Mach, K.J.; Mastrandrea, M.D.; van Aalst, M.K.; Adger, W.N.; Arent, D.J.; Barnett, J.; Betts, R.; Bilir, T.E.; et al. Technical Summary. In *Climate Change 2014: Impacts, Adaptation, and Vulnerability. Part A: Global and Sectoral Aspects. Contribution of Working Group II to the Fifth Assessment Report of the Intergovernmental Panel on Climate Change*; Field, C.B., Barros, V.R., Dokken, D.J., Mach, K.J., Mastrandrea, M.D., Bilir, T.E., Chatterjee, M., Ebi, K.L., Estrada, Y.O., Genova, R.C., et al., Eds.; Cambridge University Press: Cambridge, UK; New York, NY, USA, 2014; pp. 35–94.

37. Basett, T.J.; Fogelman, C. Déjà vu or something new? The adaptation concept in the climate change literature. *Geoforum* **2013**, *48*, 42–53.

38. State Statistical Office (Statistisches Landesamt) Baden-Wuerttemberg (Stuttgart, Germany). Available online: http://www.statistik.baden-wuerttemberg.de/ (accessed on 24 November 2016).

39. State Office for the Environment, Measurements and Nature Conservation of the Federal State of Baden-Württemberg (LUBW). *Klimaatlas Baden-Württemberg*; LUBW: Karlsruhe, Germany, 2006.

40. Zolina, O.; Simmer, C.; Kapala, A.; Bachner, S.; Gulev, S.K.; Maechel, H. Seasonally dependent changes of precipitation extremes over Germany since 1950 from a very dense observational network. *J. Geophys. Res.* **2008**, *113*, D06110. [CrossRef]

41. Albrecht, F.M.; Dietze, B. Langzeitverhalten der Starkniederschläge in Baden-Württemberg und Bayern. In *KLIWA-Berichte 8*; KLIWA: Karlsruhe, München, Offenbach, Germany, 2006.

42. Hundecha, Y.; Bárdossy, A. Trends in daily precipitation and temperature extremes across western Germany in the second half of the 20th century. *Int. J. Climatol.* **2005**, *25*, 1189–1202. [CrossRef]

43. Fink, A.H.; Brücher, T.; Krüger, A.; Leckebusch, G.C.; Pinto, J.G.; Ulbrich, U. The 2003 European summer heatwaves and drought—Synoptic diagnosis and impacts. *Weather* **2004**, *59*, 209–216. [CrossRef]

44. Hoy, A.; Hänsel, S.; Skalak, P.; Ustrnul, Z.; Bochníček, O. The extreme European summer of 2015 in a long-term perspective. *Int. J. Climatol.* **2016**. [CrossRef]

45. Puskeiler, M.; Kunz, M.; Schmidberger, M. Hail statistics for Germany derived from single-polarization radar data. *Atmos. Res.* **2016**, *178–179*, 459–470. [CrossRef]

46. Piper, D.; Kunz, M.; Ehmele, F.; Mohr, S.; Mühr, B.; Kron, A.; Daniell, J.E. Exceptional sequence of severe thunderstorms and related flash floods in May and June 2016 in Germany—Part I: Meteorological background. *Nat. Hazards Earth Syst. Sci.* **2016**, *16*, 2835–2850. [CrossRef]

47. Wagner, S.; Berg, P.; Schädler, G.; Kunstmann, H. High resolution regional climate model simulations for Germany: Part II—Projected climate changes. *Clim. Dyn.* **2013**, *40*, 415–427. [CrossRef]

48. Beniston, M.; Stephenson, D.B.; Christensen, O.B.; Ferro, C.A.T.; Frei, C.; Goyette, S.; Halsnaes, K.; Holt, T.; Jylhä, K.; Koffi, B.; et al. Future extreme events in European climate: An exploration of regional climate model projections. *Clim. Chang.* **2007**, *81*, 71–95. [CrossRef]

49. Die Bundesregierung. *Deutsche Anpassungsstrategie an den Klimawandel*; Bundesministerium für Umwelt, Naturschutz, Bau und Reaktorsicherheit: Berlin, Germany, 2008.

50. Die Bundesregierung. *Aktionsplan Anpassung der Deutschen Anpassungsstrategie an den Klimawandel*; Bundesministerium für Umwelt, Naturschutz, Bau und Reaktorsicherheit: Berlin, Germany, 2011.

51. Umweltbundesamt. *Monitoringbericht 2015 zur Deutschen Anpassungsstrategie an den Klimawandel*; Umweltbundesamt: Dessau-Roßlau, Germany, 2015.

52. Bubecka, P.; Klimmer, L.; Albrecht, J. Klimaanpassung in der rechtlichen Rahmensetzung des Bundes. *Nat. Recht* **2016**, *38*, 297–307. [CrossRef]

53. Ministerium für Umwelt, Klima und Energiewirtschaft. *Strategie zur Anpassung an den Klimawandel in Baden-Württemberg*; Ministerium für Umwelt, Klima und Energiewirtschaft: Stuttgart, Germany, 2014.

54. State Office for the Environment, Measurements and Nature Conservation of the Federal State of Baden-Württemberg (LUBW). *Zukünftige Klimaentwicklung in Baden-Württemberg*; LUBW: Karlsruhe, Germany, 2013.

55. Ministerium für Umwelt, Klima und Energiewirtschaft Baden-Württemberg. KLIMOPASS. Available online: https://um.baden-wuerttemberg.de/de/klima/klimawandel/klimawandel-in-baden-wuerttemberg/klimaforschung/klimopass/ (accessed on 17 January 2017).

56. Baur, N.; Blasius, J. *Handbuch Methoden der empirischen Sozialforschung*; Springer: Wiesbaden, Germany, 2014.

57. Flick, U.; von Kardorff, E.; Keupp, H.; von Rosenstiel, L.; Wolff, S. *Handbuch qualitative Sozialforschung: Grundlagen, Konzepte, Methoden und Anwendungen*; Beltz Psychologie-Verlag-Union: Weinheim, Germany, 2012; Volume 3.

58. Lorenz, S.; Dessai, S.; Forster, P.M.; Paavola, J. Adaptation planning and the use of climate change projections in local government in England and Germany. *Reg. Environ. Chang.* **2016**. [CrossRef]

59. Eliasson, I. The use of climate knowledge in urban planning. *Landsc. Urban Plan.* **2000**, *48*, 31–44. [CrossRef]

60. Wamsler, C.; Pauleit, S. Making headway in climate policy mainstreaming and ecosystem-based adaptation: Two pioneering countries, different pathways, one goal. *Clim. Chang.* **2016**, *137*, 71–87. [CrossRef]

61. Roberts, D.; Boon, R.; Diederichs, N.; Douwes, E.; Govender, N.; Mcinnes, A.; Mclean, C.; O'Donoghue, S.; Spires, M. Exploring ecosystem-based adaptation in Durban, South Africa: "Learning-by-doing" at the local government coal face. *Environ. Urban.* **2011**, *24*, 167–195. [CrossRef]

62. Adger, W.N.; Arnella, N.W.; Tompkins, E.L. Successful adaptation to climate change across scales. *Glob. Environ. Chang.* **2005**, *15*, 77–86. [CrossRef]

63. Robine, J.-M.; Cheung, S.L.K.; Le Roya, S.; van Oyen, H.; Griffiths, C.; Michel, J.-P.; Herrmann, F.R. Death toll exceeded 70,000 in Europe during the summer of 2003. *C. R. Biol.* **2008**, *331*, 171–178. [CrossRef] [PubMed]

64. Karlsruhe, S.; Arbeitsschutz, U. *Anpassung an den Klimawandel*; Stadt Karlsruhe, Umwelt-und Arbeitsschutz: Karlsruhe, Germany, 2008.

65. Karlsruhe, S.; Arbeitsschutz, U. *Anpassung an den Klimawandel—Bestandsaufnahme und Strategie für die Stadt Karlsruhe*; Stadt Karlsruhe, Umwelt-und Arbeitsschutz: Karlsruhe, Germany, 2013.

66. Meinke, I.; Gerstner, E.M.; von Storch, H.; Marx, A.; Schipper, H.; Kottmeier, C.H.; Treffeisen, R.; Lemke, P. Regionaler Klimaatlas Deutschland der Helmholtz-Gemeinschaft informiert im Internet über möglichen künftigen Klimawandel. *Mitt. DMG* **2010**, *2*, 5–7.

67. Fröhlich, J. Klimaanpassung im administrativen Diskurs—Das Verhältnis von Verwaltungsakteuren zu unsicherem wissenschaftlichen Wissen. *Z. Umweltpolit. Umweltr.* **2009**, *32*, 325–350.

68. Rieger, D.; Bangert, M.; Bischoff-Gauss, I.; Förstner, J.; Lundgren, K.; Reinert, D.; Schröter, J.; Vogel, H.; Zängl, G.; Ruhnke, R.; et al. ICON—ART 1.0—A new online-coupled model system from the global to regional scale. *Geosci. Model Dev.* **2015**, *8*, 1659–1676. [CrossRef]

69. Vogel, B.; Vogel, H.; Bäumer, D.; Bangert, M.; Lundgren, K.; Rinke, R.; Stanelle, T. The comprehensive model system COSMO-ART—Radiative impact of aerosol on the state of the atmosphere on the regional scale. *Atmos. Chem. Phys.* **2009**, *9*, 8661–8680. [CrossRef]

70. Flato, G.; Marotzke, J.; Abiodun, B.; Braconnot, P.; Chou, S.C.; Collins, W.; Cox, P.; Driouech, F.; Emori, S.; Eyring, V.; et al. Evaluation of climate models. In *Climate Change 2013: The Physical Science Basis. Contribution of Working Group I to the Fifth Assessment Report of the Intergovernmental Panel on Climate Change*; Stocker, T.F., Qin, D., Plattner, G.-K., Tignor, M., Allen, S.K., Boschung, J., Nauels, A., Xia, Y., Bex, V., Midgley, P.M., Eds.; Cambridge University Press: Cambridge, UK; New York, NY, USA, 2013; pp. 741–866.

71. Füssel, H.M. Adaptation planning for climate change: Concepts, assessment approaches, and key lessons. *Sustain. Sci.* **2007**, *2*, 265–275. [CrossRef]

72. Climate-ADAPT–Sharing Adaptation Information across Europe, European Climate Adaption Platform. Available online: http://climate-adapt.eea.europa.eu/ (accessed on 24 January 2017).

73. RAMSES. Science for Cities in Transition Home Page. Available online: http://www.ramses-cities.eu/ (accessed on 24 January 2017).

74. Umweltbundesamt. Tatenbank. Available online: https://www.umweltbundesamt.de/themen/klima-energie/klimafolgen-anpassung/werkzeuge-der-anpassung/tatenbank (accessed on 24 January 2017).

75. Umweltbundesamt. Klimalotse. Available online: https://www.umweltbundesamt.de/themen/klima-energie/klimafolgen-anpassung/werkzeuge-der-anpassung/klimalotse (accessed on 24 January 2017).

76. Dilling, L.; Lemos, M.C. Creating usable science: Opportunities and constraints for climate knowledge use and their implications for science policy. *Glob. Environ. Chang.* **2011**, *21*, 680–689. [CrossRef]

5

Observed Regional Climate Variability during the Last 50 Years in Reindeer Herding Cooperatives of Finnish Fell Lapland

Élise Lépy [1],* [iD] and **Leena Pasanen** [2]

[1] Faculty of Humanities, University of Oulu, 90014 Oulu, Finland
[2] Research Unit of Mathematical Sciences, University of Oulu, 90014 Oulu, Finland; leena.ruha@oulu.fi
* Correspondence: elise.lepy@oulu.fi

Abstract: In Finnish Lapland, reindeer herders' activity is strongly dependent on the surrounding natural environment, which is directly exposed to environmental changes and climatic variations. By assessing whether there is any evidence of change in climate in Fell Lapland over the last 50 years, this paper attempts to link global climatic trends with local conditions and respond to the need of information at the local level. It aims at assessing the changes in temperature, precipitation and snow cover at a regional and local scale, as well as determining the climatic trends for the period 1960–2011. Statistical methods were used to conduct analyses of the regional homogeneity, the annual and seasonal variability, and the cold intensity. The results show that the regional climate is not homogeneous and differences exist between locations. Nevertheless, it can be concluded that, in general, a warming trend is discernible for the period 1960–2011, frost and thaw cycles slightly increase, and variations in mean temperatures are more important in the winter. Precipitation is more variable according to the site but, in general, precipitation is increasing with time, especially in the winter, and the snow cover does not seem to contain any discernible trend.

Keywords: climate variability; correlations; Lapland; precipitation; reindeer herding; snow cover; temperatures; trends

1. Introduction

The living conditions of Arctic and sub-Arctic communities are affected by climate variability and environmental changes [1–3], with direct impacts on health and quality of life [4]. In Northern Finland, where the population is scattered and land use needs can lead to conflict, climate change seems to be the most prominent driver of change, even though other environmental and socioeconomic pressures have an undeniable impact on socio-ecological systems too [5,6]. While the higher sensitivity of northern latitudes, compared to most other parts of the world, to climate change has been shown [1,7,8], few multidisciplinary studies have undertaken quantitative climate research in specific remote and vulnerable areas of Finnish Lapland [9]. Also, most of the studies conducted in Finland focus on the long-term variability of one single climate variable at local-scale [10] or national-scale [11,12], whereas the present one attempts to investigate the long-term and local-scale variability of all climate variables that are specifically crucial for reindeer herding.

The present study downscales the global approach to climate to the regional and local levels, and focuses on all the seasons with a specific focus on the winter season. In Northern Finland, winter is a critical season for reindeer herding [13–15] as ice and snow conditions remain particularly important environmental factors affecting reindeer populations dynamics, as well as the daily life of local communities [16,17]; in addition to the large natural variability of winter weather conditions, the Finnish Meteorological Institute predicts that warming and increased precipitation will be stronger in the winter (respectively by 3–9 °C and by 10–40%) than in the summer (by 1–5 °C and 0–20%), and winter changes will affect the north more than the south [18]. Moreover, the number of freezing point days (with a daily minimum temperature below zero and maximum temperature above zero) is another critical point for local livelihoods that are predicted to be larger than it is currently in the north. In that sense, the variability and change in winter weather conditions, and also the increasing occurrence of extreme winter weather events, will produce many public safety and economic problems, disrupting local socioeconomic activities.

A change in climate and weather conditions will affect local people and their livelihoods. In Finnish Lapland, reindeer herders' activity is strongly dependent on the surrounding natural environment. Reindeer herding systems can be considered as nature-based cultural livelihood systems [19] whose practitioners are both Sámi and non-Sámi people [20], and it is an excellent indicator of livelihood welfare, especially as the activity is directly exposed to environmental changes and climatic variations [21,22]. Indeed, the seasonal migrations of reindeer, the availability of the ground and tree lichen [23] and the calving rate and period [13] can be affected by weather conditions, particularly in the wintertime.

The study area is situated between 67 and 69° N in the Arctic region. Beyond the effects of high latitude, the regional climate is influenced by various geographical factors such as the proximity of the Scandinavian Alps that participates in the formation of the Foehn wind, which affects the distribution of temperatures and precipitation on both sides of the mountain range; the warm waters of the North Atlantic Current, which has a moderate maritime effect on temperatures; the continental landmass located south and east, which allows high pressure systems to drive warm air in summer and cold air in winter [10]; and the spatial distribution of waterbodies, including the Baltic Sea [24]. The study area comprises three reindeer herding cooperatives of Fell Lapland in the North of Finland (Table 1 and Figure 1). More precisely, the Käsivarsi reindeer herding cooperative is partly located in the Scandinavian Alps and offers a tundra landscape where Sámi herders breed around 10,000 reindeer (Reindeer Herder's Association, n.d.) around the fells under, sometimes, very hazardous weather conditions [19]. Fell mountains are still present in the northern part of Näkkälä reindeer herding cooperative, which is located in the taiga-tundra transition zone. Further south, the boreal forest covers the partially hilly surface of Muonio reindeer herding cooperative, which includes about 6000 reindeer (Reindeer Herder's Association, n.d.) that migrate around the pine and spruce forests and peat lands of the Muonionjoki River valley [19]. The whole study area is a region where the sustainability of reindeer herding and pasture conditions is at the heart of many debates involving a "complex set of ecological, political, cultural and socioeconomic issues" [25] (p. 141). The vulnerability of reindeer herding livelihood is prone to combined social-ecological pressures [19], whose starting point is often linked to climate change.

Table 1. Description of the three meteorological stations.

Weather Stations	Station Reference Number	Coordinates	Elevation (m.a.s.l.)	Years Recorded
Enontekiö Kilpisjärvi	9001	69°03'00" N, 20°48'00" E	483	1951–1978
Enontekiö Kilpisjärvi kyläkeskus	9003	69°03'00" N, 20°47'24" E	480	
Enontekiö Palojärvi	9202	68°34'12" N, 23°19'48" E	356	1972–2000
Enontekiö Näkkälä	9201	68°36'00" N, 23°34'48" E	374	1960–present
Muonio Alamuonio	8201	67°57'36" N, 23°40'48" E	236	1946–present

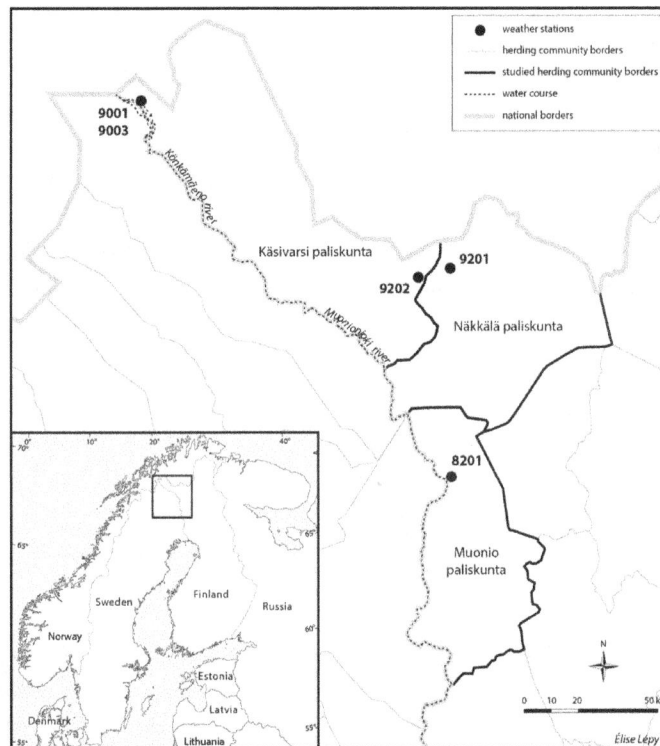

Figure 1. Location map.

By assessing whether there is any evidence of change in climate in Fell Lapland over the last 50 years, the present paper attempts to link global climatic trends with local conditions and respond to the need of information at the local level. In addition, the obtained results from climate and weather data treatment and analysis were used in semi-structured interviews with Finnish and Sámi reindeer herders of Fell Lapland [14,19]. This paper finally aims at assessing the changes in temperature, precipitation and snow cover in Fell Lapland at a regional and local scale, and at determining the climatic trends over half a century in order to anticipate the future evolution.

2. Materials and Methods

The climate datasets used in this study are extracted from three meteorological stations that belonged to the national meteorological network of Finland (the Finnish Meteorological Institute, see Table 1). The stations of Kilpisjärvi, Enontekiö and Muonio were chosen due to their location in the reindeer herding cooperatives of Käsivarsi, Näkkälä and Muonio (Figure 1).

This study focuses on three meteorological variables—the mean air temperatures, total precipitation and snow cover properties—whose variations have considerable effects on the reindeer life cycle and, therefore, on reindeer herding activities. In fact, air temperature is a crucial parameter for reindeer herders as warm springs affect the end of the hibernation period of predators, warm summers and autumns make reindeer physiologically stressed, and frost and thaw cycles alter the reindeer digging for lichen and thereby feeding [13,14,19]. The combination of low air temperatures and precipitation can have a very negative effect on winter-feeding conditions; on the other hand, high moisture rate is essential in autumn for the growth of mushrooms, part of the reindeer diet [13]. Finally, snow cover properties are a key element for reindeer seasonal migrations [26], for the accessibility to ground lichen [22,23,27–29] and for the calving period [13].

Daily data for each variable for the common reference period 1960–2011 (year 1982 is missing) were analysed for determining the regional homogeneity, the interannual and seasonal variability, and the cold intensity.

2.1. Analysing the Regional Homogeneity

The Pearson correlation coefficients was calculated to analyse the regional homogeneity of the three meteorological variables: the mean air temperatures (annual, winter, spring, summer and autumn mean temperatures), the precipitation (total precipitation in the year, winter, spring, summer and autumn precipitation), the duration of permanent snow cover in number of days (longest snow cover with a minimum thickness of one centimetre), and the maximal snow depth in centimetres. This analysis is an indicator to determine if data is homogeneous within the study area and therefore will be treated at the regional level (regional climate variability); or heterogeneous and will be treated at the local level (local climate variability).

2.2. Analysing the Interannual and Seasonal Variability

The analysis of the interannual and seasonal variability of the three selected meteorological variables at a regional or local scale was conducted in order to outline some climatic trends. Anomaly data were calculated and used to easily compare the different meteorological stations. The anomalies represent the deviations from the 1960–2011 average temperature, average of total precipitation and average length (number of days) of permanent snow cover with a thickness of at least one centimetre, for each site.

We also inspected the existence and location of a significant change point in the annual temperature variation utilising the Lepage test [30], that is, a nonparametric location-scale test that detects significant differences between two samples. Here, the first sample consists of the temperatures before or at the change point and the second sample consists of the temperatures after the change point. As the location of the change point is not known in advance, the Lepage test statistic is computed for each year, and the maximum of the absolute values of the standardised statistics is used as test statistics whose distribution is estimated by simulation [31]. The test was conducted using the R software [32] package cpm [31].

At the regional scale, the climate variability is studied in correlation with two atmospheric circulation patterns, the North Atlantic Oscillation (NAO) and the Arctic Oscillation (AO), in order to show the possible impacts of larger scale atmospheric phenomena on the variability of the mean air temperatures in Fell Lapland. Whereas the NAO is characterised by a dipole in the atmospheric pressure at sea level between the Azores high and Iceland low [33–35], the AO "can be interpreted as the surface signature of modulations in the strength of the polar vortex aloft" [36] (p. 1297). Both patterns can be measured through the NAO and AO indices. The annual NAO and AO indices have been calculated from the monthly mean NAO and AO indices obtained from the online data sets of the Climate Prediction Centre of NOAA [37]. Correlations between the monthly average temperature anomalies and the monthly NAO and AO indices were calculated for the period 1960–2011. At the local scale, the climate variability is examined at interannual and seasonal time scales by displaying the variations in anomalies of variables for Kilpisjärvi, Enontekiö and Muonio. For some of the charts, the first (Q1) and third (Q3) quartiles were calculated in order to detect extreme cases.

2.3. Analysing the Cold Intensity

The negative sum of degree-days and the number of frost and thaw cycles were determined to analyse the intensity of the cold, which is an essential parameter for reindeer herding. The season starts on 1 July and ends on 30 June. The first quartile (Q1) was calculated for each of the three stations.

When analysing data by seasons, the four seasons are referred as follow: December, January and February for the winter; March, April and May for the spring; June, July and August for the summer; September, October and November for the autumn. Data is analysed for the available years; some years are missing sporadically.

3. Results

3.1. Regional Climate Homogeneity

In order to determine if the regional climate of Fell Lapland is spatially homogeneous, the relationship between the three meteorological stations of Kilpisjärvi, Enontekiö and Muonio is analysed by performing various regression analyses. Shown in Table 2, the Pearson correlation coefficients were calculated for the three major variables of the local climate (the numbers with the highest correlation coefficients are in bold).

Table 2. Pearson correlation coefficients (r) between the different meteorological stations concerning the average temperatures, precipitation and snow conditions.

Mean Temperatures							
Winter	K [1]	E [2]	M [3]	Summer	K	E	M
K	1				1		
E	0.959	1			0.929	1	
M	0.943	0.989	1		0.932	0.984	1
Spring				Autumn			
K	1				1		
E	0.943	1			0.957	1	
M	0.947	0.988	1		0.901	0.976	1
Annual							
K	1						
E	0.977	1					
M	0.967	0.982	1				

Precipitation							
Winter	K	E	M	Summer	K	E	M
K	1				1		
E	0.206	1			0.629	1	
M	0.383	0.639	1		0.717	0.827	1
Spring				Autumn			
K	1				1		
E	0.413	1			0.402	1	
M	0.486	0.583	1		0.570	0.604	1
Total							
K	1						
E	0.31	1					
M	0.568	0.6	1				

Snow							
Permanent snow cover >1	K	E	M	Maximal snow depth	K	E	M
K	1			K	1		
E	0.6	1		E	−0.053	1	
M	0.581	0.77	1	M	0.296	0.631	1

[1] Kilpisjärvi, [2] Enontekiö, [3] Muonio.

Explained by the latitudinal location of the station, it was found that Kilpisjärvi is slightly colder than Enontekiö and Muonio with a mean annual temperature anomaly between Kilpisjärvi and Enontekiö of −0.05 °C and between Kilpisjärvi and Muonio of −0.91 °C. Nevertheless, the lower degree of continentality of Kilpisjärvi (24.48 °C) indicates that the presence of the warm waters of the North Atlantic Current is felt a bit more prominently there than in Enontekiö (27.33 °C) and Muonio (28.73 °C), although the observed average temperature is colder. Overall, although small disparities in temperatures exist between the three sites, the temperature varies similarly seasonally and annually (Table 2) and mean temperatures at the three sites are highly correlated (r > 0.9).

There was a modest regional heterogeneity in the distribution of precipitation (455.81 mm/year of average total precipitation for the period 1960–2011 at Kilpisjärvi, 443.14 mm/year at Enontekiö, and 489.42 mm/year at Muonio) and the existence of disparities in the precipitation variability. Table 2 shows that the three sites are moderately correlated annually and seasonally with the exception of the summertime (0.63 < r < 0.83). Again, Muonio and Enontekiö show the closest relationship.

Permanent snow cover revealed fairly strong correlations (r > 0.58) though not as high as the mean temperatures, it was found that Kilpisjärvi was poorly correlated to the two other sites concerning the maximal snow depth (r = −0.053 with Enontekiö and r = 0.296 with Muonio).

Overall, the results show a very close relationship between the stations for the temperatures and a weaker one for precipitation and snow conditions.

Based on those results, the temperature variability can be studied at a regional scale by using the mean temperatures of Kilpisjärvi, Enontekiö and Muonio meteorological stations (Section 3.2). On the other hand, disparities in precipitation and snow cover between the three sites are great enough that their variability must be examined at a local scale, that is, site by site (Section 3.3).

3.2. Regional Climate Variability

3.2.1. Annual Temperature Variability

An examination of the annual temperature variability and anomalies of the three sites shows that there is a slight discernible warming trend from 1960 to 2011 (Figure 2), the average annual temperature for 1960–2011 being −1.82 °C. More specifically, there appears to be a temperature shift from colder to warmer temperatures at the end of the 1980s. To confirm this shift, we performed a test based on the Lepage statistic that detected a statistically significant change point in the time series of temperature. Indeed the absolute values of standardised Lepage statistics show a peak in the year 1988 on a significance level of 0.01 (Figure 2a). The mean temperatures of the colder (from 1960 to 1988) and the warmer (from 1989 to 2011) periods are, respectively, −2.37 °C and −1.17 °C. The two periods are also discernible by the number of extreme years represented by the 25% of coldest (<Q1) and warmest years (>Q3). Indeed, the period 1960–1988 recorded the coldest annual temperatures (nine years < Q1 and none for the period 1989–2011) and the period 1989–2011 most of the warmest ones (10 years > Q3 and only two for the period 1960–1988).

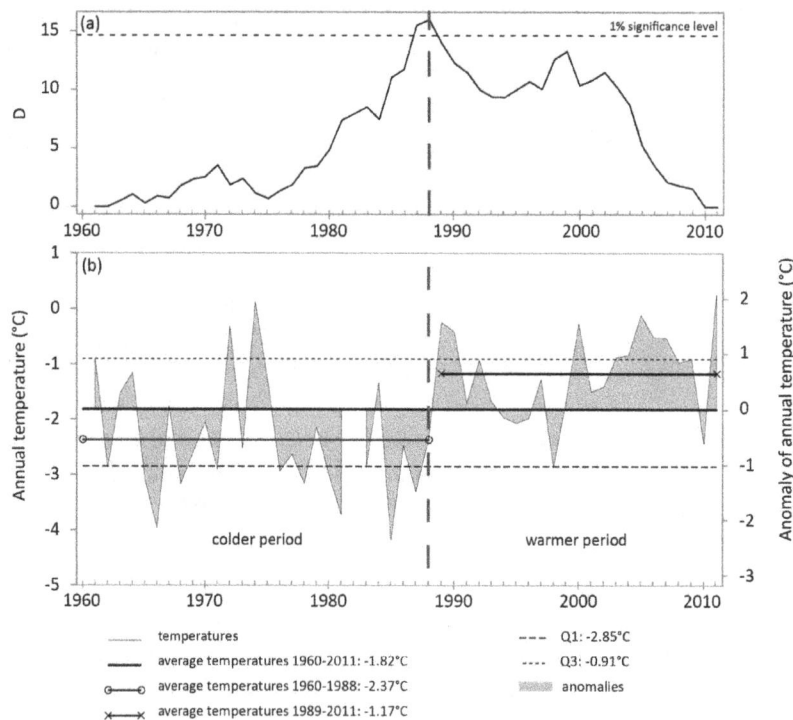

Figure 2. (**a**) Time series of the absolute values of standardised Lepage test statistics, denoted D. The year the maximum is attained is shown as vertical dashed line; (**b**) Regional average temperatures (°C) and anomalies for 1960–2011 (combination of the three weather stations). Note that year 1982 is missing.

3.2.2. Seasonal Temperature Variability

Results show that the temperature anomalies are greater during the coldest season (from November to March), that is, a quite high variability with a maximum in December, January and February (Figure 3). On the other hand, the warmest season (from April to October) is represented by weaker fluctuations.

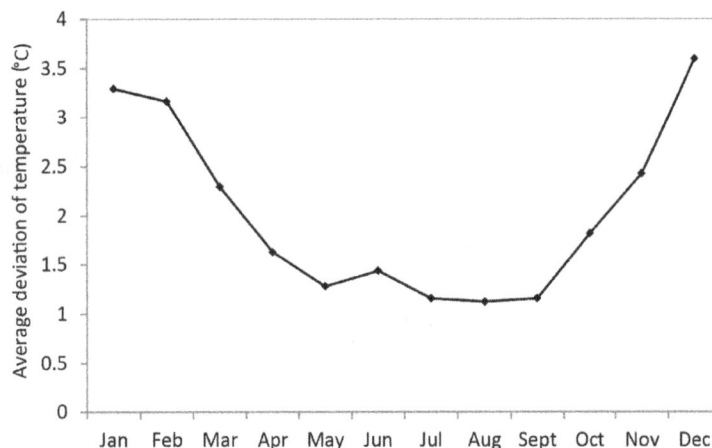

Figure 3. Monthly variability of temperature anomalies for the period 1960–2011 (combination of the three weather stations).

Both atmospheric patterns have a greater influence on temperature variations during the cold months in Fell Lapland with a maximum reached in March (Figure 4). Nevertheless, the AO index is slightly more associated with temperature variations than the NAO in winter.

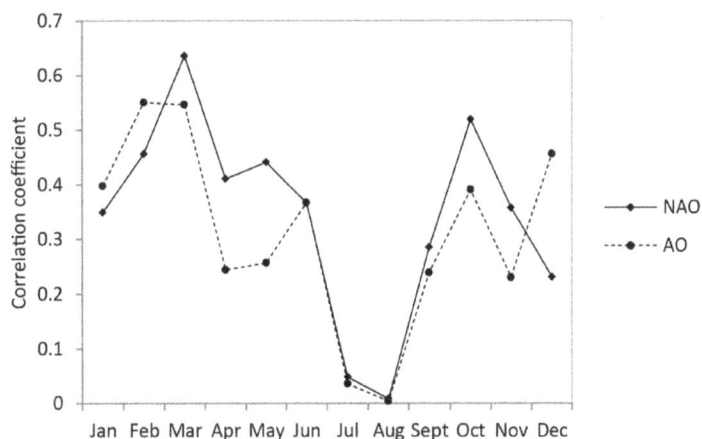

Figure 4. Monthly variation of the correlation between the monthly average temperature anomalies and the monthly North Atlantic Oscillation (NAO) and Arctic Oscillation (AO) indices for the period 1960–2011.

3.3. Local Climate Variability

3.3.1. Interannual Variability of Precipitation and Snow Cover

With the lowest average amount of precipitation for the period 1960–2011, Enontekiö has the particularity of showing certain variability without a clear trend (Figure 5). On the other hand, the annual anomalies of total precipitation of the sites of Kilpisjärvi and Muonio show a clear trend: most of the negative anomalies, meaning that it rained less than the average, occurred before

1988; and after that most of the years have shown a net tendency to receive more precipitation. The calculations of the first and third quartiles also confirm these tendencies: the period 1961–1988 recorded the driest years (11 years < Q1 for Kilpisjärvi and only two years for the period 1989–2011; 10 years < Q1 for Muonio and only three years for the period 1989–2011) and 1989–2011 the wettest ones (10 years > Q3 for Kilpisjärvi and only three years for the period 1960–1988; 10 years > Q3 for Muonio and only three years for the period 1960–1988).

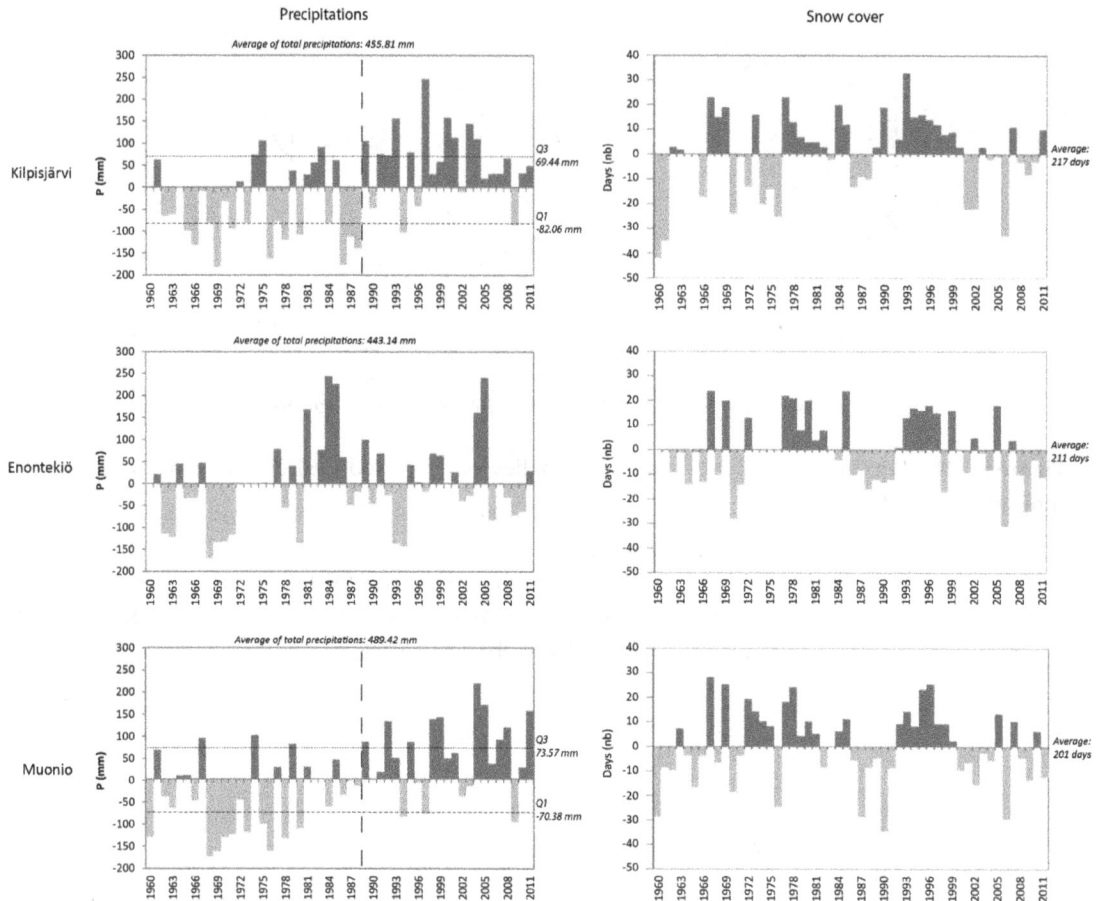

Figure 5. Annual anomaly of the total precipitation (mm) and permanent snow cover >1 cm (number of days) at Kilpisjärvi, Enontekiö and Muonio, 1960–2011. The shift point is shown as vertical dashed line.

Concerning the number of days of permanent snow cover, the average number of days gets higher northwards, namely, 217 days in Kilpisjärvi against 201 days in Muonio (Figure 5). The latitudinal positions clearly explain the differences in duration of snow cover. None of the sites shows a discernible trend but instead an irregular succession of positive and negative anomalies. At Kilpisjärvi, Enontekiö and Muonio, the snow cover has had the tendency to last, respectively, 8, 7 and 9 days shorter since the beginning of the 21st century compared to the period 1960–2000.

3.3.2. Seasonality of Interannual Variability in Precipitation

Results from the annual anomaly of precipitation for the four seasons at Kilpisjärvi, Enontekiö and Muonio reveal two main points (Figure 6). First, the site of Muonio is the one which records the largest amount of precipitation for all seasons except for the winter (Kilpisjärvi). Then, the summer precipitation is the most important for the three sites for the period 1960–2011: on average 154.91 mm for Kilpisjärvi, 178.81 mm for Enontekiö and 195.88 mm for Muonio; the driest season is the spring with, respectively, 74.54 mm, 70.65 mm and 83.64 mm.

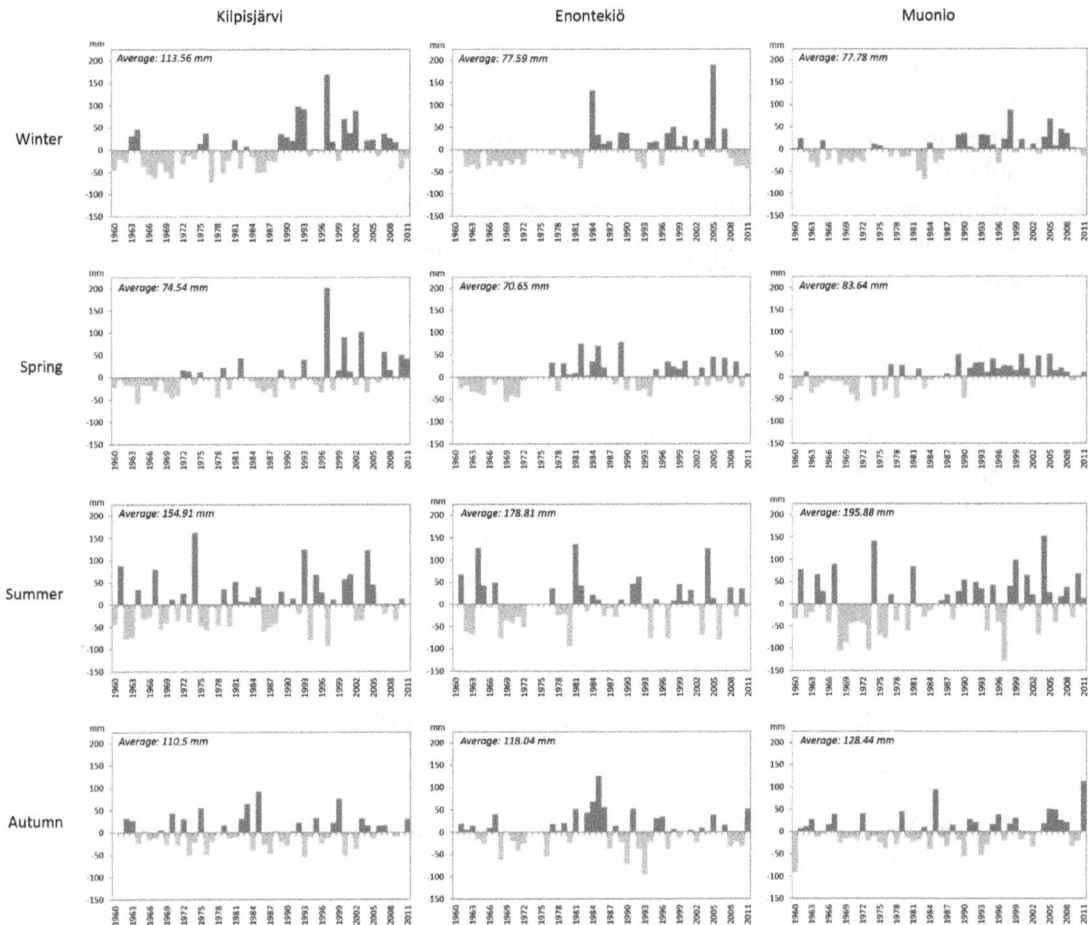

Figure 6. Annual anomaly of seasonal precipitation (mm) at Kilpisjärvi, Enontekiö and Muonio, 1960–2011.

Concerning the seasonal variability of precipitation, it can be observed that the amount and amplitude of fluctuations differ from site to site and from season to season:

1. In the winter, the amplitude of the anomalies, especially positive ones, has increased with the time for all cases, even though it is more visible for Kilpisjärvi. For the three sites, two distinct periods are discernible: the first one from the 1960s to the end of the 1980s (a bit earlier for Enontekiö, mid-1980s), annual anomalies were mostly negative meaning less precipitation; and the second one from the late 1980s onwards, annual anomalies have been positive at the exception of the late 2000s. This increase in average amount of precipitation needs to be interpreted in concordance with temperature trends meaning that the share of rainy precipitation has probably grown at the expense of snowfall.

2. In general, the variability of spring precipitation is quite low with the exception of the last decade in Kilpisjärvi. However, for the three sites, the first positive anomalies have been recorded in the 1970s and, since then, they have been much more frequent than the negative ones, meaning that springs get slightly wetter.

3. The summer is the most variable season in terms of precipitation, and this is the case in the three case studies. The similarity of the variability of the three stations was already shown in Table 2. No specific trends can be noticed.

4. Since the 1960s, the autumn season has recorded a quite low annual variability of precipitation for the three weather stations without any specific trend.

3.4. Cold Intensity

One of the major concerns of reindeer herders is the variability of temperatures. Combined with a high rate of precipitation, it can be harmful for their herds. Thus, negative temperatures are welcomed in the autumn only if the ground is dry; otherwise, reindeer lichen can be trapped under a solid layer of ice. In the winter, what reindeer herders fear the most is the variability of temperatures around zero, which affects the phase changes of the snow cover. During the birth season in the spring, cold temperatures are not preferable for reindeer calves. These are the reason why analysing degree-days and frost and thaw cycles are relevant for this study.

3.4.1. Degree-Days

The cold intensity can be represented by the sum of the degree-days below 0 °C. According to Jaagus [38], it is the most important indicator of the severity of a winter. This indicator is calculated by summing up all the negative mean daily temperatures of a cold season and by discarding the positive ones. In the case of Kilpisjärvi and Muonio (Figure 7), it is clear that the cold was more intense before 1990 in view of the number of years that recorded a negative degree-days sum inferior to -2072.75 and -2190.15, respectively. The winter of 1966 was definitely the coldest one. Since the beginning of the 1990s, none of the years has recorded a sum inferior to the first quartile limit. For Enontekiö, the data range was smaller, but every year before 1990 recorded a sum of negative degree-days below the first quartile limit set at -2067.25. Finally, the cold intensity is not always related with the length of the winter season. The interannual and long-term variability of degree-days is logically correlated to temperature variability ($r > 0.67$ for each of the three stations): the shift point appears at the same time. Moreover, it is also more strongly correlated to the AO index variability ($r > 0.55$ for each of the three stations) than the NAO index variability ($r < 0.43$ for each of the three stations), as winter temperatures depend more on the AO (see Figure 3).

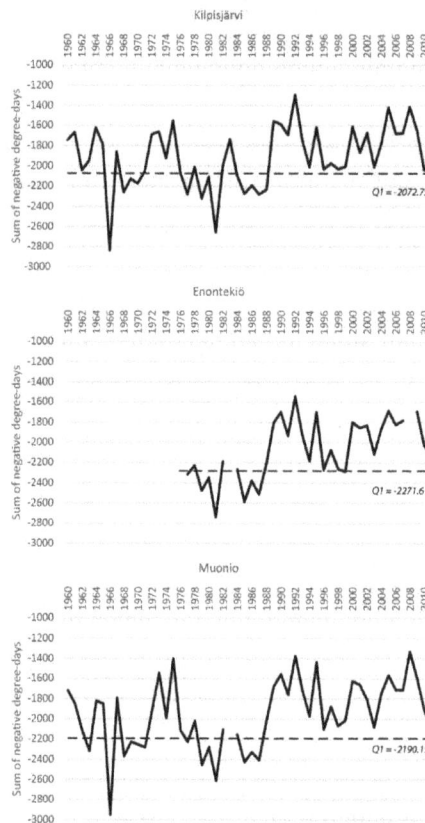

Figure 7. Sum of negative degree-days at Kilpisjärvi Enontekiö and Muonio, 1960–2011.

3.4.2. Frost and Thaw Cycles

Frost and thaw cycles during the autumn and winter seasons are badly perceived by reindeer herders as they greatly affect lichen digging. Frost and thaw cycles mean the periods when temperatures pass from positive to negative, and vice versa. The more frost and thaw cycles there are, the more difficult it is to access the food that might be covered by a hard layer of ice. Figure 8 shows the number of frost and thaw cycles based on daily average temperatures from 1 July to 30 June for the three case stations. Each case represents a distinct trend:

- at Kilpisjärvi weather station, the average of cycles is approx. 11 per year. Among the three stations, it is the one that normally records the most of cycles every year, up to 19 in 2003. Since the 1990s, a slight increase of those cycles has occurred.
- at Enontekiö, whose average is about nine cycles per year, the increase is very clear even though the station does not record as many frost and thaw cycles as the two other stations.
- at Muonio, the number of cycles flows around the average of 10 per year without any discernible trend even for recent years.

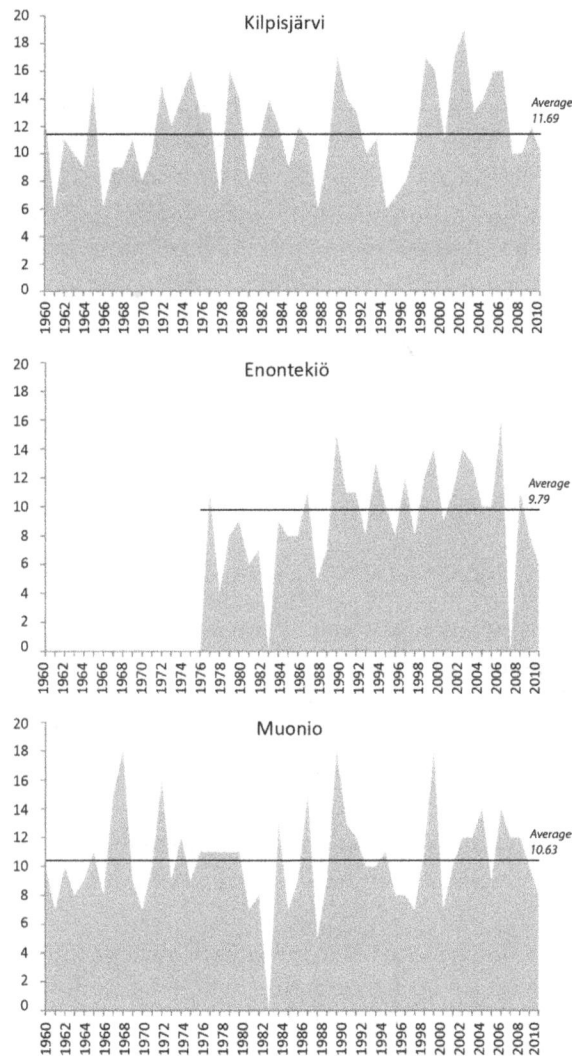

Figure 8. Number of frost and thaw cycles based on the daily average temperatures at Kilpisjärvi, Enontekiö and Muonio, 1960–2011.

4. Discussion

Spatial and temporal variations of climate of Finnish Fell Lapland in Northern Finland has been analysed over the last 50 years and have revealed changes in temperatures, precipitation and snow cover for three weather stations located in the Finnish reindeer husbandry area. The following conclusions can be drawn:

- It is difficult to speak about full climate regional homogeneity when only mean temperatures have revealed quite good correlation coefficients (r > 0.97), unlike precipitation and snow cover. Nevertheless, for all cases, annual and seasonal, Enontekiö and Muonio weather stations are the closest ones.

- For the annual temperature data, a warming trend is discernible for the period 1960–2011 with two distinct periods: a colder period between 1960 and 1988 with greater negative anomalies and strong cold intensity (lowest negative peaks of degree-days recorded before 1990); and a warmer period from 1989 to 2011 with greater positive anomalies and +1.20 °C of average temperatures compared to the previous period. It can also be seen that this most recent period has recorded slightly more frost and thaw cycles whose high occurrence is strongly linked to the warming of the climate at the high latitudes. As for seasonal data, variations in mean temperatures have been found to be more important during wintertime, partly due to the strong correlation with the AO index.

- Muonio is the weather station that has recorded the largest amount of precipitation. The annual data of precipitation have shown that a drier period occurred before 1988 for Kilpisjärvi and Muonio. Most of the positive anomalies have been recorded after 1988, meaning that the general trend is about receiving more precipitation. Nevertheless, no clear trend has appeared for Enontekiö. Concerning the seasonal variability of precipitation, the results have shown that summer is the season with the highest amplitude and number of variations for the three cases. As for the autumn, no specific trend is discernible. Winter and spring precipitation, however, has evolved differently, and Kilpisjärvi has stood out from the other weather stations. For both seasons, Kilpisjärvi has recorded larger amplitude of positive anomalies compared to Enontekiö and Muonio despite the quite low variability of precipitation. For both seasons and for the three sites, precipitation is increasing with time with more and more rainy winter precipitation, especially from the beginning of the 1990s. A slight decrease has been recorded from the beginning of the 2000s.

- None of the sites shows a discernible trend in the snow cover.

Besides the undeniable correlation between temperature and precipitation type and snow cover properties, the climate variability and shifts in Finnish Lapland is explained by the local geography and also by the impacts of the ocean-atmosphere coupling.

Indeed, with a sub-meridian orientation, the Scandinavian Mountains, which peak at 2469 m above sea level, induce important climatic contrasts on both sides of the mountain range. While the oceanic climate is confined at the Norwegian mountainside and, therefore, brings abundant precipitation, the drying effect of the Foehn wind plays an essential role in precipitation distribution on the eastern mountainside, especially in the Swedish coasts and in Lapland where precipitation is lower.

Moreover, the large-scale atmospheric phenomena have a great influence on the local and regional climate variability. Fell Lapland is located at relatively high latitudes and, as most of the northern regions, it experiences "the largest fluctuations due to [its] position in the transition zone between the west wind belt and the Arctic climate" [39] (p. 508). In fact, the position of Northern Europe between the ocean and the continent and between the tropical and polar regions favours the confluence of multiple air masses (Figure 9). Thus, this is a meeting place for the maritime polar air masses, cold and moist air from south of Greenland; the maritime tropical air masses, warm and moist air that is important for the energy transfer from the warm Atlantic waters to Northern Europe; the continental polar air masses that originate from the thermal high stabilised in Russia; the maritime Arctic air masses responsible for the coldness from February to the arrival of the spring; and the continental Arctic air masses that get formed in Lapland starting in the autumn [40]. Therefore, the climate variability is strongly governed by the position and the dynamic of the action centres. In Northern Europe, their distribution is affected by oscillations in pressure patterns that lead to change in temperature patterns. Even though it has been demonstrated that the NAO "is the most prominent and recurrent pattern of atmospheric variability over the middle and high latitude of the Northern Hemisphere, especially during the cold season" [41] (p. 113), and therefore has a major influence on winter temperatures as Hurrell [28] has pointed out, those results are in line with findings of previous research studies led in Finland. In fact, Irannezhad et al. [11] showed the significant influence of the AO in the winter in most areas of Finland and of the NAO in the spring in Northern Finland. Furthermore, the time series of the NAO and AO indices show a higher peak at the end of the 1980s that could be linked to the 1988 shift in temperature and precipitation time series. Also, some studies have demonstrated the linkages between the sea surface temperature (SST) variability of the North Atlantic [42] and the Baltic Sea [43] and the NAO, making the SST variability of both water bodies be a possible factor for the climate variability of Finnish Lapland. For instance, Buchan et al. [44] have shown the impacts of the North Atlantic SST anomalies on Northern cold weather events.

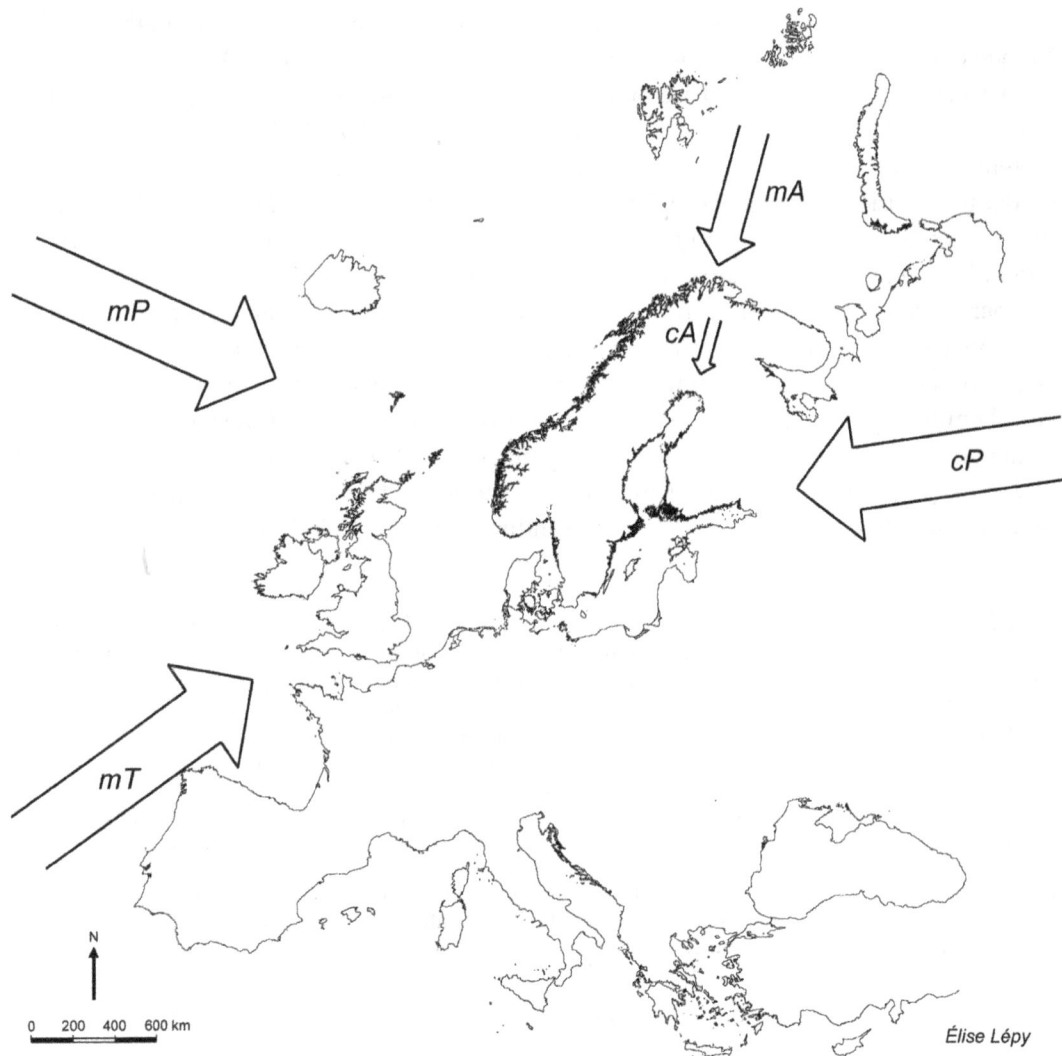

Figure 9. Air masses influencing the climate of Northern Europe (mP: maritime polar; mT: maritime tropical; mA: maritime arctic; cA: continental Arctic; cP: continental polar).

A time span of 50 years is too short to get an overview of long-term changes in the past for predicting the future climates, but this study can be considered as a complement to the works led by Lee et al. [9] in the eastern part of Finnish Lapland for the period 1876–1993. Interpretations of data might differ since great changes have occurred at the beginning of the 1990s emphasising warming trends, more precipitation and a slight diminution of snow cover in general. Indeed, Lee et al. [9] have not noticed any significant warming or cooling period between 1876 and 1993. As for the precipitation, they concluded that there was a significant increase for the winter, spring and autumn seasons, while this study confirms only the two first seasons.

Even though future climate changes will manifest themselves differently in different locations, we can attempt to draw a parallel between this study and others conducted in other Finnish reindeer management areas. Thus, Turunen et al. [15] have carried out climate model simulations for the period 2035–2064 in Sodankylä, and they have concluded that winters "will be characterized by ephemeral snow cover formation and melting several times during the winter" with a delay of the seasonal snow cover formation and earliness of snow cover melting. However, these results are also consistent with other studies conducted in Northern Fennoscandia, especially the ones of Kivinen et al. [45] reporting warming trends with an increase of extreme warm events and a decrease of extreme cold events (Figure 2), among other studies [46,47]. Moreover, the high variability of temperatures during the winter found in Figure 3 might explain the conclusion of Kivinen et al. [45] on the fact that any significant warming trends were detected in winter. The present results also confirm the conclusions of Lehtonen et al. [48] predicting an increase in winter precipitation in Northern Europe.

Disparities within Fell Lapland are maybe not huge but reindeer herders still have to cope with different weather situations according to their geographical location. Indeed, this study has shown, by different means, the particularity of Kilpisjärvi weather station that can be explained by the very close proximity of the Lake Kilpisjärvi. In their study on the influence of natural conditions on the spatial variation of climate in Lapland, Vajda and Venäläinen [24] have shown that lakes situated in Lapland influence the microclimate of their surroundings. In general, the parallel between reindeer herders' perceptions and climatic trends is interesting. In fact, reindeer herders have estimated that shorter snow season and thinner snow cover in the winter would be advantageous for reindeer to access ground lichen, and more summer precipitation would also boost the vegetation growth and particularly mushrooms that reindeer eat [13,14]. However, an increase of frost and thaw cycles and heavy rain on snow in the winter are very badly perceived by herders, as it would have implications on snow cover properties and ground-icing blocking the access to winter food [29]. These changes in climate would also have other increasing adverse indirect effects on reindeer herding as the behaviour of predator populations and the development of competing land uses [19].

Acknowledgments: This study is part of the contribution to the WP8: Traditional livelihoods of the CLICHE (Impacts of climate change on Arctic environment, ecosystem services and society) project, part of the research programme on Climate Change (FICCA) launched by the Academy of Finland in 2011. Dr. Élise Lépy's work was supported by the PITCH project (*Primary Industries and Transformational Change*) funded by the Norwegian Research Council and the project "Understanding the cultural impacts and issues of Lapland mining: a long-term perspective on sustainable mining policies in the North" part of the Arctic Academy Programme (ARKTIKO) launched by the Academy of Finland in 2014. Dr. Leena Pasanen's work was supported by the EBOR project (*Ecological history and long-term dynamics of the Boreal forest ecosystem*) funded by the Academy of Finland. The authors also thank the two anonymous reviewers for their insightful comments and suggestions.

Author Contributions: Élise Lépy is responsible for the acquisition of data, the conception and design of the work, and the statistical analyses and interpretations of climate data at the exception of the Lepage test; she also drafted the manuscript as well as all tables and figures. Leena Pasanen has conducted and interpreted the results from the Lepage test. Both authors were involved in the final editorial revisions for publication.

Conflicts of Interest: The authors declare no conflict of interest.

References

1.   ACIA. *Arctic Climate Impact Assessment Scientific Report*; Cambridge University Press: Cambridge, UK, 2005.

2.   Barber, D.; Lukovich, J.V.; Keogak, J.; Baryluk, S.; Fortier, L.; Henry, G.H.R. The changing climate of the Arctic. *Arctic* **2008**, *61*, 7–26. [CrossRef]

3.   Larsen, J.N.; Anisimov, A.; Constable, A.; Hollowed, A.B.; Maynard, N.; Prestrud, P.; Prowse, T.D.; Stone, J.M.R. Polar regions. In *Climate Change 2014: Impacts, Adaptation, and Vulnerability. Part B: Regional Aspects. Contribution of Working Group II to the Fifth Assessment Report of the Intergovernmental Panel on Climate Change*; Barros, V.R., Field, C.B., Dokken, D.J., Mastrandrea, M.D., Mach, K.J., Bilir, T.E., Chatterjee, M., Ebi, K.L., Estrada, Y.O., Genova, R.C., et al., Eds.; Cambridge University Press: Cambridge, UK; New York, NY, USA, 2014; pp. 1567–1612.

4.   Rautio, A.; Poppel, B.; Young, K. Human health and well-being. In *Arctic Human Development Report. Regional Processes and Global Linkages*; Nymand, L., Fondahl, G., Eds.; Nordic Council of Ministers: Copenhagen, Denmark, 2015; pp. 297–346.

5.   Adger, W.N. Vulnerability. *Glob. Environ. Chang.* **2006**, *16*, 268–281. [CrossRef]

6.   Folke, C. Resilience: The emergence of a perspective for social-ecological systems analyses. *Glob. Environ. Chang.* **2006**, *16*, 253–267. [CrossRef]

7.   Callaghan, T.V.; Bergholm, F.; Christensen, T.R.; Jonasson, C.; Kokfelt, U.; Johansson, M. A new climate era in the sub-Arctic: Accelerating climate changes and multiple impacts. *Geophys. Res. Lett.* **2010**, *37*, L14705. [CrossRef]

8.   Stocker, T.F.; Qin, D.; Plattner, G.K.; Tignor, M.; Allen, S.K.; Boschung, J.; Nauels, A.; Xia, Y.; Bex, V.; Midgley, P.M. (Eds.) *IPCC Climate Change 2013: The Physical Science Basis. Contribution of Working Group I to the Fifth Assessment Report of the Intergovernmental Panel on Climate Change*; Cambridge University Press: Cambridge, UK; New York, NY, USA, 2013.

9.   Lee, S.E.; Press, M.C.; Lee, J.A. Observed climate variations during the last 100 years in Lapland, Northern Finland. *Int. J. Climatol.* **2000**, *20*, 329–346. [CrossRef]

10. Pike, G.; Pepin, N.C.; Schaefer, M. High latitude local scale temperature complexity: The example of Kevo Valley, Finnish Lapland. *Int. J. Climatol.* **2013**, *33*, 2050–2067. [CrossRef]

11. Irannezhad, M.; Marttila, H.; Kløve, B. Long-term variations and trends in precipitation in Finland. *Int. J. Climatol.* **2014**, *34*, 3139–3153. [CrossRef]

12. Irannezhad, M.; Chen, D.; Kløve, B. Interannual variations and trends in surface air temperature in Finland in relation to atmospheric circulation patterns, 1961–2011. *Int. J. Climatol.* **2014**, *35*, 3078–3092. [CrossRef]

13. Heikkinen, H.I.; Kasanen, M.; Lépy, É. Resilience, vulnerability and adaptation in reindeer herding communities in the Finnish-Swedish border area. *Nordia Geogr. Publ.* **2013**, *41*, 107–122.

14. Lépy, É. Perceptions des éleveurs de rennes de Laponie finlandaise. *Communications* **2017**, *2*, 47–61.

15. Turunen, M.T.; Rasmus, S.; Bavay, M.; Ruosteenoja, K.; Heiskanen, J. Coping with difficult weather and snow conditions: Reindeer herders' views on climate change impacts and coping strategies. *Clim. Risk Manag.* **2016**, *11*, 15–36. [CrossRef]

16. Rasmus, S.; Kumpula, J.; Jylhä, K. Suomen poronhoitoalueen muuttuvat talviset sää- ja lumiolosuhteet. *Terra 126* **2014**, *4*, 69–185.

17. Riseth, J.A.; Tømmervik, H.; Bjerke, J.W. 175 years of adaptation: North Scandinavian Sámi reindeer herding between government policies and winter climate variability (1835–2010). *J. For. Econ.* **2016**, *24*, 186–204. [CrossRef]

18. Jylhä, K.; Ruosteenoja, K.; Räisänen, J.; Venäläinen, A.; Tuomenvirta, H.; Ruokolainen, L.; Saku, S.; Seitola, T. *The Changing Climate in Finland: Estimates for Adaptation Studies. ACCLIM Project Report 2009*; Finnish Meteorological Institute: Helsinki, Finland, 2009.

19. Lépy, É.; Heikkinen, H.; Komu, T.; Sarkki, S. Participatory meaning-making of environmental and cultural changes in reindeer herding in the northernmost border area of Sweden and Finland. *Int. J. Bus. Glob.* **2017**, in press.

20. Jernsletten, J.L.; Klokov, K. *Sustainable Reindeer Husbandry*; Centre for Saami Studies: Tromsø, Norway, 2002.

21. Forbes, B.; Bölter, M.; Müller-Wille, L.; Hukkinen, J.; Müller, F.; Gunslay, N.; Konstatinov, Y. (Eds.) *Reindeer Management in Northernmost Europe*; Ecological Studies 184; Springer: Berlin, Germany, 2006.

22. Helle, T.; Kojola, I. Demographics in an alpine reindeer herd: Effects of density and winter weather. *Ecography* **2008**, *31*, 221–230. [CrossRef]

23. Riseth, J.A.; Tømmervik, H.; Helander-Renvall, E.; Labba, N.; Johansson, C.; Malnes, E.; Bjerke, J.W.; Jonsson, C.; Pohjola, V.; Sarri, L.E.; et al. Sámi traditional ecological knowledge as a guide to science: Snow, ice and reindeer pasture facing climate change. *Polar Rec.* **2011**, *47*, 202–217. [CrossRef]

24. Vajda, A.; Venäläinen, A. The influence of natural conditions on the spatial variation of climate in Lapland, Northern Finland. *Int. J. Climatol.* **2003**, *23*, 1011–1022. [CrossRef]

25. Kitti, H.; Gunslay, N.; Forbes, B.C. Defining the quality of reindeer pastures: The perspectives of Sámi reindeer herders. In *Reindeer Management in Northernmost Europe*; Ecological Studies 184; Forbes, B.C., Bölter, M., Müller-Wille, L., Hukkinen, J., Müller, F., Gunslay, N., Konstatinov, Y., Eds.; Springer: Berlin, Germany, 2006; pp. 141–165.

26. Côté, S.; Festa-Bianchet, M.; Dussault, C.; Tremblay, J.P.; Brodeur, V.; Simard, M.; Taillon, J.; Hins, C.; Le Corre, M.; Sharma, S. Caribou herd dynamics: Impacts of climate change on traditional and sport harvesting. In *Nunavik and Nunatsiavut: From Science to Policy. An Integrated Regional Impact Study (IRIS) of Climate Change and Modernization*; Allard, M., Lemay, M., Eds.; ArcticNet Inc.: Quebec City, QC, Canada, 2012; pp. 249–269.

27. Turunen, M.; Soppela, P.; Kinnunen, H.; Sutinen, M.L.; Martz, F. Does climate change influence the availability and quality of reindeer forage plants? *Polar Biol.* **2009**, *32*, 813–832. [CrossRef]

28. Bartsch, A.; Kumpula, T.; Forbes, B.; Stammler, F. Detection of snow surface thawing and refreezing in the Eurasian Arctic with QuikSCAT: Implications for reindeer herding. *Ecol. Appl.* **2010**, *20*, 2346–2358. [CrossRef] [PubMed]

29. Hansen, B.B.; Aanes, R.; Herfindal, I.; Kohler, J.; Sæther, B.E. Climate, icing, and wild arctic reindeer: Past relationships and future prospects. *Ecology* **2011**, *92*, 1917–1923. [CrossRef] [PubMed]

30. Lepage, Y. A combination of Wilcoxon's and Ansari-Bradley's statistics. *Biometrika* **1971**, *58*, 213–217. [CrossRef]

31. Ross, G.J. Parametric and nonparametric sequential change detection in R: The cpm package. *J. Stat. Softw.* **2015**, *66*, 1–20.

32. R Core Team. *R: A Language and Environment for Statistical Computing*; R Foundation for Statistical Computing: Vienna, Austria, 2014.

33. Rogers, J.C. The association between the North Atlantic Oscillation and the South Oscillation in the Northern Hemisphere. *Mon. Weather Rev.* **1984**, *112*, 1999–2015. [CrossRef]

34. Hurrell, J.W. Decadal trends in the North Atlantic Oscillation: Regional temperatures and precipitation. *Science* **1995**, *269*, 676–679. [CrossRef] [PubMed]

35. Cohen, J.; Barlow, M. The NAO, the AO, and global warming: How closely related? *J. Clim.* **2005**, *18*, 4498–4513. [CrossRef]

36. Thompson, D.W.J.; Wallace, J.M. The Arctic Oscillation signature in the wintertime geopotential height and temperature fields. *Geophys. Res. Lett.* **1998**, *25*, 1297–1300. [CrossRef]

37. NOAA. Available online: http://www.cpc.ncep.noaa.gov/products/precip/CWlink/pna/nao.shtml (accessed on 30 June 2017).

38. Jaagus, J. Trends in sea ice conditions in the Baltic Sea near the Estonian coast during the period 1949/1950–2003/2004 and their relationships to large-scale atmospheric circulation. *Boreal Environ. Res.* **2006**, *11*, 169–183.

39. Chen, D.; Hellström, C. The influence of the North Atlantic Oscillation on the regional temperature variability in Sweden: Spatial and temporal variations. *Tellus* **1999**, *51A*, 505–516. [CrossRef]

40. Lépy, É. Les Glaces de mer en mer Baltique. Étude Géographique et Implications Environnementales et Sociétales à Partir de l'étude Comparée de la Baie de Botnie (Oulu, Finlande) et du Golfe de Riga (Lettonie). Doctor's Thesis, University of Caen-Basse-Normandie, Caen, France, 2009.

41. Hurrell, J.W.; Deser, C. Northern hemisphere climate variability during winter: Looking back on the work of Felix Exner. *Meteorol. Z.* **2015**, *24*, 113–118. [CrossRef]

42. Gastineau, G.; Frankignoul, C. Influence of the North Atlantic SST variability on the atmospheric circulation during the twentieth century. *J. Clim.* **2015**, *28*, 1396–1416. [CrossRef]

43. Stramska, M.; Bialogrodzka, J. Spatial and temporal variability of sea surface temperature in the Baltic Sea based on 32-years (1982–2013) of satellite data. *Oceanologia* **2015**, *57*, 223–235. [CrossRef]

44. Buchan, J.; Hirschi, J.J-M.; Blaker, A.T.; Sinha, B. North Atlantic SST anomalies and the cold North European weather events of winter 2009/10 and December 2010. *Mon. Weather Rev.* **2014**, *142*, 922–932. [CrossRef]

45. Kivinen, S.; Rasmus, S.; Jylhä, K.; Laapas, M. Long-term climate trends and extreme events in Northern Fennoscandia (1914–2013). *Climate* **2017**, *5*, 16. [CrossRef]

46. Beniston, M.; Stephenson, D.B.; Christensen, O.B.; Ferro, C.A.; Frei, C.; Goyette, S.; Halsnaes, K.; Holt, T.; Jylhä, K.; Koffi, B.; et al. Future extreme events in European climate: An exploration of regional climate model projections. *Clim. Chang.* **2007**, *81*, 71–95. [CrossRef]

47. Vikhamar-Schuler, D.; Isaksen, K.; Haugen, J.E.; Tømmervik, H.; Lucks, B.; Schuler, T.V.; Bjerke, J.W. Changes in winter warming events in the Nordic Arctic Region. *J. Clim.* **2016**, *29*, 6223–6244. [CrossRef]

48. Lehtonen, I.; Ruosteenoja, K.; Jylhä, K. Projected changes in European extreme precipitation indices on the basis of global and regional climate model ensembles. *Int. J. Climatol.* **2014**, *34*, 1208–1222. [CrossRef]

6

Deconstructing Global Temperature Anomalies: An Hypothesis

Norman C. Treloar

540 First Avenue West, Qualicum Beach, BC V9K 1J8, Canada; norman.treloar@gmail.com

Abstract: This paper evaluates contributions to global temperature anomalies from greenhouse gas concentrations and from a source of natural variability. There is no accepted causation for the apparent interrelationships between multidecadal oscillations and regime changes in atmospheric circulation, upwelling, and the slowdowns in global surface temperatures associated with a ~60-year oscillation. Exogenous tidal forcing is hypothesized as a major causal agent for these elements, with orthogonal components in tidal forcing generating zonal and meridional regime-dependent processes in the climate system. Climate oscillations are simulated at quasi-biennial to multidecadal timescales by tidal periodicities determined by close approaches of new or full moon to the earth. Subtracting a tidal analog of the ~60-year oscillation from global mean surface temperatures reveals an exponential component comparable with greenhouse gas emission scenarios, and which is responsible for almost 90% or contemporary global temperature increases. Residual subdecadal temperature anomalies correlate with the subdecadal variability of evolved carbon dioxide (CO_2), ENSO activity and tidal components, and indicate a causal sequence from tidal forcing to greenhouse gas (GHG) release to temperature increase. Tidal periodicities can all be expressed in terms of four fundamental frequencies. Because of the potential importance of this formulation, tests are urged using general circulation models.

Keywords: tidal forcing; temperature slowdowns; atmospheric circulation; zonal and meridional regimes; greenhouse gas emission scenarios

1. Introduction

1.1. Background

Global temperatures and sea surface temperatures (SSTs) vary over ranges that include multidecadal to quasi-biennial. A current controversy concerns the presence and nature of decadal-scale slowdowns in global temperature anomalies, particularly the recent slowdown from about 1998 to the present [1,2]. This slowdown "has provided the scientific community with a valuable opportunity to advance understanding of internal variability and external forcing ... " [3]. This paper will suggest a unifying and parsimonious physical (tidal) hypothesis to drive ocean/atmosphere variability and simulate the temporal variability of global temperature anomalies on timescales ranging from multidecadal to quasi-biennial. It is hoped that documenting the hypothesis would enable subsequent comparisons with other treatments.

Global temperature and its slowdowns have temporal patterns shared in other fields, for example the following: The Atmospheric Circulation Index (ACI) is a measure of winter wind-flow regimes from the Atlantic to West Siberia. Based on earlier work, Klyashtorin [4] classified the state of the ACI at a chosen time in terms of the accumulated difference between its zonal and meridional properties (herein referred to as Z-M). Using this formulation, the ACI exhibits a ~60-year oscillation, and regime change is signaled by a reversal in the accumulated ACI curve at a maximum or minimum. The temporal variation of detrended global temperature (and length of day, LOD) curves resemble the ACI Z-M parameter [4], although with differences of a few years in lead or lag times. The LOD varies over decades by milliseconds and is strongly correlated [5] with atmospheric angular momentum (AAM). In zonal wind regimes, the LOD increases when the earth's rotation is slowed by a change in the earth's mass distribution and an exchange of angular momentum between the earth's surface and its atmosphere [6], a process promoted by precipitation or by evaporation over the oceans [7].

Dickey et al. [8] noted the common decadal variability of the LOD, and the angular momentum of the earth's core and surface air temperatures, and found significant correlations with LOD leading model-corrected temperatures by eight years. They concluded that either oscillations in the core's magnetic field modulated atmospheric factors, or that some other indirect effect of another fundamental process affected climate, or that there was another process that affects both.

In this journal, Oviatt et al. [9] reviewed some of these decadal patterns, drawing attention to zonal and meridional regimes that affect global temperatures, and described the effects on temperature of ocean upwelling. Global temperatures rise during zonal ACI regimes, and "pause" during meridional regimes. Following discussions at a recent workshop on decadal variability [10], the word "slowdown" will be used herein instead of the words "pause" or "hiatus" that have often been used to describe the arrested temperature rises in meridional regimes.

Oviatt et al. [9] listed three hypotheses to explain decadal shifts: (1) atmospheric–ocean interaction dynamics; (2) the rate of Atlantic meridional overturning circulation (AMOC); and (3) a statistical combination of climate indices and Arctic sea ice variability. They concluded there was no accepted causation to explain the decadal changes.

The first hypothesis includes the suggestion that the hiatus is associated with the negative phase of the Interdecadal Pacific Oscillation (IPO), and the accompanying cooling and strong easterly winds over the equatorial Pacific. The second hypothesis includes the idea that the warming hiatus is a result of large heat uptake by the deep ocean. From large ensemble simulations, Liu et al. [11] showed that, in hiatus decades, the Indian Ocean shows anomalous warming and accelerated ocean heat content increase below 50 m. This is associated with a La Niña-like climate shift and enhanced heat transport of the Indonesian Throughflow, and the warming occurs concurrently with Pacific cooling. Meridional overturning and wind-driven decadal variability in ocean basins were possible proximate causes. For further discussion of ocean heat uptake and global temperatures, see Whitmarsh et al. [12].

Several authors have noted the apparent relationship between the successive rises and pauses in global temperatures on the one hand and the temperatures and phases in Pacific-basin oscillations on the other (e.g., [1,2,13]). However, just ten of 262 model simulations with the Coupled Model Intercomparison Project Phase 5 (CMIP5) produced the early-2000s slowdown in global surface temperature rise from the decadal modulation associated with an IPO negative phase [14]).

Such empirical or computer-modeled relationships beg the question: What is the ultimate driver for ocean and climate variability? The seemingly close relationships between such oceanic, geophysical and climatic parameters may derive either from internally-generated and teleconnected terrestrial endogenous processes or through an external forcing mechanism. The possible sources of influences external to the climate system have generally focused on aerosols, volcanoes and solar irradiance, but such natural sources of variability seem to be inadequately represented in models; see for instance Trenberth [1] and Santer et al. [3].

This paper presents the hypothesis that exogenous tidal forcing from the sun and moon is a major cause of climate variability. The tidal hypothesis leads to the following propositions:

(1)   tidal forces from the sun and moon vary predictably;

(2)   these external tidal forces exist in alternately dominating meridional (approximately north–south) and zonal west–east) directions;

(3)   these tidal forces provide an exogenous driver of global ocean and atmospheric variability, on timescales from subdecadal to multidecadal, in a manner to some degree deterministic, predictable and testable; and

(4)   this exogenous forcing engenders decadal-scale slowdowns in global mean surface temperatures.

Relationships between tidal forces and climate variability may useful in long-term risk assessment for natural resource management, including fisheries applications described by Klyashtorin [4] and Oviatt et al. [9]. However, here, the hypothesis will be related only oscillations present in global mean surface temperatures and in some major climate systems on timescales from quasi-biennial to multidecadal.

1.2. The Tidal Hypothesis

The understanding of tidal effects from the sun and moon has progressed as a result of seminal studies [15–17], in the course of which the parameterization has become complex as more spectral components and terms have been added to the tidal potential; see for instance Kantha and Clayson [18].

Doodson [15] derived six fundamental astronomical frequencies governing the tides, with corresponding periods covering the range from a lunar day to almost 21,000 years. The three that seem most likely candidates to assist in explaining annual-to-multidecadal climate processes had periods of 18.847 and 18.613 years, corresponding to intervals involving the sun's mean longitude, the longitude of the moon's perigee and the longitude of the moon's ascending node. Doodson expanded the number of tidal frequencies f by adding or subtracting these frequencies or their harmonics, such as with $f_1 + 2f_2 - f_3$ and so on. Adding and subtracting two frequencies (reciprocals of periods) generates two more frequencies. This process of frequency combination (sometimes called frequency demultiplication, e.g., [19]) has been invoked (for example) by Keeling and Whorf [20] and by two studies of the quasi-biennial oscillation to be described later. Such frequency combinations are a common feature of systems with interacting oscillators, as with intermodulation in electrical systems and vibration-rotation bands in spectroscopy. Treloar [21] chose an approach that has become accessible in recent decades. As the tidal potential depends on the distances and directions of the sun and moon in relation to the earth, the latter parameters can be computed from astronomical polynomial algorithms [22], which are reasonably accurate over several centuries. Using this source, and with a simple physical model reflecting the combined mass/distance^3 contributions from sun and moon, three-dimensional tidal forces varying over time were partitioned into components parallel and perpendicular to the plane of the moon's orbit, approximately equivalent to zonal and meridional (or east–west and north–south) earth-based directions respectively. There is therefore a distinction between zonal and meridional tidal regimes in the understanding of the climate oscillations discussed here. Previous tidal approaches to the climate system have often focused on meridional forcing associated with the 18.6-year lunar nodal cycle. As developed, the formulation [21] identified the same, but found more prominent meridional components, and even more prominent zonal components. This study will suggest that the tidal components found in the earlier study, and others added in the present one, unlock many puzzles surrounding oscillations in the climate system. However, given the complexity of the topic and the simplicity of the physical model used, the hypothesis developed here is inevitably exploratory.

From time series analysis, high-frequency components in both directional senses were derived, showing that tidal maxima corresponded to events of close perigee coinciding with new moon. Some high-frequency components had nearly coincident maxima ("beats") at decadal or multidecadal intervals. This beating method follows a procedure by Keeling and Whorf [20], which it must be said has been the subject of critiques by Munk et al. [23] and Ray [24], and has previously shown limited success in correlating with oscillations in the climate system. Tidal components derived from the present formulation show more success. The beating defined periods, phase angles (peak-and-valley timing) and amplitudes of multidecadal and decadal oscillations of "parent" and "daughter" tidal components (see for example [21], Figure 1), some of which seem not to have been previously identified or examined in relation to oscillations in atmospheric and oceanic components of the climate system. The defining feature of the approach is that the climate parameters discussed apparently respond to the coincidence of close perigee with new (or sometimes full) moon, which can be described generally as "close syzygy".

The Appendix summarizes important features of the prior study [21] and the way they are extended to the present study. For example, it:

- describes the zonal or meridional characteristics of the components;
- explains the assumption that tidal periods are expected to be time-averaged;
- describes small amendments to period and timing from the original data [21] based on tests for close syzygy with online Fourmilab software [25]; and
- distinguishes between the treatment of the 18.60-year lunar nodal cycle in this and some previous studies of tidal forcing

It was suggested [21] that the 18.60-year oscillation could be represented as the 18.02-year saros cycle averaged over the 186.0-year "parent" cycle. The relationship is shown in the Appendix A to the present paper, and resembles a pattern (Keeling and Whorf, [20]) in which 18.02-year cycles fall within 186-year arcs. It should be noted that Ray [24] has expressed reservations about this and other features of the Keeling and Whorf paper.

1.3. Upwelling and Ocean Temperatures

Upwelling appears mainly off the west coasts of the continents (see [8]) or in the middle of the equatorial area of the oceans. Upwelling is induced by strong winds blowing over the ocean surface [26] and by tidal forcing of vertical mixing. Munk and Wunsch [27] concluded that the meridional overturning circulation may be mainly determined by the relatively small power of vertical mixing available to return the fluid to the surface layers. They considered four sources of the power needed to sustain abyssal mixing, and found that surface buoyancy forcing and geothermal heating were relatively unimportant, but that winds and tides accounted for most of the mixing: The sun and moon provide a total of 3.7 terawatts of tidal power, more than half the power needed for vertical mixing in the ocean. Keeling and Whorf [20,28] suggested that tidal periodicities produced climate effects through upwelling of cold water and consequent SST variability, and it is a focus of this paper to assess the degree of concordance between tidal periodicities and the temporal variability of ocean and global temperatures.

If tidal forcing drives zonal or meridional upwelling, then we expect that tidal forcing would be correlated with ocean temperatures. One might anticipate that the deduced pattern of tidal periods, phase angles and amplitudes would be most manifest in globally-averaged temperature data, but that individual ocean oscillations would experience different degrees of zonal or meridional forcing, and respond differently to the tidal components listed.

2. Materials and Methods

The only results carried forward from the Appendix A summary are the period and timing of the tidal components, the "beating amplitude" of the three multidecadal components (see below), and the assumption that tidal components and the climate oscillations responding to them vary in cosine form with the latter oscillation amplitudes being tidal-regime-dependent but constant during a regime. The number (seven) of periodicities generating tidal maxima at configurations of close perigee at new moon [21] is expanded in this paper. Although tidal component amplitudes are adjusted empirically to match the climate data, the periods and phases (timing of peaks and valleys) of the tidal components are essentially fixed by the prior analysis.

This paper denotes tidal frequencies by ν, and treats components derived by the close syzygy approach in a similar manner to the above: as fundamental frequencies, their harmonics and combinations. This approach leads to four fundamental frequencies which are apparently able, with their harmonics and frequency combinations, to simulate much of the quasi-biennial to multidecadal variability in temperature and ocean data. The four fundamental frequencies, denoted by ν_1, ν_2, ν_3 and ν_4, correspond to the 59.75-, 86.81-, 186.0- and 5.778-year periodicities derived [21]; findings from this source are summarized and slightly updated in the Appendix A.

Testing the tidal hypothesis depends crucially on the timing of extrema (maxima and minima) in the cycles, which is in their various phase angles or, as characterized here, in their "reference times". Working within this limit, tidal components derived from analysis generally have a significant statistical presence ($p \leq 0.05$) in at least some of the oscillations to be described.

A causal relationship is possible if the period and phase (the timing of peaks and valleys) of a tidal oscillation coincides with the period and phase of an oscillation in temperature or another parameter. However, in cases with small differences in phase, one oscillation will lead and the other lag in time. If a temperature or other variable closely resembling a tidal analog, lags the analog by a small amount, then it is possible that the tidal oscillation causes the other oscillation. It is commonly found in research of this type that the lag time is small in relation to the period of the oscillation (such as a lag time of two months between two oscillations having a period of ten years). In such cases, the two oscillations are in virtual synchrony, allowing the possibility of a causal relationship between the two. This circumstance arises in this paper. A relatively small lag time may represent a reasonable interval for a stimulus to have its response. In an early part of this paper, a lag time of several years is proposed for a ~60-year oscillation, this lag reflecting the time taken for a proposed process of migration of released ocean gases to the upper atmosphere. The process will the explained below.

Climate-related oscillation data at decimal year time t can be compared with the tidal oscillations as captured by cosine relationships, in which the amplitude of a tidal oscillation at time t in decimal years is proportional to $\cos(2\pi[t + t_{lag} - t_0]/P)$ or $\cos(2\pi\nu[t + t_{lag} - t_0])$, where P is the component period in years, ν is the component frequency in reciprocal years, t_0 (as described in the Appendix A) is the component reference time or "date-stamp" (1918.20 or 2039.96), and t_{lag} the time in decimal years that the climate response lags the tidal stimulus. For example, the correlation of SST data can be tested against the 18.60-year oscillation for an SST time lag of 0.1 years when SST data are expressed in the form: $\cos(2\pi[t + 0.1 - 1918.20]/18.60)$, where t represents the decimal year corresponding to an SST data point, and so on for other pairs of tidal and climate oscillations.

Groups of tidal components are introduced progressively in three sections covering multidecadal to quasi-biennial periods to simulate climate oscillations operating on corresponding timescales. These sections begin with components derived by the above beating process, and progress to harmonics and frequency combinations, all components related to close syzygy events. Prior to later discussion, a fourth section summarizes the relationships found between tidal components and the other climate oscillations considered.

The first three sections are discussed in the following terms:

(i) Multidecadal scale (Section 3.1):

The aim is to examine the degree to which a combination of multidecadal tidal components can simulate the ~60-year oscillation implicated as a cause of multidecadal fluctuations and slowdowns in global temperature [4]. The temperature data employed are HadCRUT4 (gridded monthly mean near-surface air temperatures from the Hadley Climate Research Unit, version 4) annual global mean surface temperature (GMST) anomalies, decadally smoothed with a 21-point binomial filter [29].

(ii) Intermediate-period scale (Section 3.2):

Following the multidecadal-scale simulation of smoothed anomaly data, this section examines the degree to which, after removing the ~60-year oscillation, tidal components having periods between the multidecadal set and a period of about 5 years can simulate the *residual* decadally smoothed GMST anomalies.

(iii) Short-period scale (Section 3.3):

This section examines the degree to which adding short-period tidal components can simulate *unsmoothed* climate datasets. The short-period tidal set is compared with: (a) HADCRUT4 unsmoothed annual GMST residual anomalies [29] from 1959 to 2016; (b) the University of Alabama in Huntsville (UAH) lower tropospheric (LT) temperatures generated from composite satellite data for the period 1979 to 2016 [30] and converted to annual data; and (c) the January to December annual increments in ppm carbon dioxide levels measured at Mauna Loa [31] for the period 1959 to 2016. Inter-correlations between these datasets are also examined. Datasets involving differences between sites may attenuate evidence for the ~60-year oscillation or effects from greenhouse gases, but they would also attenuate the response from climate drivers (tidal or otherwise) common to those sites. Such attenuation may be expected with the Southern Oscillation Index (SOI) and North Atlantic Oscillation (NAO), with Indian Ocean Dipole datasets, and with tree-ring, coral or other datasets involving multiple widely separated monitoring sites, and these are excluded from this study. Newman et al. [32] suggested that the Pacific Decadal Oscillation (PDO) is a combination of three geographically-separated eigenmodes (in the North, Central and East Pacific; for similar reasons, the PDO has also been excluded.

However, further simulations with short-period tidal components are analyzed between locations that may reflect single or uniform responses to forcing. The data sources are respectively 69 years (1948 to 2016) of monthly Quasi-Biennial Oscillation (QBO) data from the NOAA/ESRL (Earth System Research Laboratory) [33], 67 years (1950 to 2016) of monthly Oceanic Niño Index (ONI) data representing the three-month running mean of ERSST.v4 SST anomalies (Extended Reconstructed SST anomalies, version 4) in the Niño 3.4 region from National Oceanic and Atmospheric Administration (NOAA) Physical Sciences Division [34], and 161 years (1856 to 2016) of monthly Atlantic Multidecadal Oscillation (AMO) data based on the Kaplan SST dataset and using United States rainfall and Mississippi River outflow [35].

The QBO is an oscillation of equatorial zonal winds in the stratosphere, and is measured by the 30mb zonal wind at the equator in m/s. It is characterized by downward propagating easterly and westerly wind regimes in the equatorial stratosphere, but which affects pole-to-pole stratospheric flow. It is the major pattern of variability in the equatorial stratosphere, and "although several GCMs [general circulation models] have produced simulations of the QBO, there is no simple set of criteria that guarantees a successful simulation" [36]. Only four of 30 models submitted to the Coupled Model Intercomparison Project 5 (CMIP5) [37], and only five out of 15 submitted to the Chemistry-Climate Model Validation Activity [38], have a QBO signal.

NOAA uses the ONI operationally to classify the presence of a full-fledged El Niño or La Niña, according as the three-month Niño 3.4 mean is greater than +0.5 or less than −0.5 °C. [39]. The three-monthly smoothed data are compiled monthly but, in this case, some smoothing is present.

As compiled [35], the AMO data were defined and detrended by subtracting the global mean SST anomalies from the North Atlantic SST anomalies but retain the ~60-year oscillation as well as decadal and subdecadal contributions. The AMO has exhibited the ~60-year variability over the last 8000 years, and this oscillation is not likely attributable to forcing via the Gleissberg solar cycle; however, coupling from the AMO to regional climate conditions appears to be modulated by orbitally induced shifts in large-scale ocean-atmosphere circulation [40]. There is evidence that the AMO is driven by the Atlantic Meridional Overturning Circulation (AMOC) [41,42], which invites a comparison of its possible meridional nature with meridional character derived from the tidal formulation in contrast to the zonal character of the QBO and ONI.

(iv) Summary and comparison of the degree to which tidal components contribute to, and may ultimately explain, the climate oscillations mentioned.

It is important to note here the variation in results that may stem from the choice of HadCRUT4 data in relation to other data sources. Although there is often little to choose between the various standard sources, Karl et al. [43] have amended GMST data from the National Oceanic and Atmospheric Administration (NOAA) in the light of past biases especially with SST data from ships and buoys. Their corrections had their greatest effect over recent years and do not support the presence of a perceived slowdown since 1998 when decadal temperature changes over the slowdown are compared to those over the last half of the 20th century. Yan et al. [44] say that the presence or absence of the recent slowdown depends on the year chosen as its start when the slowdown is judged more in the light of yearly temperature data, so that decadal change may be a more appropriate criterion. Prior to this slowdown, trends in the new annual NOAA data are robust with other datasets. Respecting the HadCRUT4 data used in this paper, Cowtan and Way [45] found biases in the HadCRUT4 data from unobserved regions particularly over the poles and in Africa, and corrected the biases by two methods: kriging and a hybrid method. This reconstruction slightly raised global temperatures over the recent slowdown, but a comparison showed [46] HadCRUT4, Cowtan and Way and several other sources of global temperature data with temperatures leveling off (but not declining) since around 1998 or 2000.

Importantly, Yan et al. suggest that the slowdown phenomenon does not represent a change in climate warming but rather a manifestation of the way that the ocean redistributes heat, that ocean heat content is a better measure of our warming planet, and that the term "global warming hiatus" should be replaced with "global surface warming slowdown".

3. Results

3.1. Multidecadal Scale

The three previously derived [21] multidecadal tidal components, with periods 59.75, 86.81 and 186.0 years, are used for simulations described in this section, and Table 1 lists their properties relevant to the simulations. The derived reference time t_0 of tidal extrema (maxima or minima) can be used to generate other corresponding times of extrema by adding or subtracting integer multiples of the cycle periods. For example, another maximum for the 86.81-year cycle occurs in 2039.96-86.81, and so on. The 59.75- and 86.795-year events listed in Tables I and II [21] can be generated by integer-multiple differences of the 59.75- and (now amended) 86.81-year periods from 2039.96. The t_0 parameter represents an effective "time-stamp" for each cyclic component, and the tidal hypothesis stands or falls if this timing is found to be incompatible with patterns in climate or ocean oscillations. This table also lists the "beating amplitudes" described in the Appendix A.

Table 1. The nature of multidecadal tidal components in this parameterization. Beating amplitudes for the multidecadal cycles are the respective heights of the envelopes shown in the Appendix A.

Period P, Years	Frequency Designation	Reference Time, t_0	Beating Amplitude
59.75	ν_1	2039.96	1700
86.81	ν_2	2039.96	500
186.0	ν_3	1918.20	160

Following the procedure of Klyashtorin [4] with the ACI, the difference is accumulated between the cosinusoidal amplitudes of multidecadal zonal and meridional components over time. A simple zonal-minus-meridional (Z-M) curve would serve to indicate a regime according as the curve is above or below a central horizontal axis. However, the Z-M parameter is framed [4] in terms of the accumulated or integrated difference between zonal and meridional components. This formulation produces a ~60-year oscillation that resembles that seen in detrended global temperatures [4].

As a test of sensitivity and robustness, the accumulated Z-M differences over time were derived from the tidal hypothesis and compared with the ACI in three different ways. For the 59.75-, 86.81- and 186.0-year tidal components, the respective cosine amplitudes were taken to be proportional to the Table 1 beating, and the zonal-meridional differences were defined using the amplitudes A of:

(1)   the 59.75-year zonal cycle minus the meridional 86.81-year cycle only, i.e., A59.75-A86.81;
(2)   59.75-year cycle minus the sum of the meridional 86.81- and 186.0-year cycles, i.e., A59.75-(A86.81 + A186.0); and
(3)   the 59.75-year cycle minus the difference between the 86.81- and 186.0-year cycles, i.e., A59.75-(A86.81 − A186.0) .

Regressions of the three Z-M differences against ACI data over time gave similar results, the analogs accounting for 83–86% of the variance of the ACI, allowing small lead or lag times. The chief reason for this similarity is that the 186.0-year cycle makes a relatively small contribution.

While the theoretically most appropriate analog is currently unclear, the regression result suggests that the subsequent tidal simulation of global temperatures will not differ crucially if a different zonal-meridional difference formulation is chosen among these three.

The scaled accumulated Z-M difference generates a rather irregular ~60-year tidal oscillation. Of the above three tidal analogs, the first formulation, accounting for the greatest fraction (86%) of the ACI variance, is compared with the ACI in Figure 1. The amplitude at time t (in decimal years) is given in the first instance by:

$$1700 \, \cos(2\pi[t - 2039.96]/59.75) \; - \; 500 \, \cos(2\pi[t - 2039.96]/86.81) \tag{1}$$

In Figure 1, the tidal Z-M calculation begins in 1841.5; the timing of annual data is placed at mid-year. The Z-M difference is then accumulated over the following years and the successive differences scaled to the ACI as $0.01 \times [\text{Accumulated } (Z - M)] - 300$. Although the two parameters are virtually synchronous over this range, 87% of the ACI variance is captured when the ACI lags the tidal Z-M component by one year.

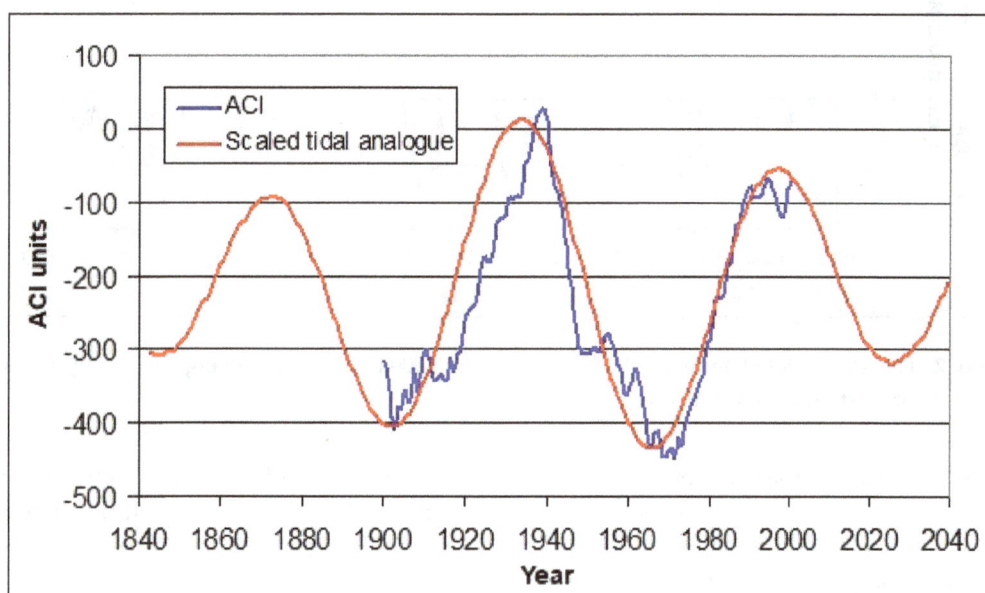

Figure 1. The accumulated balance over time of zonal minus meridional components in the Atmospheric Circulation Index (ACI), compared with the tidal zonal (59.75-year) minus meridional (86.81-year) differences. The tidal curve is vertically scaled to the ACI anomaly in Leonid Klyashtorin's [4] zonal units.

The exogenous tidal mechanism should be essentially unchanged over centuries. The cosine components enable the analog to be projected into the past and future, as shown. The tidal turning points in Figure 1 occur in the mid-years of 1845, 1872, 1902, 1934, 1966, 1997 and 2026. The rising and falling segments of the two curves define zonal and meridional regimes respectively. The zonal-minus-meridional difference appears to be important, and implies an opposition between forces in the two orthogonal directions.

The accumulated Z-M difference in Equation (1) (unscaled to the ACI) is testable against long-term climate reconstructions and datasets. Over the last century or so, the interval between successive maxima or minima has varied between 62 and 64 years, but the oscillation is slightly irregular, with a mean period of 59.75 years, since it is produced by the 59.75-year cycle in combination with the lower amplitude 86.81-year cycle.

A zero to one-year lag time will be shown to apply to near-surface global and ocean oscillation data at subdecadal to multidecadal periods and relatively small amplitude, and define the zonal or meridional character of the responses by the tidal components. However, an eight-year lag applies to the large-amplitude ~60-year multidecadal oscillation in global temperatures, as deduced in the following. Ocean temperatures will be shown to respond to a combination of these two factors. The accumulated Z-M difference between 1500 and 2100 is shown in Figure 2, incorporating this eight-year lag to link tidal forcing to correspond to the temperature response. From 1500 to 2100, successive extrema occur around the middle of years of 1521, 1553, 1585, 1617, 1646, 1672, 1701, 1731, 1763, 1795, 1826, 1853, 1880, 1910, 1942, 1974, 2005 and 2034. While the Figure 1 results are taken to apply to surface phenomena in near-real-time, the eight-year lagged results will be suggested to reflect processes occurring higher in the atmosphere.

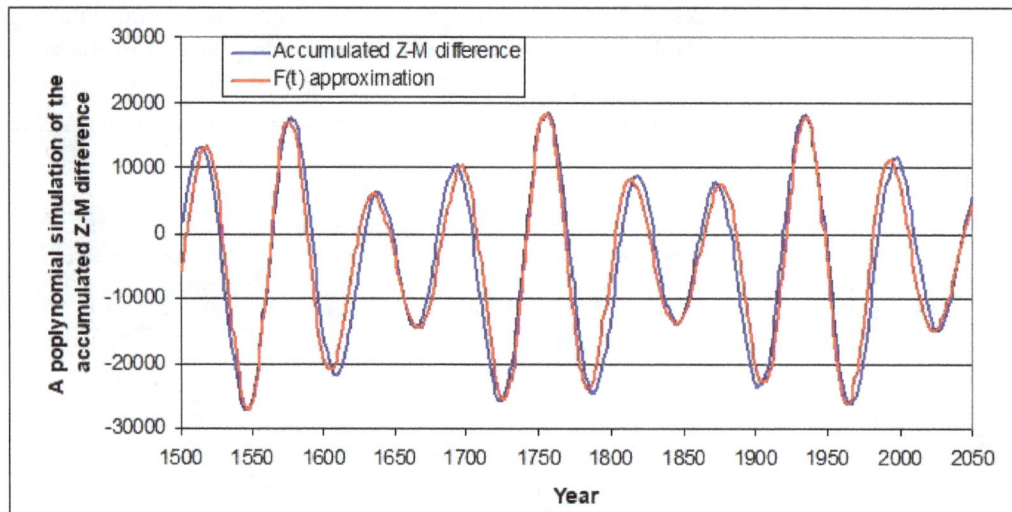

Figure 2. The accumulated annual Z-M difference from 1500 to 2100 with an eight-year lag, and an analog incorporating the F(t) function.

Figure 2 shows that the curve of accumulated differences is amplitude-modulated by a long-period oscillation. Analysis shows this latter oscillation to have a maximum at 1944.1 and a period of 191.7 years. This period is attributable to the frequency combination process described earlier, viz. $1 / 59.75 - 1 / 86.81 = 1 / 191.7$.

The maximum and period of the oscillation means that a minimum coincides with the 2039.96 reference time for the other two oscillations. Shown in Figure 2, the ~60-year oscillation over 600 years can be approximated in the form $aF(t) + b$, where $a = 6900$, $b = -4650$ and

$$F(t) = \{\cos(2\pi[t-1943.1]/59.75)\} \{\cos(2\pi[t-1952.1]/191.7) + 2.35\} \tag{2}$$

F(t) is a convenient approximation that avoids the annual accumulation process, and allows the multidecadal tidal formulation to be tested against long-term climate data containing the ~60-year oscillation, whether data are annual or monthly. Comparing the accumulated Z-M difference with the F(t) approximation between 1500 and 2050, the mean difference in the times of corresponding extrema is 2.8 years. The F(t) approximation will be applied in Section 3.3 in relation to the ~60-year oscillation present in measured monthly and reconstructed annual Atlantic Multidecadal Oscillation data.

Tidal forcing parallel and perpendicular to the plane of the moon's orbit were labeled [21] as "zonal" and "meridional", respectively, but the terms are directionally different from their east–west and north–south sense on our planet, because of the tilt of the Earth in relation to the lunar orbit. The rotational axis of the earth is tilted by about 23.5° to the plane of the ecliptic, and the plane of the lunar orbit is tilted a further 5° to the ecliptic, suggesting that meridional tidal forcing should be aligned in a direction 28.5° west of north, perhaps fortuitously resembling the anti-phase pattern in equatorial and North Pacific temperatures (see for example Mantua et al. [47]).

Figure 1 shows that the ACI has additional decadal-scale elements, and a similar pattern appears [4] in detrended global temperatures. The decadally smoothed HadCRUT4 annual GMST anomalies [29] were simulated by a procedure by incorporating the sum of:

(1)   the ~60-year tidal oscillation, iterating its vertical scaling and displacement, and its lead time with respect to the GMST anomalies' and

(2)   an exponential rise in background temperatures, iterating its asymptotic starting year, a multiplier for the exponential, and the exponent itself.

The least-squares errors (LSEs) between the sum of these factors and the GMST anomalies were found for each of the Z-M options examined above for the ACI case. The LSEs were approximately 0.28 °C for the smoothed dataset and 1.6 °C for the unsmoothed set.

In all cases, the Z-M analogs produced matches to the GMST anomalies with the ~60-year tidal oscillation leading the ~60-year surface temperature oscillation by eight years. Thus, recent maxima in the tidal ~60-year oscillation are during the years 1880, 1942 and 2005, although the last is indistinct in the temperature record because of rising background temperatures. Recent minima occurred during 1853, 1910 and 1974. In relation to detrended global temperatures, Klyashtorin [4] states that the LOD leads by six years, and the ACI leads by four years. The result of the tidal analog leading global temperatures by eight years and the ACI by one year (in Figure 1) is not consistent with Klyashtorin's statement, but the general progression is that the tidal Z-M oscillation leads, and the ACI, LOD and detrended global temperatures follow, so that tidal forcing is prior in time and perhaps drives the other processes.

The sum of the exponential and tidal factors accounted for 91–98% of the variance in the smoothed GMST anomalies, the higher number from the Z-M analog chosen for Figures 1 and 2 with the unsmoothed set, the three Z-M analogs all account for about 88% of the variance. Figure 3 shows the iterated result with the Z-M analog for the unsmoothed GMST data: the exponential, the multidecadal tidal ~60-year oscillation with an eight-year lead time, the sum of the two, and the unsmoothed GMST anomalies. The ~60-year oscillation has a mean valley-to-peak height of about 0.2 °C, or a mean cosine amplitude of about 0.1 °C.

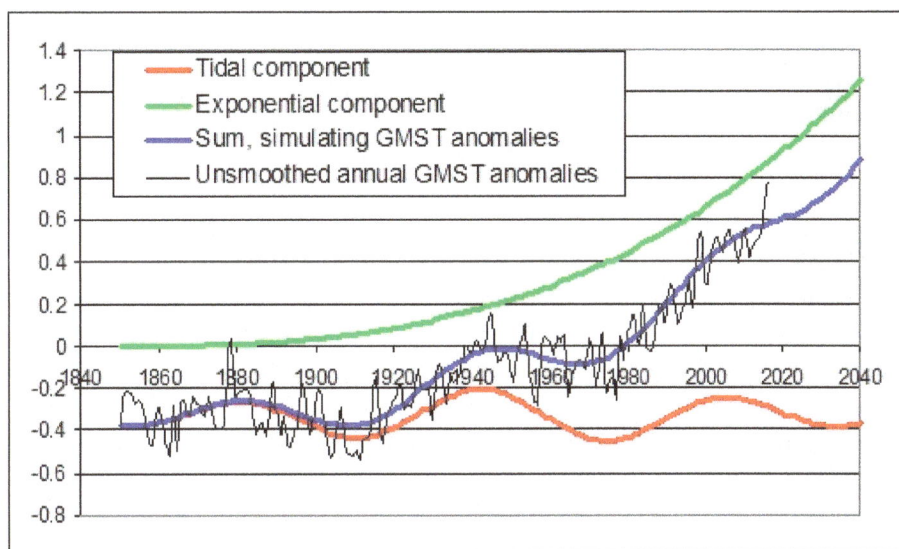

Figure 3. Adding tidal and exponential components to simulate annual unsmoothed HadCRUT4 global mean surface temperature (GMST) anomalies, the curves projected to 2040.

Annual data were assigned a mid-year timing, in keeping with the decimal year timing of tidal extrema. The generating factors to determine GMST anomalies at decimal year t from 1850.5 to 2016.5 by the iteration process were:

(1) The tidal ~60-year oscillation for the unsmoothed case, incorporating Equation (1) above:

$$a \{1700 \cos(2\pi[t - 2039.96]/59.75) - 500 \cos(2\pi[t - 2039.96]/86.81)\} + b \qquad (3)$$

where a = 5.48×10^{-6}; b = $-0.37\,°C$; and with this Z-M difference accumulated after 1841.5. The series of accumulated tidal values from 1842.5 onward are reassigned such that values at decimal year t become values at t + 8, to co-vary with the eight-year lag in ~60-year global temperature oscillations;

(2) The exponential had the asymptote 1851.5, so that the exponential was assigned a value of zero for both 1850.5 (at the first GMST datum) and 1851.5; the iterated function had the form

$$c\,(t - 1851.5)^d \text{ where } c = 5.2 \times 10^{-7} \text{ and } d = 2.80 \qquad (4)$$

(3) These two factors are then summed for each decimal year t.

The same procedure was adopted for the decadally smoothed data from 1850.5 to 2016.5. The results for the asymptote, a, b, c and d, were quite similar: 1851.5, 5.36×10^{-6}, $-0.37\,°C$, 5.2×10^{-7} and 2.80, respectively.

The beating Z and M amplitudes in Table 1, obtained by applying Lorentzian envelopes as described in the Appendix A, are accumulated as shown in Figure 2 and have a mean peak to valley accumulated amplitude nearly 40,000 units. The 5.48 or 5.36×10^{-6} factors are multiples of the accumulated Z-M differences, producing the mean peak to valley ~60-year oscillation temperature range in Figure 3 of about 0.2 °C.

The exponential removed from 1850 to 2016 HadCRUT4 GMST data is not expected to be greatly affected by small corrections over the recent slowdown (see Section 2)—noting that the slowdown itself is simulated by the current tidal meridional regime.

The mean valley-to-peak mid-heights for the three ~60-year peaks can be compared to the contribution of the exponential component to GMST anomalies over the same three intervals. If the formulation for the ~60-year oscillation is valid, then it defines an additional exponential or quasi-exponential component with an asymptote in the region of 1850 like that shown in Figure 3, whatever GMST contributions there may be from other sources affecting radiative forcing from atmospheric greenhouse gases, such as solar irradiance, volcanic eruptions or sulfate aerosols. With the two components shown in Figure 3, the percent contribution of the exponential component for unsmoothed or decadally smoothed GMST anomalies, averaged over successive ~60-year oscillations (from valley to valley for the tidal component), is shown in Figure 4.

Extrapolating the exponential to 2050.5 and 2100.5 produces an anomaly for the unsmoothed and smoothed cases of 1.4 and 2.7 °C, respectively. While such an extrapolation should be treated with caution, the projection is of the same order as 2100 "best estimates" for A1T and B2 [48] and RCP6.0 [49] greenhouse gas emission scenarios, with temperature change relative to 1980–1999 and 1886–2005, respectively. Figure 4 implies that, in this formulation, the percent contribution of atmospheric greenhouse gas concentrations to GMST anomalies has risen markedly since 1850, and that the contemporary contribution of GHGs to global temperature rise is almost 90% of the sum of GHG and ~60-year tidal component contributions.

When the accumulated Z-M difference rises, the zonal component dominates; when the difference falls, the meridional component dominates. Temperatures tend to rise in zonal regimes, and exhibit a slowdown in meridional regimes. The implication is that the directional tidal property (zonal or meridional) influences global temperatures through physical effects on oceanic and atmospheric circulation.

On the parameterization in Figure 3, increased radiative forcing from rising greenhouse gas concentrations in the atmosphere during ~30-year-long zonal regimes is later reversed in the following ~30-year-long meridional regimes. Thus, tidal forcing is hypothesized to modulate the radiative effects of GHGs on global temperature. The slowdown in unsmoothed temperature data after 1998 may reflect the timing of near-real-time (Figure 1) and lagged (Figure 2) meridional regimes beginning in

1997 and 2005 respectively and the significant 1998 El Niño. The slowdown is absent in the decadally smoothed HadCRUT4 anomalies, and the simulated anomaly curve in Figure 3 suggests that no more slowdowns will be seen in the smooth "Sum" curve in future, because of the overriding influence of the exponential radiative forcing component.

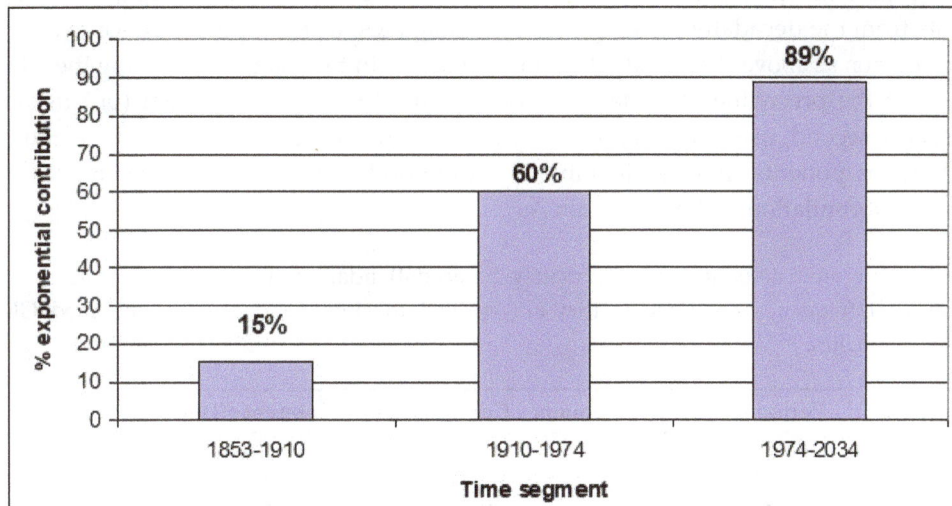

Figure 4. The mean percent contribution of the exponential component to GMST anomalies on the multidecadal scale after removing a ~60-year tidal oscillation.

The multidecadal pattern similarity indicates an interdependence between core and atmospheric angular momenta, length of day, surface temperature, zonal or meridional circulation regimes and tidal forcing. Since multidecadal tidal components lead global temperatures by eight years—greater than lead times for the other processes mentioned by Klyashtorin [4] and Dickey et al. [8]—exogenous tidal forcing from the sun and moon may be a prima facie proximate cause for those processes. If so, then the inevitable question is whether there are feasible physical associations to explain that causality.

There are indicators of interactions between some of these physical components. For example:

(1) Oceanic processes can affect angular momenta and LOD. During El Niños, easterly winds along the equatorial Pacific decrease, which increases atmospheric angular momentum. However, the earth's total angular momentum must stay constant, so the speed of rotation of the solid earth slows down, and LOD increases [50]. In addition, when evaporation or precipitation occurs over the oceans, mass is redistributed, producing changes in the earth's atmospheric angular moment [51].

(2) Atmospheric processes can affect angular momenta and LOD. The characteristic time for vertical transport of gases from the surface to the stratosphere is 5–10 years [52], and water vapor and carbon dioxide are present in atmospheric layers up to the stratosphere. Conceivably, evaporation and migration of these greenhouse gases from oceans to stratosphere occurs during the eight years' lead time deduced above, during which the LOD and AAM change due to mass re-distribution, and global temperatures increase from the evolved greenhouse gases.

(3) The evaporation process may be initiated by directional properties of tidal forcing. For example, a coupled model [53] simulated a significant increase in global AAM, the increase contributed by an acceleration of zonal mean zonal wind in the tropical-subtropical upper troposphere. Consistent with this, global temperatures increase most in zonal circulation regimes after 1844, 1903 and 1966 (Figure 3), in response to zonal tidal regimes that lead global mean surface temperatures by eight years.

(4) The quasi-biennial oscillation (QBO) modulates the zonal mean wind and the mean meridional circulation, and induces in the tropics large-scale transport of chemical species into the

stratosphere [40]. The relationship of tidal components to the QBO and the connection with tropical processes will be discussed in Section 3.3.

3.2. Intermediate-Period Scale

This section deals with the residuals obtained after subtracting the ~60-year and exponential components from the decadally smoothed GMST anomalies, using the Z-M choice to simulate the ~60-year oscillation as above. The residual simulation occurs in two stages: first, using the "daughters" (harmonics) of the preceding multidecadal components to simulate residual (bi-) decadal-scale anomalies; and, second, using higher frequency components to simulate the residuals more closely. The daughter components are listed in Table 2, with periods, frequencies and reference extrema t_0 following their formulation in the Appendix A.

Table 2. Intermediate-period (approximately bidecadal) tidal components derived from three fundamental frequency components that are applied to simulate decadally smoothed GMST residual anomalies.

Period P, Years	Frequency Combination	Reference Time, t_0
14.94	$\nu_5 = 4\nu_1$	2039.96
21.70	$\nu_6 = 4\nu_2$	2039.96
18.60	$\nu_7 = 10\nu_3$	1918.20

The decadally smoothed residual anomalies oscillate along a horizontal axis over a range generally less than about 0.1 °C. as displayed in Figure 5. The smoothed residuals were simulated by regression against the two most prominent "daughters", the 21.70-year meridional and 14.94-year zonal tidal components; the 18.60-year component made no discernible contribution.

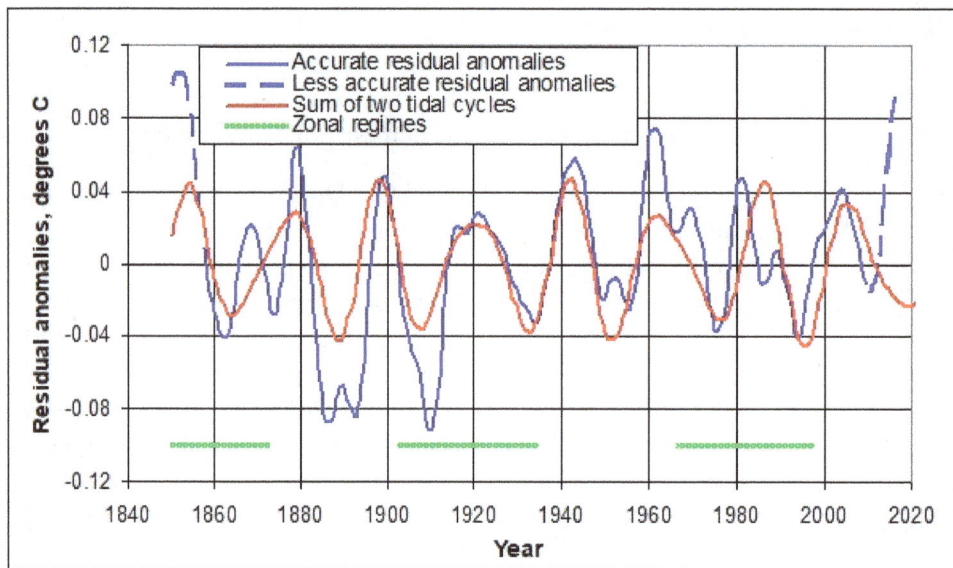

Figure 5. Residual HadCRUT4 decadally smoothed anomalies after removing a ~60-year oscillation and an exponential rise in temperature, compared with a sum of 21.70- and 14.94-year tidal components with anomalies lagging the tidal combination by 0.45 years. Residuals calculated from anomalies prone to "end effects" in the smoothing process are labeled "less accurate".

The sum of component cosine functions for the two cycles at time t in decimal years is:

$$-0.034 \, \cos(2\pi[t + 0.45 - 2039.96]/21.70) \; - \; 0.013 \, \cos(2\pi[t + 0.45 - 2039.96]/14.94) \qquad (5)$$

The possibility of a lag time between tidal stimulus and GMST response is evaluated statistically by comparing system data at each time t in decimal years with a formulation for each tidal component $\cos(2\pi[t + t_{lag} - t_0]/P)$. For example, testing a three-month lag time for GMST (or residual) data points at decimal years t in relation to the 14.94-year tidal cycle means testing the GMST data against the cosine relationship $\cos(2\pi[t + 0.25 - 2039.96] / 14.94)$.

The best-fit residual anomaly lag time was 0.45 years after omitting from the regression the first and final seven years of the residuals to avoid end effects in the smoothing [54] that may have been responsible for the high anomalies at the beginning and end of the series. The small corrections proposed to HadCRUT4 data over the last slowdown (Section 2) fall shortly before or during the latter time with end effects, so that any late-year biases in the HadCRUT4 data should have little effect on these results.

On multiple linear regression, the cosine amplitude multipliers for the 21.70- and 14.94-year harmonic components were $-0.034\ ^\circ\text{C}$ and $-0.013\ ^\circ\text{C}$; the probability t statistics were -10.6 and -3.8, and the p statistics were 5.5×10^{-2} and 1.3×10^{-4} respectively; the multiple R was 0.68. The negative multipliers and t statistics imply that tidal maxima from the two components were accompanied by lower GMSTs, consistent with tidal forcing promoting ocean upwelling. The 18.60-year harmonic component made no significant contribution ($p = 0.58$).

The residual curve shows lower temperatures than the tidal curve for minima around 1890, 1910 and the late 20th century. Volcanic eruptions in 1883 (Krakatau), 1902 (Santa Maria and Mt. Pelée), 1912 (Novarupta) and 1982 (El Chichón) may account for some of the difference by reduction in solar insolation; however, an effect from the 1991 eruption of Mount Pinatubo is not evident. Figure 5 indicates that higher frequency components are needed to improve agreement with these anomalies.

Volcanic eruptions eject dust particles and gases. Some of these gases, such as water vapor and carbon dioxide, are greenhouse gases that cause warming, while the emitted particulates block the sun and cause cooling. The cooling effect is usually most noticeable in temperature curves like the above, because the effect is relatively large and occurs over a short period. Ejected greenhouse gases disperse in the atmosphere and their warming effect is longer-lasting and can be less noticeable. The cooling effect from volcanic particulates is the feature most closely evaluated in climate studies. Greenhouse gas contributions from volcanoes are almost negligible compared to those from human sources.

The second stage in simulating the decadally smoothed anomalies is to apply more intermediate-period components to address the subdecadal "fine structure"; the added tidal components are listed in Table 3. The original formulation [21] found an irregular meridional component with a mean period of 5.775 years, later amended to 5.778 as described in the Appendix A. The additional components are identified here as frequency combinations. Thus, frequencies corresponding to components with periods 8.848 and 47.96 years are generated from $\nu_5 + \nu_6$ and $\nu_5 - \nu_6$, therefore equivalently from ν_1 and ν_2. Since the generating components had reference times t_0 at 2039.96, the new components are assumed to have the same t_0. The period results were slightly corrected to means of 8.850 and 47.92 years by finding intervals from 2039.96 with frequent occurrences of close syzygy. In the case of the slightly irregular 8.850-year component, Fourmilab shows that every third interval occurs near alternately close new and moon events, but these close syzygy events are displaced from the 2039.96 reference time t_0 by half a cycle. Over many three-fold cycle intervals, the mean cycle period is 8.850 years. Despite the half-cycle displacements, when climate oscillations are compared with functions $\cos(2\pi[t + t_{lag} - 2039.96]/P)$, the same t_{lag} is found to apply to the 8.850-year component. It was unexpected to find the period (8.850 years) of the moon's apsidal precession numerically generated by a combination of ν_5 and ν_6, and equivalently of ν_1 and ν_2, components. Other new components generated are 3.495- and 16.69-year components from ν_4 and ν_8, so ultimately from ν_1, ν_2 and ν_4. Often, 3.5-, 5.8-, or ~17-year ocean cycles have been reported (e.g., [55–57]). As may be expected, higher frequency tidal components tend to show more significance in monthly than in annual datasets.

Table 3. The 5.778-year component and additional components generated by frequency combination from Table 1 components. The latter four are generated from both zonal and meridional component frequencies.

Period P, Years	Frequency Combination	Reference Time, t_0
5.778	ν_4	2039.96
8.850	$\nu_8 = \nu_5 + \nu_6 = 4\nu_1 + 4\nu_2$	2039.96
47.92	$\nu_9 = \nu_5 - \nu_6 = 4\nu_1 - 4\nu_2$	2039.96
3.495	$\nu_{10} = \nu_4 + \nu_8 = 4\nu_1 + 4\nu_2 + \nu_4$	2039.96
16.69	$\nu_{11} = \nu_4 - \nu_8 = 4\nu_1 + 4\nu_2 - \nu_4$	2039.96

With these added components, the curve for "most reliable residual anomalies" can be empirically simulated more closely than in Figure 5. Figure 6 shows one result, assigning regime-dependent amplitudes according to Table 4. Discontinuities in Figure 6 occur as regimes change. Reaching statistical significance ($p \leq 0.05$) in component correlations is affected by the small sample size [N(zonal) = 80; N(meridional) = 75]. All six components reach statistical significance in at least one regime type except the 5.778-year component, although it is the 8.850- and 5.778-year components that simulate the subdecadal features. The 18.60-year component had a less significant contribution to the simulation than the other six components, and has been omitted from Table 4 and from the simulation. The best-fit lag time for the residuals is 0.40 years.

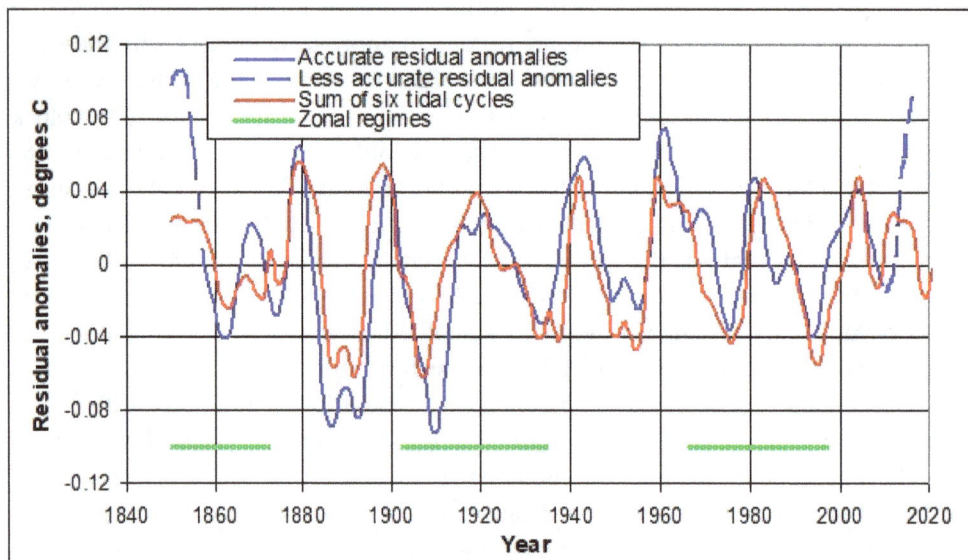

Figure 6. The result of adding four regime-dependent components to the previous analysis, compared with residual HadCRUT4 decadally smoothed GMST anomaly residuals, and with anomalies lagging the tidal combination by 0.40 years. Residuals calculated from anomalies prone to "end effects" in the smoothing process are labeled "less accurate".

Table 4. Regime-dependent amplitudes for cosine functions for tidal components with a 0.40-year lead time in relation to the decadally smoothed GMST anomaly residuals, in the Figure 6 simulation. The amplitude A entries are in °C for each component, and are given for each period P by $A\cos(2\pi[t + 0.4 - 2039.96]/P)$ with t in decimal years.

Period P, Years	21.70	14.94	8.850	47.92	16.69	5.778
A (Zonal), °C	−0.04	−0.01	−0.01	−0.005	−0.015	0
A (Meridional), °C	−0.03	−0.02	0.015	−0.015	−0.015	0.01

As mentioned above, the sum of the tidal ~60-year cycle and the exponential rise (Figure 3) accounts for 98% of the variance in the decadally-smoothed HadCRUT4 anomalies. For the residual decadal-scale anomaly in Figure 5, the accompanying two-component tidal curve captures 41% of the remaining variance, while the six-component and regime-dependent approach associated with Figure 6 and Table 4 account for 53.5%. The results suggest that tidal forcing modulates the approximately bidecadal variation in GMST anomalies.

The improved agreement shown in Figure 6 is derived empirically. However, at face value these results are consistent with the earlier rise-and-slowdown multidecadal results, that global mean surface temperatures are regime-modulated. Since global temperatures are responsive to ocean surface temperatures, this then suggests that including regime changes may improve the way that (for example) ocean oscillations could be characterized.

Table 5 displays values of correlation coefficient (R) and root mean square error (RMSE) as progressively more components have been added to simulate HadCRUT4 decadally smoothed GMST anomalies. Model 1, a single exponential, was not reported earlier in this text, but is included as a baseline reference. The least-squares exponential has the form:

$$(1) -0.34 + 4.8 \times 10^{-7} \ (\text{Year} - 1850.5)^{2.83} \text{ for the unsmoothed case; and} \qquad (6)$$

$$(2) \ 0.34 + 5.2 \times 10^{-7} \ (\text{Year} - 1850.5)^{2.81} \text{ for the decadally smoothed case.} \qquad (7)$$

Model 2 consists of the exponential plus the tidal ~60-year oscillation. For the smoothed dataset, Model 3 consists of Model 2 components plus the 21.70- and 14.94-year tidal cycles. For the smoothed dataset, Model 4 consists of Model 2 components plus six regime-dependent tidal cycles. Figure 7 shows the improvement in simulating these anomalies, as measured by RMSE values.

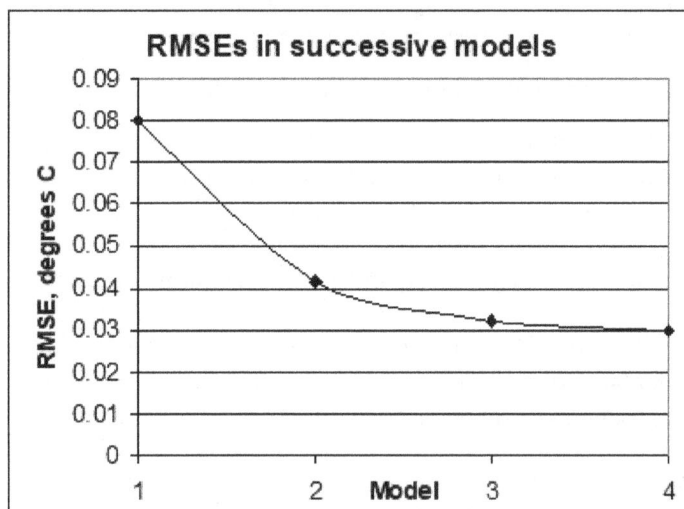

Figure 7.   Root mean square errors (RMSEs) in successive models for decadally smoothed HadCRUT4 anomalies.

Table 5.   Statistical results for succeeding models of simulating HadCRUT4 decadally smoothed GMST anomalies.

Components of Regression	Decadally Smoothed Case: R, RMSE (°C)
Model 1. A single exponential factor: $5.0 \times 10^{-6} \ (t - 1850.5)^{2.82} - 0.34$	0.957, 0.0800
Model 2. The sum of the exponential and the tidal ~60-year oscillation.	0.988, 0.0418
Model 3. As #2, plus 2 tidal cycles.	0.994, 0.0321 See Figure 5.
Model 4. As #2, plus 6 regime-dependent tidal cycles.	0.995, 0.0297 See Figure 6 and Table 4.

The remaining unassigned errors in Figure 6 may be attributable to factors such as the averaging of unsmoothed contributions and an absence of components operating on the decadal scale. Tidal components will be compared with unsmoothed data in the rest of this paper.

To summarize the hypothesis at this stage, decadally smoothed global mean surface temperatures have the following contributing sources:

(1)   an exponential component that resembles models for the effects of greenhouse gas concentrations;

(2)   a ~60-year oscillation caused by an eight-year lagged response to exogenous multidecadal tidal forcing, possibly through the delayed temperature response from vertical transport to the atmosphere of greenhouse gases during zonal regimes;

(3)   intermediate-period temperature variability caused by decadal-scale tidal forcing, with different contributions during zonal and meridional regimes, and with temperature response lagging the tidal stimuli by 0.4 or 0.45 years (about five months); and

(4)   other contributions, not quantified in this paper, that may include those from episodic volcanic emissions and man-made emissions of pollutant aerosols.

3.3. Short-Period Scale

To simulate unsmoothed oscillations in this section, two components are added (Table 6) with approximately bidecadal periods derived from Fourmilab data on syzygy events.

Table 6. Bidecadal tidal components.

Period P, Years	Frequency Designation	Reference Time, t_0
2.396	$\nu_{12} = 4\nu_8 - 3\nu_2 = 16\nu_1 + 13\nu_2$	2039.96
2.213	$\nu_{13} = 4\nu_8 = 16\nu_1 + 16\nu_2$	2039.96

3.3.1. Unsmoothed Residual Anomalies, Lower Tropospheric Temperatures and CO_2 Levels

Both the UAH LT temperature data (1979 to 2016) [30] calculated as an annual mean and the annual Mauna Loa carbon dioxide increases (1959 to 2016) [31] show linear increases over time. These increases are detrended with the transformations:

$$0.5[\text{UAH datum} - 0.01232(\text{Year}) + 24.59] \text{ and } 0.175[CO_2 \text{ datum} - 0.02865(\text{Year}) + 55.408]. \quad (8)$$

The unsmoothed GMST residuals are effectively detrended by removal of the exponential and ~60-year tidal components, and are not rescaled. Amplitudes were empirically adjusted for the shortest-period tidal components as displayed in Table 7.

Table 7. A simulation for tidal component cosine amplitudes (in °C) over the period 1959.5 to 2016.5. The amplitude A entries are in °C for each component, and are given for each period P by $A\cos(2\pi[t - 2039.96]/P)$ with t in decimal years.

Period P, Years	2.213	2.396	3.495
A (Zonal), °C	0	−0.03	0.08
A (Meridional), °C	−0.06	0	0

The oscillating components in the detrended parameters are compared in Figure 8. The size of the unsmoothed GMST residual oscillations over the period of the recent slowdown appear to be too large to show evidence of needing the small HadCRUT4 corrections described in Section 2. The CO_2 data and the two temperature datasets vary in near-synchrony. A causal relationship between evolved CO_2 and temperature might be argued either way, but the inter-connection seems undeniable.

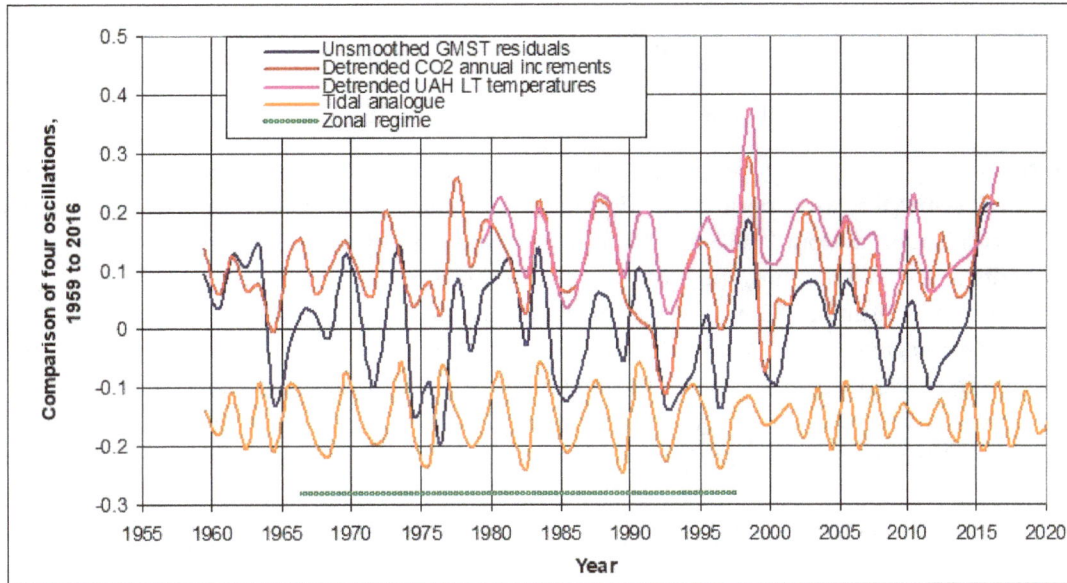

Figure 8. The scaled temporal variation from 1979 to 2016 of the unsmoothed GMST residual anomalies, detrended UAH lower tropospheric temperatures, and detrended CO_2 annual increments and tidal analog. The latter three curves are vertically offset for clarity.

For periods where these data coincide (from 1959 to 2016, or 1979 to 2016), regression p statistics are given in Tables 8 and 9.

Table 8. Probability p statistics ($N = 58$) in pair-wise regressions with higher-frequency oscillations from 1959 to 2016. The tidal data are derived from the simulation in Table 7.

CO_2 annual increments	7.9×10^{-8}	
3 tidal component simulation	0.0012	0.0026
	Unsmoothed GMST residual anomalies	CO_2 annual increments

Table 9. Probability p statistics ($N = 38$) in pair-wise regressions of UAH LT temperatures with oscillations from 1979 to 2016. The tidal data are derived from the simulation in Table 7.

Unsmoothed GMST residual anomalies	3.3×10^{-1}
CO_2 annual increments	2.3×10^{-6}
3 tidal component simulation	0.0020

For both the unsmoothed anomalies and the UAH tropospheric temperatures, the highest significance is found with a lag of 0.15 years with respect to the tidal composite. However, the highest significance for the CO_2 annual increments is obtained with a lag of 0.24 years. Whether the lag time differences are meaningful is a topic for further study.

These results show significant correlations between detrended UAH lower tropospheric temperatures, detrended CO_2 annual increments, unsmoothed GMST anomalies using a simulation with only three components. The results support a causal connection between the simultaneous increase of CO_2 and temperature on subdecadal timescales. The unsmoothed GMST anomalies and a three-component simulation derive from the present formulation of exogenous tidal forcing, this forcing appearing from Figure 8 to generate the simultaneous evolution of CO_2 and increase in temperature. The regression of unsmoothed residuals vs. the three tidal-component combination is not quite significant ($p = 0.052$), but the correlation is much higher in the zonal regime ($p = 0.0065$).

The oscillations in this figure closely resemble ENSO events, and relationships between the parameters in the figure may be worth exploring further in work on deep ocean heat content and the redistribution of heat, along the lines discussed by Yan et al. [44]. The association of oscillations of CO_2 increments with those reflecting ENSO activity will be discussed in Section 4.1.

3.3.2. The Quasi-Biennial Oscillation (QBO)

Gruzdev and Bezverkhny [58] deduced the presence of two regimes in QBO data, with periods of 2 and 2.5 years; the latter period was dominant from 1974 to 1993, the closest tidal equivalent being the zonal interval from 1966.5 to 1997.5. To compare with the suggestion [48] of 2- and 2.5-year components in the two regimes, the most significant components in meridional and zonal regimes (with periods of 2.213 and 2.396 years respectively) are displayed in Figure 9; their respective p values in the respective regimes are 1.6×10^{-38} and 3.3×10^{-59}. Again, we find a strongly regime-dependent oscillation. The highest correlation with the combination of two tidal components is with a QBO lag time of 0.15 years ($p = 9.0 \times 10^{-88}$). A smaller but statistically important contributor to the meridional regime is the 3.495-year component ($p = 9.7 \times 10^{-9}$), whose contribution is omitted from Figure 9. In the Figure, maxima in the 2.213-year oscillation and QBO coincide around 1956; taken with the coincidences in the later meridional regime, this suggests a phase match over more than 25 cycles of the 2.213-year component, and consistent with a maximum near its 2039.96 reference time.

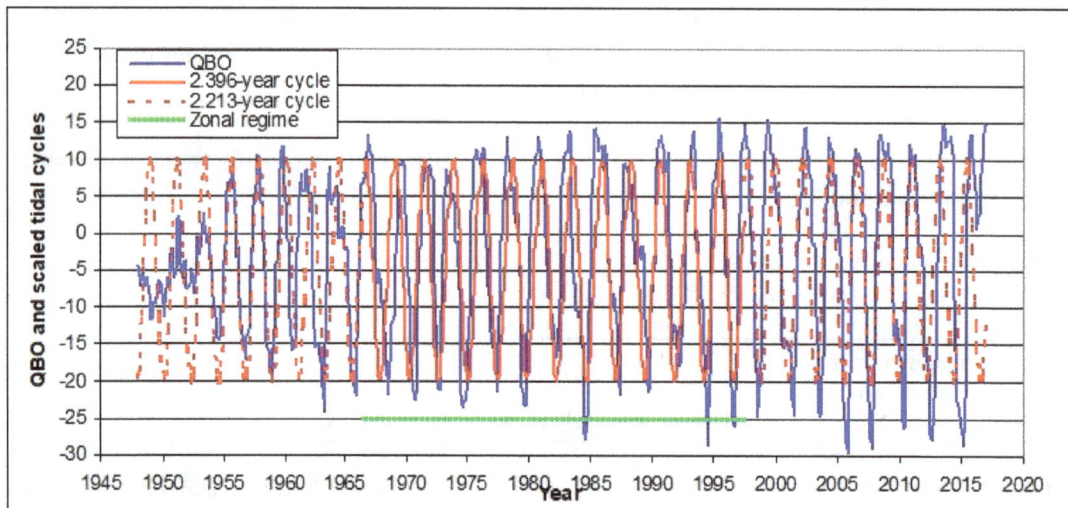

Figure 9. Comparison of the monthly QBO dataset with the vertically scaled 2.123- and 2.396-year tidal oscillations, with a QBO lag time of 0.15 years.

The QBO was mentioned in Section 3.2 concerning the transport of greenhouse gases to the upper atmosphere over a mean duration of eight years during ~30-year zonal regimes. The QBO modulates the zonal-mean wind and is associated with global circulation patterns, including the upward transport and distribution of chemical species from the tropics [35].

Baldwin and Tung [59] noted that the QBO includes not only the ~28-month period but that using the annual frequency in applying the process of frequency combination mentioned earlier produces two other periods (about 20 and 8.6 months). See also Ruzmaikin et al. [60]; their QBO study invokes the same modulation process.

3.3.3. The Oceanic Niño Index (ONI)

The ONI illustrates the effect of regime differences with subdecadal tidal components (see Figure 10). Correspondence is most marked in the zonal regime from 1966 to 1997. The highest significance is found with the ONI lagging the tidal components by 0.22 years, and Tables 10 and 11 show amplitudes (°C) in a simulation and correlation results for subdecadal tidal components in the ONI; the multiple R is 0.34.

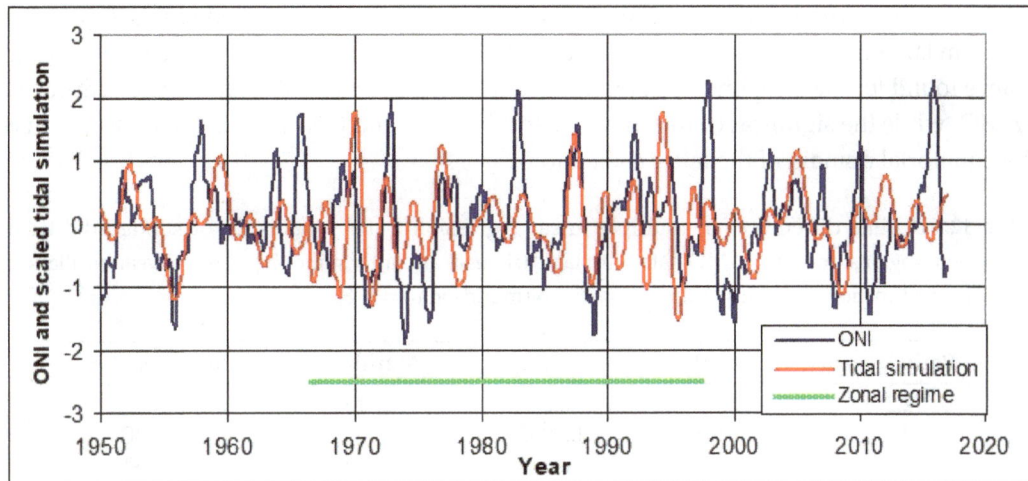

Figure 10. A comparison of the Oceanic Niño Index with a combination of five subdecadal tidal components.

Table 10. Amplitudes for the five subdecadal tidal components in a simulation of the ONI. The ONI lags the tidal components by 0.22 years.

Period, years	2.213	2.396	3.495	5.778	8.850
A (Zonal), °C	−0.6	0.5	0.5	0.3	0
A (Meridional), °C	0	−0.4	0	0.4	0.4

Table 11. Regime differences for subdecadal tidal components in multiple linear regression statistics with the ONI, for a lag time of 0.22 years.

Period, Years	Zonal t	Zonal p	Meridional t	Meridional p
2.213	−2.7	0.0076	1.2	0.25
2.396	3.8	0.00016	−1.8	0.066
3.495	8.5	6.1×10^{-16}	1.7	0.088
5.778	2.7	0.007	4.6	5.6×10^{-5}
8.850	−0.031	0.98	4.0	6.2×10^{-5}

The ONI is closely related to the SOI and other Pacific equatorial oscillations, and the peaks and valleys in this Figure, as in Figure 8, generally correspond to important ENSO events. The SOI is also an important global climate predictor [61]. The preceding analysis suggests that, to the degree that tidal forcing simulates the ONI, and that it is exogenous and deterministic, the tidal approach may eventually have merit as a climate predictor.

3.3.4. The Atlantic Multidecadal Oscillation

The AMO has exhibited a ~60-year oscillation over the last 8000 years, as mentioned in Section 2, and shorter-term components are better seen by removing that component. The ~60-year oscillation is represented by the eight-year-lagged function F(t) described in Section 3.1. Regression results to note are:

(1)   a regression of F(t) against the AMO produces a p statistic of 2×10^{-139}; and

(2)   the amplitude of the F(t) function in the AMO is 0.047 °C, compared to the 0.02 °C amplitude for the ~60-year oscillation in GMST anomalies (Figure 3).

After removing the ~60-year components in the AMO, a multiple linear regression of the residual against higher-frequency tidal components led to the regime-dependent correlation results shown in

Table 12. Regressions with lag times indicated a lag time of only 0.04 years; whether this is meaningfully different from lag times found with other oscillations discussed in this section is uncertain. The greater significance found for the components in meridional regimes is consistent with the mid-latitude nature of the AMO, while the significance found above for the QBO and ONI in zonal regimes is consistent with the equatorial domain of the latter oscillations.

Table 12. Regime differences in multiple linear regression t and p statistics of subdecadal tidal components against monthly AMO data with lag 0.04 years and after removing the ~60-year oscillation with the F(t) function. Samples: N(zonal) = 954; N(meridional) = 978.

Period, Years	Zonal t	Zonal p	Meridional t	Meridional p
2.213	2.0	0.042	1.9	0.054
2.396	−3.6	0.00031	3.6	0.00040
3.495	1.9	0.055	3.5	0.00044
5.778	−1.2	0.25	2.8	0.0050
8.850	1.6	0.10	6.9	7.8×10^{-12}
14.94	−1.9	0.064	0.47	0.64
16.69	−6.9	9.6×10^{-12}	1.1	0.26
18.60	−3.7	0.00026	4.6	5.6×10^{-6}
21.70	0.50	0.62	−2.0	−0.044

Based on the regression results after removing the ~60-year component, a simulation of the AMO is given by tidal component amplitudes A in Table 13 and the simulation is shown in Figure 11.

Table 13. Regime-dependent component amplitudes A in a simulation of the AMO after removing a ~60-year oscillation. The AMO lags tidal components by 0.04 years.

Period, Years	2.213	2.396	3.495	5.778	8.850	14.94	16.69	18.60	21.70
A (Zonal), °C	0.02	−0.03	0.03	−0.01	0.01	−0.02	−0.06	−0.05	0.00
A (Meridional), °C	0.03	0.03	0.03	0.02	0.12	0.00	0.00	0.09	−0.02

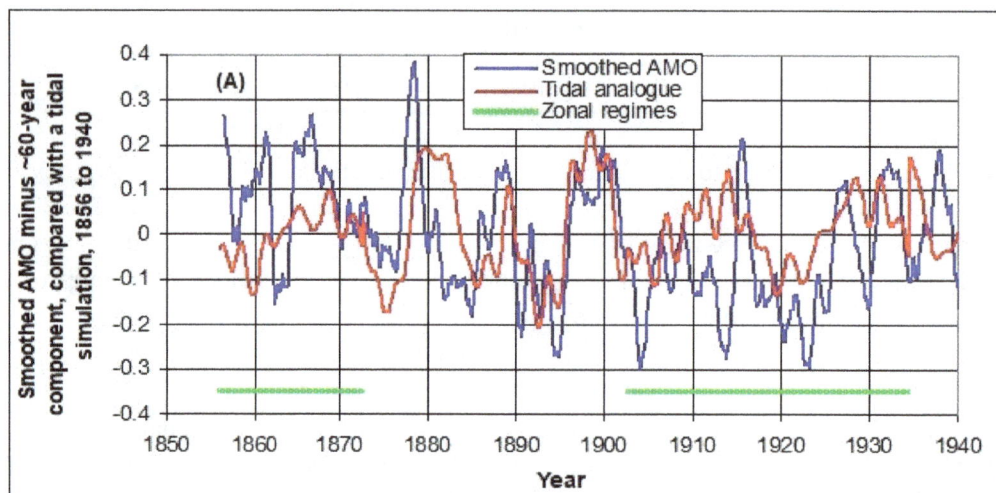

Figure 11. *Cont.*

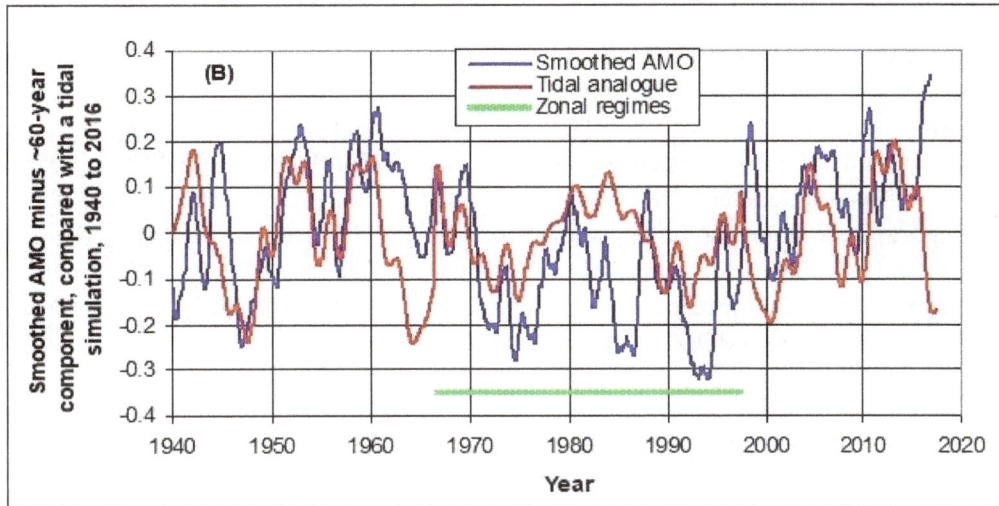

Figure 11. A simulation of the AMO over the time segments: (**A**) 1856–1940; and (**B**) 1940–2016, after removing the ~60-year oscillation captured by F(t), and then smoothing with a 13-month running mean. These are compared with a tidal analog, for an AMO lag time of 0.04 years with respect to the tidal analog. The tidal components are listed in Table 13.

The overall correlation of the simulation is rather poor (R ~0.3), and the use of multiple tidal variables renders the AMO simulation tentative at best. However, these attempts suggest that some aspects are becoming clearer, that:

(1)   while analyses above for equatorial or global oscillations showed no statistical evidence of the 18.60-year tidal component, it is a significant contributor to the AMO;

(2)   the 180-year component correlation is significant during zonal and meridional regimes; and

(3)   these aspects may be consistent with the mid-latitude and meridional (in geographic terms) nature of the AMO.

The Z-M curve in Figure 2 somewhat resembles AMO proxies derived from annual Puerto Rican $\delta^{18}O$ coral data (1751–2004) [62], Labrador algae (1365–2007) [63] and Nordic Sea ice (about 1600–1990, curve inverted) [64]. A comparison [62] showed some differences between the Puerto Rican $\delta^{18}O$ coral proxy and more land-based AMO reconstructions by Gray et al. [65] and Mann et al. [66]. In relation to a marine oscillation (the AMO), a proxy from tidal forcing might correspond more closely with marine AMO proxies from coral, algae or sea ice. The tidal formulation may be a starting point for understanding these reconstructions, including the amplitude modulation noted [64] in longer-term AMO data.

3.4. Summary and Comparison of the Degree to Which Tidal Components Contribute to, and May Ultimately Explain, Climate Oscillations.

Figure 3 and accompanying text show that the presence of slowdowns in global mean surface temperature anomalies can be simulated and explained by an exponential component mirroring atmospheric greenhouse concentration with the addition of a combination of three multidecadal tidal components producing a slightly variable ~60-year tidal oscillation. The process is an atmospheric consequence of tidally-driven regime changes that are suggested to modulate the release of greenhouse gases from the oceans. Zonal regimes provide the greater amount released, modulating the global radiative forcing and generating slowdowns as a result.

The two different tidally-generated regimes impose different temperature and CO_2 oscillations on the bidecadal to subdecadal scale, presumably influencing both surface and sub-surface ocean circulation in different ways. Decadally smoothed GMST anomalies, LT temperatures and atmospheric concentrations of greenhouse gases (using CO_2 as an indicator) can be simulated by two to six intermediate-scale tidal components (Figures 5 and 6). Subdecadal tidal components simulate GMST residual anomalies, detrended UAH LT temperatures and detrended annual CO_2. The subdecadal Figure 8 shows minima and maxima similar to the timing of major ENSO events, suggesting that ENSO modulates greenhouse gas release from the oceans. In Figure 10, the ONI is moderately well simulated by the five subdecadal tidal components; the ONI is a standard for identifying major ENSO events. In Figure 9, a combination of the two shortest period tidal components closely simulate QBO 30mb zonal winds. In Figure 11, after removing a simulation of the ~60-year oscillation, the AMO is less well simulated by intermediate and short period components, but the 18,60-year component appears to contribute to the AMO, presumably because of its meridional and mid-latitude character.

4. Discussion

4.1. General Comments

The tidal hypothesis requires the conditional adoption of a perspective and a vocabulary involving exogenous tidal forcing, phase-defined cycles, zonal and meridional regimes, and combinations from a limited number of fundamental frequencies. The tidal components given in this paper are summarized in Table 14, as combinations of four fundamental frequencies, and in a form suggested by the classical Doodson numbering system.

Table 14. Tidal frequency components in climate oscillations examined in this paper. The penultimate column lists the regime in which the components show greatest responses in the climate oscillations discussed.

Frequency Designation	Period, Years	Frequency Combination	Representation for ν_1 ν_2 ν_3 ν_4	Main Active Regime	t_0 (AD)
ν_1	59.75	ν_1	1000	Zonal	2039.96
ν_2	86.81	ν_2	0100	Meridional	2039.96
ν_3	186.0	ν_3	0010	Meridional	1918.20
ν_4	5.778	ν_4	0001	Meridional	2039.96
ν_5	14.94	$4\nu_1$	4000	Both	2039.96
ν_6	21.70	$4\nu_2$	0400	Both	2039.96
ν_7	18.60	$10\nu_3$	00(10)0	Meridional	1918.20
ν_8	8.850	$4\nu_1 + 4\nu_2$	4400	Meridional	2039.96
ν_9	47.92	$4\nu_1 - 4\nu_2$	4-400	Both	2039.96
ν_{10}	3.495	$4\nu_1 + 4\nu_2 + \nu_4$	4401	Both	2039.96
ν_{11}	16.69	$4\nu_1 + 4\nu_2 - \nu_4$	440-1	Zonal	2039.96
ν_{12}	2.396	$16\nu_1 + 13\nu_2$	(16)(13)00	Zonal	2039.96
ν_{13}	2.213	$16\nu_1 + 16\nu_2$	(16)(16)00	Meridional	2039.96

The tidal periods listed in Table 14 are supported by Fourmilab intervals between syzygy events and by empirical comparison with climate oscillations. The contributions to the climate oscillations discussed have come from four fundamental frequency components or their combinations, for which most have the same reference time t_0 (2039.96). The multiples for ν_{12} may be somewhat at odds with the others, but selection rules governing the choice of frequency multiples for this system are not known to this author. Apart from the eight-year lag time postulated with reference to slowdowns in global temperature anomalies, the decadally smoothed GMST residuals lag the tidal stimuli by about five months and the unsmoothed residuals and other unsmoothed oscillations examined lag the tidal stimuli in times ranging from less than three months down to about two weeks.

In attributing solar and lunar components to the spectrum of the NAO, Berger [67] observes that "... for the cycles to be both well-defined and anything other than whole numbers (that is, not tied to seasons), they have to rely on outside forcing." This paper suggests an emerging structure and exogenous explanation for zonal and meridional modes of variability in the global ocean and atmosphere, as an alternative to the view that these modes of variability are internally generated in the oceans.

The hypothesis leads to an interpretation: of oscillations in zonal and meridional senses; of ocean variability in turn affecting global temperatures; of slowdowns in meridional regimes; and of the global temperature baseline rising in an exponential manner consistent with the increase over time of greenhouse gas emissions. Lower-frequency tidal components simulate the alternate rises and slowdowns in global temperatures; intermediate-frequency components simulate the detrended oscillation of decadally smoothed global temperatures; and higher-frequency components simulate the detrended oscillation of unsmoothed global temperatures, lower troposphere temperatures and annual carbon dioxide increments.

This study suggests the possibility of a ~60-year cyclic process in the oceans involving the dissolution and evolution of greenhouse gases this being promoted by zonal and meridional regimes created by tidal forcing and the regime-dependent differences in amounts of greenhouse gases released from the oceans. The ~60-year cycle in regime-dependent upwelling is a possible causal agent for the corresponding cycle in the transport of nutrients to the upper ocean layer and for corresponding cycles in fish catch; see Klyashtorin [4]. The lags in AAM, LOD and global temperature may conceivably be responses to evolved gases rising in the atmosphere; their similar temporal profiles [4] are understandable in relation to the profile of multidecadal components in tidal forcing (Figure 1). Because the temporal pattern of multidecadal tidal forcing is prior to the corresponding patterns for ACI, LOD and AAM, and is also exogenous, tidal forcing is construed to be a prior cause of those phenomena.

The ~60-year ACI oscillation, mirrored in (lagged) global temperatures, is formulated as an accumulated excess of zonal over meridional circulation, implying that the two circulation patterns are in some way opposed. Global temperatures have tended to rise more in zonal than in meridional regimes, the latter producing temperature slowdowns. The slowdowns, the consistent rise and oscillations in HadCRUT4 global temperature anomalies can be largely represented as a composite of the eight-year lagged accumulated difference of multidecadal tidal cycles, the near-real-time effect of higher-frequency tidal cycles, and an exponential rise in temperature. The global temperature composite projection to 2100 is comparable to published anthropogenic greenhouse gas emission scenarios.

Following a recent workshop on decadal climate variability [10], the question was posed: How long will the current slowdown last? Tom Knutson of NOAA estimated the upper bound to be 2030, while John Fyfe of the Canadian Centre for Climate Modeling and Analysis (CCCma) concluded that the slowdown is near its end if not already ended. The tidal approach presented here suggests that, in a sense, both may be right. Figure 1 shows the present meridional phase ending in 2025, agreeing with Knutson, while Figure 3 indicates that rising greenhouse gas levels would eliminate slowdowns in the future, agreeing with Fyfe.

There is a clear contrast between the response of atmospheric and oceanic oscillations to the tidal formulation: The atmospheric oscillations discussed—the ACI, smoothed and unsmoothed GMST residuals, detrended lower tropospheric temperatures, detrended atmospheric carbon dioxide levels, the QBO—all closely follow temporal patterns encompassed by the tidal formulation. However, the arguably simple ocean oscillations—the ONI and AMO—show a much poorer relationship to the tidal formulation, which suggests there is at least one more driver acting on the oceans.

White et al. [68] found that variability in Australian three- to five-year drought episodes were associated with the spatio-temporal evolution of global standing modes and travelling waves in co-varying sea surface temperature and sea-level pressure anomalies. A physical understanding of the non-stationarity of global SST/SLP (sea level pressure) modes and waves, and the thermodynamics of ocean-atmosphere coupling and teleconnections, required a global physical model.

The recent global temperature slowdown has been associated with cool eastern Pacific SSTs, which in turn stem from strengthening Pacific trade winds and a resulting increase in equatorial upwelling in the region [9]. McGregor et al. [69] linked the strengthening trade winds and Walker circulation, and decreased SSTs in the eastern Pacific to a warming trend in Atlantic SSTs and trans-basin coupled atmosphere/ocean variability. They noted previous modeling studies revealing a physical linkage between Atlantic and Pacific climate variability on decadal timescales, and found statistical evidence that Atlantic sea surface temperature anomalies play the dominant role in controlling Atlantic-Pacific trans-basin variability. They conclude that ocean basin coupling resulted in tropical Pacific cooling which in turn contributed to a decadal slowdown in global temperatures, but that the low correlation between trans-basin variability and detrended global temperatures meant that such coupling "is clearly not the main driver for the multidecadal [changes] in global warming".

Tidal forcing is a candidate to explain the multidecadal changes referred to, along with some higher frequency changes in the ocean oscillations described in the text. However, the rather moderate results obtained with ocean oscillations show that the tidal formulation is insufficient to explain all temporal variability in these oscillations, and that factors such as teleconnections and travelling waves linking ocean basins are also required. The tidal formulation may disclose the portion of ocean variability needing more study—for example, during zonal regimes for oscillations with meridional character such as the AMO and during meridional regimes for oscillations with zonal character such as the ONI.

The role of pollutant aerosols and solar irradiance on temperature has not been evaluated here, since the close simulation of temperature data by components already proposed limit the ability to define contributions from additional sources. McGregor et al. [70] examined the notion that radiative forcing from solar irradiance and volcanic emissions modulate the variance of the El Niño Southern Oscillation (ENSO). They were unable to reject the null hypothesis that solar variability has no influence, but could reject the null hypothesis that volcanic forcing had no effect.

Volcanic activity of a body requires internal heating. Interestingly, volcanism on Jupiter's moon Io is provided by the internal heat from tidal dissipation caused by Jupiter's large mass and the varying pull on Io due to Io's orbital eccentricity [71]. For volcanic activity on earth, this internal heating is provided by radioactivity [72]. The possible contribution of volcanoes was mentioned relating to Figure 5, but the results show little evidence of a residual pattern consistent with solar irradiance. Notwithstanding the potential effects of non-tidal forces on climate, this study suggests that cyclic patterns in climate, including those with periodicities 80–90 and 22 years that are often ascribed to Gleissberg and Hale variations in solar irradiance, are attributable predominantly and parsimoniously to tidal forcing.

The above coupling of radiative forcing with ENSO was taken up by Spencer and Braswell [73], who attributed multidecadal temperature variations to ENSO. Using ten adjustable parameters, they developed three models: the first involved a forcing pair, namely radiative forcing from anthropogenic GHGs and volcanic eruptions; the second model added to the first pair a non-radiative internal forcing proportional to an ENSO-related dataset; and the third model added to the first pair a radiative forcing proportional to the ENSO-related dataset. They found that the agreement of the models with monthly upper-level ocean temperature variations improved through the succession of models, to give rather good agreement. Their interpretation was that stronger El Niño activity caused internal radiative forcing of the climate system. The paper has attracted criticism on various grounds (e.g., [74]), but the association of ENSO, radiative forcing and global temperatures is a fruitful avenue to explore, and their paper provokes the question: What drives variations in ENSO activity?

Here, tidal forcing is hypothesized to drive global temperature variables on timescales from multidecadal to subdecadal by modulating the ocean release of CO_2 and other GHGs, the resulting pattern of radiative forcing, and the variations observed in ENSO activity. A recent paper [75] found nonlinear linkage between radiative forcing and global temperature anomalies, supporting a view that global warming may affect oscillations such as ENSO through feedback effects.

Overall, it is proposed that multidecadal timescale GMST anomalies are a combination of an exponential increase reflecting atmospheric greenhouse concentrations added to tidally-generated regime changes that modulate atmospheric GHG concentrations and their resulting radiative effect, and one result is the formation of temperature slowdowns. The multidecadal regimes determine the nature of climate oscillations on the bidecadal to subdecadal timescale. On that scale, and with the exception of likely contributions by episodes of volcanic eruptions that reduce insolation and radiative forcing by GHGs, tidal forces modulate atmospheric and presumably both surface and subsurface ocean movement that in turn affect GMST anomalies, ENSO activity and atmospheric concentrations of, and radiative forcing by, greenhouse gases.

Figure 12 is a schematic diagram of major steps in hypothesized chains of causality operating on the two timescales. It is hoped that this outline will stimulate and assist future tests of mechanisms and climate consequences of this hypothesis.

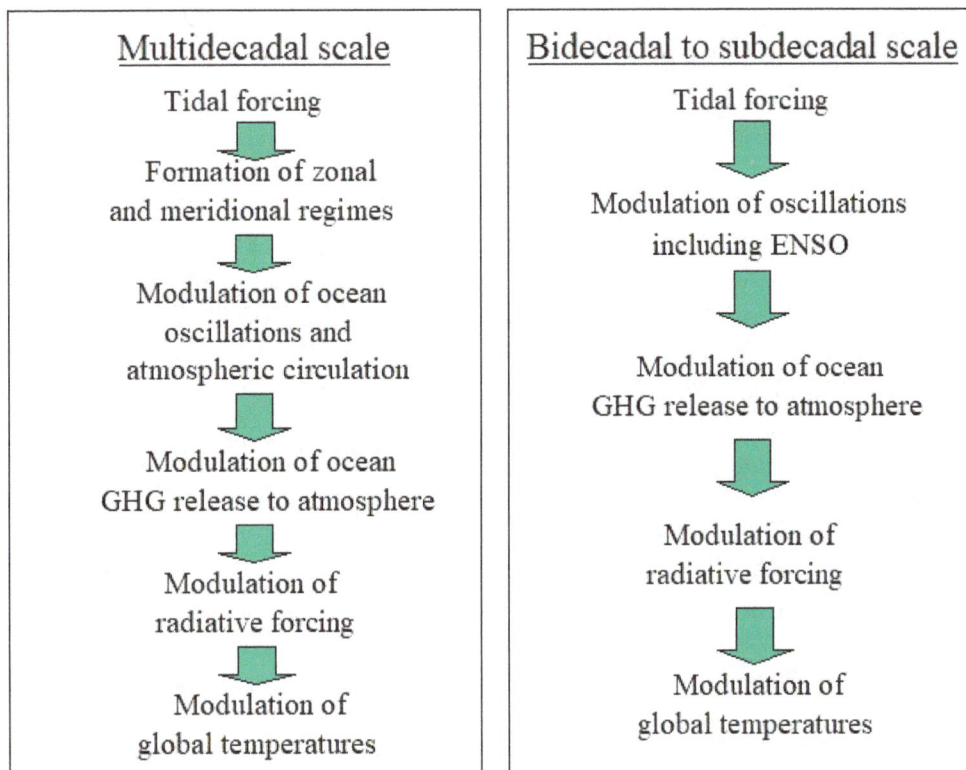

Figure 12. Schematic diagram of suggested steps linking tidal forcing to global climate effects.

4.2. Future Examination of the Tidal Hypothesis

In view of the potential that the tidal hypothesis offers to resolve questions on the pressing issue of anthropogenic climate change, it is urged that the hypothesis be examined through appropriate multidisciplinary studies. The original study [21] was based on a physical approach to orthogonal components in exogenous tidal forcing from the sun and moon. Many elements of the formulation were effectively fixed by the prior study, such as component reference times, the four fundamental tidal frequencies, and the nature and timing of zonal and meridional regimes. As such, the original formulation is inherently deterministic, predictive and substantially inflexible, and the developments in this paper are open to testing and falsifying in ways that include the following:

(i) The formulation, from time-averaged intervals of close syzygy determined from astronomical algorithms, varies from classical methods, and generates tidal components that are novel in terms of their periodicities, phases, and zonal or meridional regime-dependence. These qualitative features of the formulation need theoretical scrutiny. Tests could also be made for the presence and zonal or meridional nature of derived cycles over an extended timescale such as that shown in Figure 2, and for any proxy or other evidence of additional multidecadal changes in atmospheric or oceanic data over several centuries that coincide in timing with tidal regime changes.

(ii) Amplitudes for tidal contributions are specified in °C. Such quantitative features require examination by physical and GCM modeling and by empirical tests against other oscillations in the climate system.

(iii) It is speculated that greenhouse gases are released from the oceans by exogenous tidal forcing during zonal regimes, and generate a ~60-year oscillation in global temperatures with an average delay of eight years. The physical basis should be examined for this, in conjunction with assessing the physical likelihood that the tidal process causes lagged ~60-year oscillations in the ACI and LOD.

(iv) Without major mitigation of the increase in greenhouse gas concentrations, the implication that there should be no more decadal-scale slowdowns in global temperatures would become testable over the longer term.

In his book "Consilience", E.O. Wilson writes [76]: "The greatest challenge today … in all of science … is the accurate and complete description of complex systems. … . They think they know most of the elements and forces. The next task is to reassemble them, at least in mathematical models that capture the key properties of the entire ensembles." The climate system may require novel elements and methods to explore its complexity, and this paper suggests an element that may reward closer scrutiny. The empirical results look promising, and much may be gained by including tidal forcing in mathematical models intended to capture key properties of the climate system.

5. Conclusions

This paper may resolve some complexities surrounding the most pressing climate issue of our time: anthropogenic global warming. Following the original tidal hypothesis for exogenous climate forcing based on a simple three-dimensional physical analysis of the tidal influence on the earth by the sun and moon with all three bodies considered as rigid masses [21], this paper develops the hypothesis to derive tidal components that empirically but parsimoniously simulate several features related to the global climate system. These features include:

(i) a proximate cause for a regime-driven ~60-year oscillation in atmospheric circulation, the LOD and AAM;

(ii) the alternating rise and slowdown in global mean surface temperatures resulting from regime modulation and the ~60-year oscillation;

(iii) an exponential component in background global temperatures consistent with existing greenhouse gas emission scenarios; and

(iv) changes in atmospheric carbon dioxide concentrations, ENSO activity and global temperatures in response to subdecadal tidal components.

The tidal formulation represents a forcing that is exogenous, deterministic and therefore predictable. If the tidal parameterization can be physically validated and appropriately parameterized in GCMs, it would contribute to a greater and more comprehensive understanding of the past, present and future nature of temperature slowdowns and other features of the global climate system.

Appendix A

The original study [21] was not directly concerned with, nor did it address, fluid dynamics or tidal heights in the ocean surface. There is no disputing that high ocean tides are associated with full as well as new moon, but the study approached the tidal three-body problem involving the earth, moon and sun very simply as three interacting rigid spheres, with their separations defined in terms of their centers. (While gravitational force is proportional to the inverse square of distance to the perturbing body, tidal force is proportional to the inverse cube.)

The general three-body problem cannot be solved analytically, at least by practicable methods, so astronomical algorithms [22] were used to define the positions and directions over time of both moon and sun relative to the earth. With the known masses of moon and sun, the resultant three-dimensional combined forcing from the two bodies was resolved along orthogonal axes, parallel and perpendicular to the moon's orbit, analogous to zonal and meridional directions on earth. After the analysis was complete, it was found, as might be expected, that the greatest tidal force on the earth (considered as a vector) occurs close to new moon and a lunar eclipse—so that the forces of sun and moon on the earth act in the same direction—and when the lunar distance is close to perigee, where "close" means "less than or in the region of 357,000 km".

Treloar identified an irregular tidal component with mean period of 5.775 years, having meridional character. In addition, parent and daughter tidal pairs were identified with periods 86.795 and 20.295, 186 and 18.02, and 59.75 and 13.53 years. The first two pairs were directionally meridional, and the latter pair zonal, a characterization that allows comparison with circulation patterns as in the ACI. Daughters within a parent "envelope" were amplitude-modulated by the parent and phase-shifted from daughters around a neighboring parent. Figure 1 in Treloar [18] shows an overlapping pattern in the 59.75/13.53 pair, and similar features are shown in figures in Keeling and Whorf [28].

Wood [77] listed periods virtually identical to the above parent components, viz. at 86.834, several around 186 years, and 59.75 years. He also listed periods almost identical to those of the above daughter components, viz. at 20.294, 18.03 and 13.502 years, respectively; interestingly, the 13.502- and 20.294-year periods are in the ratio 2:3. The ~20.3-year tidal component may have been noted only by Wood and Treloar.

The daughter periods (e.g., ~20.3 years) do not exactly subdivide the parent periods (e.g., 86.795 years) so that there is an intervening phase shift (see figures in Treloar [21]). It is anticipated that a systemic response to such "interrupted" forcing would be as a time-averaged cycle, in the example just quoted as a cycle period of 86.795/4 or 21.70 years. The results in this paper support the idea that daughter components are manifested as time-averaged periodicities of about 21.70 (= 86.795/4), 18.60 (= 186.0/10) and 14.94 (= 59.75/4) years, respectively.

Using Fourmilab online software [25], the timing of tidal events involving close perigee at syzygy (new or full moon) were examined over multiple centuries, where "close" indicates "less than or in the region of 357,000 km". The tidal components in Treloar [21] had extrema (maxima or minima) at a reference time or "date-stamp" t_0 of 1918.20 for the 186.0/18.60 pair and 2039.96 for the other components. Close syzygy events apparently occur frequently enough in the sequence of cycles to establish a significant response by the ocean/atmosphere system. Stepping back intervals from the reference extrema times over several centuries led to virtually insignificant amendments to the seven periods mentioned: 5.778, 86.81, 21.70, 186.0, 18.60, 59.75 and 14.94 years, respectively. The agreement of Fourmilab results with cycle periods in Treloar [18] supports the accuracy of the algorithm and beating procedures. While small differences exist between the derived tidal components and those established from other sources, the resulting components have the advantage of being calibrated from a single, independent source. The amended tidal periods are listed in Tables 1–3. In the present paper, these derived early tidal components are expanded in number by harmonic and frequency combination methods described in the text.

Low frequency climate data can have Lorentzian band shapes [78] and sources therein), and envelopes for the beating envelopes are well described by overlapping Lorentzian functions, as the Equations for L below as a function of time t, where L_0 (the central amplitude) and H (the half band width at half maximum) are determined empirically from the fit to the beating events:

$$L(t) = L_0 \, (1/\pi) \, H/((t - T + zP)^2 + H^2) \tag{A1}$$

The amplitudes of parent and daughter cycles can be contained within Lorentzian envelopes, summed over integer values of z, as:

$$A(t) = \Sigma \, [\, L(t) \, (1 + \cos (2\pi \, (t - T + zP) \, / \, P\prime))] \tag{A2}$$

where (for example, for the 20.295-year cycle), P' is the daughter period (20.295 years), T is a parent cycle maximum such as 2039.96, P is the parent period (86.81 years), and z is an integer. The values empirically determined for L_0 and H were: For the 86.81-year cycle, 10,000 and 13; for the 186.0-year cycle, 4000 and 16; and for the 59.75-year cycle, 12,000 and 4.5.

Figure A1 displays Lorentzian envelopes for beats generated for the three pairs of tidal cycles. The Lorentzian L_0 and H parameters for the successive pairs are 10,000 and 13, 4000 and 16, and 12,000 and 4.5. The first two pairs are meridional, and the third pair is zonal. The figure displays "beating amplitudes" of 500, 160 and 1700, which, after testing other options, are used in the text as provisional measures of the respective component contributions.

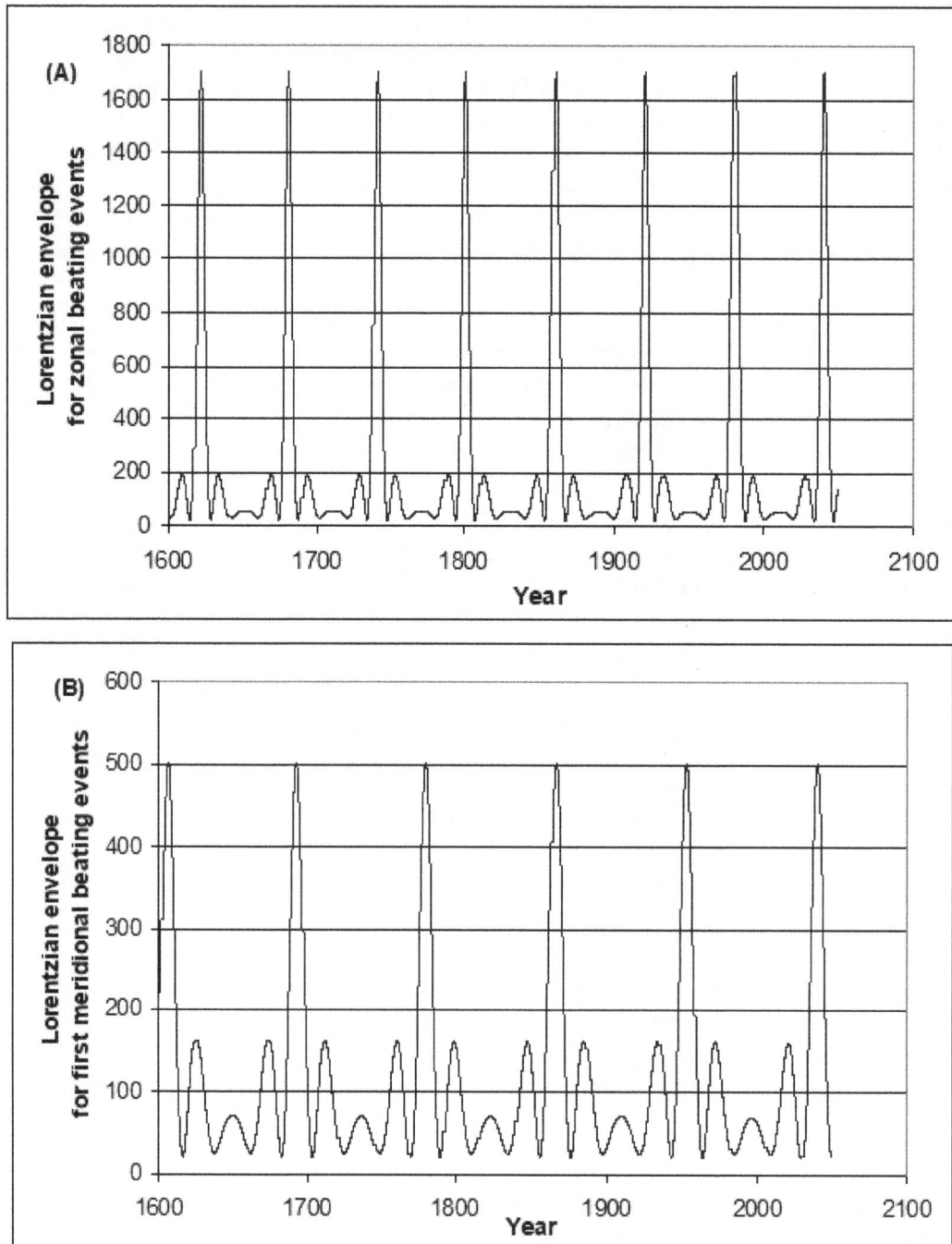

Figure A1. *Cont.*

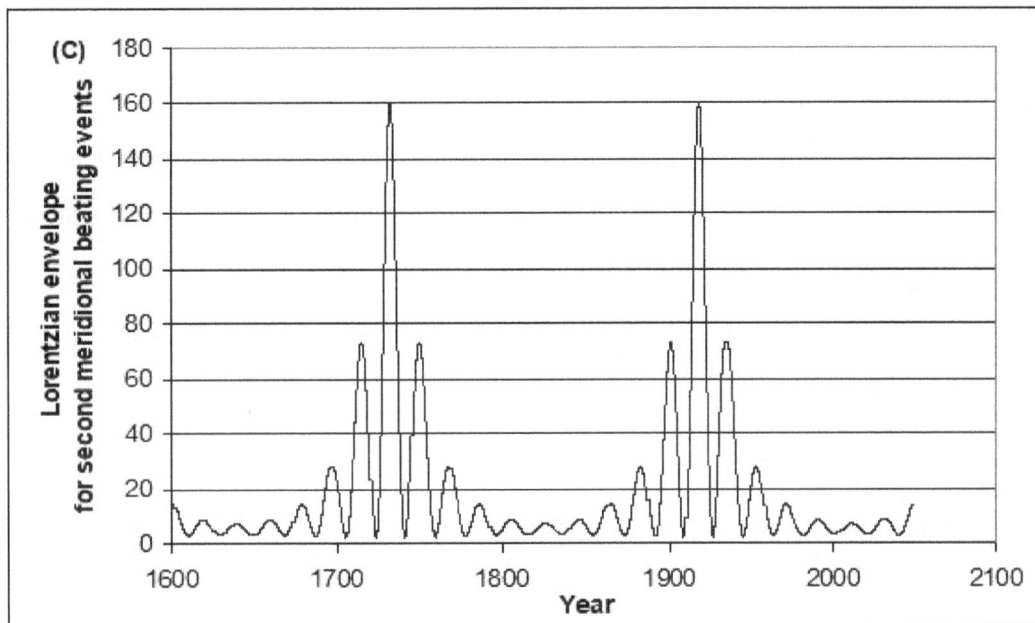

Figure A1. Lorentzian envelopes of beating events in tidal forcing for: (**A**) the 59.75-/14.94-year zonal pair; (**B**) the 86.81-/21.70-year meridional pair; and (**C**) the 186.0-/18.60-year meridional pair.

While regional climate data show non-uniform response to tidal parameters, it may be anticipated that global data should respond to the tidal parameters derived from Figure A1, and to the amplitudes of the component cycles. To test this, we tentatively identify the "beating amplitudes" of 86.81-, 186.0- and 59.75-year components by their central peak heights in Figure A1. These amplitudes supplement cycle reference times and periods for testing against climate data.

Figure A1 indicates that the 18.60-year component is not the most prominent decadal or bidecadal tidal component derived using this formulation. Tidal effects from the 18.60-year lunar nodal cycle have often focused on times of greatest positive or negative lunar declination, with no necessary relationship to the sun's position in relation to the moon's; the primary focus in the present formulation is on close syzygy, an orientation involving both bodies, termed "lunisolar" as opposed to the alternative "lunar" case involving only the tidal effect of the moon. However, the lunisolar term applies to the 186.0/18.60 case since the beating process in the first instance identified the major 186.0-year "close new moon" events centered on 1918.20 (centered around March 12) and at 186.0-year intervals, each such event surrounded before and after by intervals centering on 18.03-year saros events, from which 18.60-year time-averaged intervals were then defined. The 18.03-year saros is an eclipse cycle, though the 18.60-year cycle can be considered as a period that averages a succession of ten saros cycles with an intervening phase shift, in the same manner as described for several time-averaged tidal cycles in Treloar [18] and shown in Figure A1C. Between parent 186.0-year cycle maxima, the 18.03-year intervals of close perigee at new moon get rapidly out of synchrony with the time-averaged 18.60-year cycle over time, perhaps limiting climate responses to 18.60-year periodicities.

Acknowledgments: This paper would not have appeared without the unstinting advice and encouragement of Greg McKeon AM, to whom many thanks are offered. Grateful thanks are also due to the late Leonid Klyashtorin for information on the ACI, to Pieter Tans for permission to use NOAA/ESRL data on CO_2 annual increments, and to several anonymous referees for their invaluable counsel. This study was unfunded.

Conflicts of Interest: The author declares no conflict of interest.

References

1. Trenberth, K.E. Has there been a hiatus? *Science* **2015**, *349*, 691–692. [CrossRef] [PubMed]
2. Fyfe, J.C.; Meehl, G.A.; England, M.H.; Mann, M.E.; Santer, B.D.; Flato, G.M.; Hawkins, E.; Gillett, N.P.; Xie, S.-P.; Kosaka, Y.; et al. Making sense of the early-2000 warming slowdown. *Nat. Clim. Chang.* **2016**, *6*, 224–228. [CrossRef]
3. Santer, B.D.; Fyfe, J.C.; Pallotta, G.; Flato, G.M.; Meehl, G.A.; England, M.H.; Hawkins, E.; Mann, M.E.; Painter, J.F.; Bonfils, C.; et al. Causes of differences in model and satellite tropospheric warming rates. *Nat. Geosci.* **2017**, *10*, 478–485. [CrossRef]
4. Klyashtorin, L.B. *Climate Change and Long-Term Fluctuations of Commercial Catches: The Possibility of Forecasting*; UN FAO Fisheries Technical Paper 410; Food and Agriculture Organization of the United Nations: Rome, Italy, 2001.
5. Hide, R.; Birch, N.T.; Morrison, L.V.; Shea, D.J.; White, A.A. Atmospheric angular momentum fluctuations and changes in the length of the day. *Nature* **1980**, *286*, 114–117. [CrossRef]
6. Volland, H. Atmosphere and Earth's rotation. *Surv. Geophys.* **1996**, *17*, 101–144. [CrossRef]
7. Gonella, J.A. Ocean-atmosphere coupling and short-term fluctuations of earth rotation. *Oceanol. Acta* **1987**, *10*, 123–127.
8. Dickey, J.O.; Marcus, S.L.; de Viron, O. Air temperature and anthropogenic forcing: Insights from the solid Earth. *J. Clim.* **2011**, *24*, 569–574. [CrossRef]
9. Oviatt, C.; Smith, L.; McManus, M.C.; Hyde, K. Decadal patterns of westerly winds, temperatures, ocean gyre circulations and fish abundance: A review. *Climate* **2015**, *3*, 833–857. [CrossRef]
10. Purcell, A.; Huddleston, N. *Frontiers in Decadal Climate Variability: Proceedings of a Workshop*; The National Academies Press: Washington, DC, USA, 2016.
11. Liu, W.; Xie, S.-P.; Lu, J. Tracking ocean heat uptake during the surface warming hiatus. *Nat. Commun.* **2016**, *7*, 10926. [CrossRef] [PubMed]
12. Whitmarsh, F.; Zika, J.; Czaja, A. *Ocean Heat Uptake and the Global Surface Temperature Record*; Grantham Institute Briefing Paper No. 14; Imperial College London: London, UK, 2015.
13. England, M.H.; McGregor, S.; Spence, P.; Meehl, G.A.; Timmerman, A.; Cai, W.; Gupta, A.S.; McPhaden, M.J.; Purich, A.; Santoso, A. Recent intensification of wind-driven circulation in the Pacific and the ongoing warming hiatus. *Nat. Clim. Chang.* **2014**, *4*, 222–227. [CrossRef]
14. Meehl, G.A.; Arblaster, J.M.; Bitz, C.M.; Chung, T.Y.; Teng, H. Antarctic sea-ice expansion between 2000 and 2014 driven by tropical Pacific decadal climate variability. *Nat. Geosci.* **2016**, *9*, 590–595. [CrossRef]
15. Doodson, A.T. The Harmonic Development of the Tide-Generating Potential. *Proc. Roy. Soc. A Math. Phys. Eng. Sci.* **1921**, *100*, 305–329. [CrossRef]
16. Cartwright, D.E.; Taylor, R.J. New computations of the tide generating potential. *Geophys. J. R. Astron. Soc.* **1971**, *23*, 45–74. [CrossRef]
17. Tamura, Y. A harmonic development of the tide-generating potential. *Bull. Inf. Marées Terrestres* **1987**, *99*, 6813–6855.
18. Kantha, L.K.; Clayson, C.A. *Numerical Models of Oceans and Oceanic Processes*; Academic Press: New York NY, USA, 2000.
19. Tsien, H.S. *Engineering Cybernetics*; McGraw-Hill: New York, NY, USA, 1954.
20. Keeling, C.D.; Whorf, T.P. Possible forcing of global temperatures by the oceanic tides. *Proc. Nat. Acad. Sci. USA* **1997**, *94*, 8321–8328. [CrossRef] [PubMed]
21. Treloar, N.C. Luni-solar influences on climate variability. *Int. J. Climatol.* **2002**, *22*, 1527–1542. [CrossRef]
22. Meeus, J. *Astronomical Algorithms*; Willman-Bell: Richmond, VA, USA, 1991.
23. Munk, W.H.; Dzieciuch, M.; Jayne, S. Millennial climate variability: Is there a tidal connection? *J. Clim.* **2002**, *15*, 370–384. [CrossRef]
24. Ray, R.D. Decadal climate variability: Is there a tidal connection? *J. Clim.* **2007**, *20*, 3542–3560. [CrossRef]
25. Fourmilab Software. Available online: www.fourmilab.ch/earthview/pacalc.html (accessed on 24 April 2017).
26. Toggweiler, J.R.; Samuels, B. Effect of Drake Passage on the global thermohaline circulation. *Deep-Sea Res.* **1995**, *42*, 477–500. [CrossRef]
27. Munk, W.H.; Wunsch, C. Abyssal recipes II: Energetics of tidal and wind mixing. *Deep Sea Res. I* **1998**, *45*, 1977–2010.

28. Keeling, C.D.; Whorf, T.P. The 1,800-year oceanic tidal cycle: A possible cause of rapid climate change. *Proc. Nat. Acad. Sci. USA* **2000**, *97*, 3814–3819. [CrossRef] [PubMed]

29. Morice, C.P.; Kennedy, J.J.; Rayner, N.A.; Jones, P.D. Quantifying uncertainties in global and regional temperature change using an ensemble of observational estimates: The HadCRUT4 dataset. *J. Geophys. Res.* **2012**, *117*, D08101. [CrossRef]

30. University of Alabama in Huntsville (UAH). Lower Troposphere Temperature. Available online: http://www.nsstc.uah.edu/data/msu/v6.0beta/tlt/uahncdc_lt_6.0beat5.txt (accessed on 4 January 2017).

31. NOAA/ESRL Mauna Loa Data on CO_2 Annual Increments. Available online: ftp://aftp.cmdl.noaa.gov/products/trends/co2/co2_gr_mlo.txt (accessed on 28 February 2017).

32. Newman, M.; Alexander, M.A.; Ault, T.R.; Cobb, K.M.; Deser, C.; DiLorenzo, E.; Mantua, N.; Miller, A.J.; Minobe, S.; Nakamura, H.; et al. The Pacific Decadal Oscillation, Revisited. *J. Clim.* **2016**, *29*, 4399–4427. [CrossRef]

33. NOAA/ESRL QBO: Physical Sciences Division of NOAA/ESRL. Available online: https://www.esrl.noaa.gov/psd/data/correlation/qbo.data (accessed on 15 January 2017).

34. NOAA/PSD ONI: Physical Sciences Division of NOAA/ESRLE. Available online: https://www.esrl.noaa.gov/psd/data/correlation/oni.data (accessed on 3 February 2017).

35. Enfield, D.B.; Mestas-Nunez, A.M.; Trimble, P.J. The Atlantic multidecadal oscillation and its relationship to rainfall and river flows in the continental U.S. *Geophys. Res. Lett.* **2001**, *28*, 2077–2080. [CrossRef]

36. Baldwin, M.P.; Gray, L.J.; Dunkerton, T.J.; Hamilton, K.; Haynes, P.H.; Randel, W.J.; Holton, J.R.; Alexander, M.J.; Hirota, I.; Horinouchi, T.; et al. The quasi-biennial oscillation. *Rev. Geophys.* **2001**, *39*, 179–229. [CrossRef]

37. Lott, F.; Denvil, S.; Butchart, N.; Cagnazzo, C.; Giorgetta, M.A.; Hardiman, S.C.; Manzini, E.; Krismer, T.; Duvel, J.-P.; Maury, P.; et al. Kelvin and Rossby-gravity wave packets in the lower stratosphere of some high-top CMIP5 models. *J. Geophys. Res. Atmos.* **2014**, *119*, 2156–2173. [CrossRef]

38. Butchart, N.; Charlton-Perez, A.J.; Cionni, I.; Hardiman, S.C.; Haynes, P.H.; Krüger, K.; Kushner, P.J.; Newman, P.A.; Osprey, S.M.; Perlwitz, J.; et al. Multimodel climate and variability of the stratosphere. *J. Geophys. Res. Atmos.* **2011**, *116*, D05102. [CrossRef]

39. NOAA Temperature Record Updates and the 'Hiatus'. Available online: https://climatedataguide.ucar.edu/climate-data/nino-sst-indices-nino-12-3-34-4-oni-and-tni (accessed on 23 August 2017).

40. Knudsen, M.F.; Seidenkrantz, M.-S.; Jacobsen, B.H.; Kuijpers, A. Tracking the Atlantic Multidecadal Oscillation through the last 8000 years. *Nat. Commun.* **2011**, *2*. [CrossRef] [PubMed]

41. Knight, J.R.; Allan, R.J.; Folland, C.K.; Vellinga, M.; Mann, M.E. A signature of persistent natural thermohaline circulation cycles in observed climate. *Geophys. Res. Lett.* **2005**, *32*, 475–480. [CrossRef]

42. Rahmstorf, S.; Box, J.E.; Feulner, G.; Mann, M.E.; Robinson, A.; Rutherford, S.; Schaffernicht, E.J. Exceptional twentieth-century slowdown in Atlantic Ocean overturning circulation. *Nat. Clim. Chang.* **2015**, *5*, 475–480. [CrossRef]

43. Karl, T.R.; Arguez, A.; Huang, B.; Lawrimore, J.H.; McMahon, J.R.; Menne, M.J.; Peterson, T.C.; Vose, R.S.; Zhang, H.-M. Possible artifacts of data biases in the recent global surface warming hiatus. *Science* **2015**, *348*, 1469–1472. [CrossRef] [PubMed]

44. Yan, X.-H.; Boyer, T.; Trenberth, K.; Karl, T.R.; Xie, S.-P.; Nieves, V.; Tung, K.-K.; Roemmich, D. The global warming hiatus: Slowdown or redistribution? *Earth's Future* **2016**, *4*, 472–482. [CrossRef]

45. Cowtan, K.; Way, R.G. Coverage bias in the HadCRUT4 temperature series and its impact on recent temperature trends. *Quart. J. R. Meteorol. Soc.* **2014**, *140*, 1935–1944. [CrossRef]

46. Climate Data Sets. Available online: http://www.realclimate.org/index.php/archives/2015/06/noaa-temperature-record-updates-and-the-hiatus/ (accessed on 22 October 2017).

47. Mantua, N.J.; Hare, S.R.; Zhang, Y.; Wallace, J.M.; Francis, R.C. A Pacific interdecadal climate oscillation with impacts on salmon production. *Bull. Am. Meteorol. Soc.* **1997**, *78*, 1069–1079. [CrossRef]

48. IPCC (Intergovernmental Panel on Climate Change). *Climate Change 2007: Synthesis Report. Summary for Policymakers*; Fourth Assessment Report; IPCC: Geneva, Switzerland, 2007.

49. IPCC (Intergovernmental Panel on Climate Change). *Contribution of Working Groups I, II and III to the Fifth Assessment Report of the Intergovernmental Panel on Climate Change*; Climate Change 2014: Synthesis Report; Core Writing Team, Pachauri, R.K., Meyer, L.A., Eds.; IPCC: Geneva, Switzerland, 2014; 151p.

50. Mann, K.H.; Lazier, J.R.N. *Dynamics of Marine Ecosystems: Biological-Physical Interactions in the Oceans*, 3rd ed.; Blackwell: Oxford, UK, 2006.

51. Egger, J. The moisture torque. *Meteorol. Z.* **2006**, *15*, 671–673. [CrossRef]

52. Jacob, D.J. *Introduction to Atmospheric Chemistry*; Princeton University Press: Princeton, NJ, USA, 1999.

53. Huang, H.P.; Weickmann, K.M.; Hsu, C.J. Trend in atmospheric angular momentum in a transient climate change simulation with greenhouse gas and aerosol forcing. *J. Clim.* **2001**, *14*, 1525–1534. [CrossRef]

54. How We Calculate the Time Series of Smoothed Annual Average Temperature. Available online: http://www.metoffice.gov.uk/hadobs/hadcrut3/smoothing.html (accessed on 26 February 2017).

55. Tourre, Y.M.; Rajagopalan, B.; Kushnir, Y. Dominant patterns of climate variability in the Atlantic Ocean during the last 136 years. *J. Clim.* **1999**, *12*, 2285–2299. [CrossRef]

56. White, W.B.; Tourre, Y.M. Global SST/SLP waves during the 20th century. *Geophys. Res. Lett.* **2003**, *30*, 53–57. [CrossRef]

57. Chen, G.; Shao, B.; Han, Y.; Ma, J.; Chapron, B. Modality of semiannual to multidecadal oscillations in global sea surface temperature variability. *J. Geophys. Res. Oceans* **2010**, *115*, C03005. [CrossRef]

58. Gruzdev, A.N.; Bezverkhny, V.A. Two regimes of the quasi-biennial oscillation in the equatorial stratospheric wind. *J. Geophys. Res.* **2000**, *105*, 29435–29443. [CrossRef]

59. Baldwin, M.P.; Tung, K.-K. Extratropical QBO signals in angular momentum and wave forcing. *Geophys. Res. Lett.* **1994**, *21*, 2717–2720. [CrossRef]

60. Ruzmaikin, A.; Feynman, J.; Jiang, X.; Yung, Y.L. Extratropical signature of the quasi-biennial oscillation. *J. Geophys. Res.* **2005**, *110*, D11111. [CrossRef]

61. Stone, R.C.; Hammer, G.L.; Marcussen, T. Prediction of global rainfall probabilities using phases of the Southern Oscillation Index. *Nature* **1996**, *384*, 252–255. [CrossRef]

62. Kilbourne, K.H.; Quinn, T.M.; Webb, R.; Guilderson, T.; Nyberg, J. Paleoclimate proxy perspective on Caribbean climate since the year 1751: Evidence of cooler temperatures and multidecadal variability. *Paleoceanography* **2008**, *23*. [CrossRef]

63. Moore, G.W.K.; Halfar, J.; Majeed, H.; Adey, W.; Kronz, A. Amplification of the Atlantic Multidecadal Oscillation associated with the onset of the industrial-era warming. *Nat. Sci. Rep.* **2017**, *7*. [CrossRef] [PubMed]

64. Miles, W.M.; Divine, V.D.; Furevik, T.; Jansen, E.; Moros, M.; Ogilvie, A.E.J. A signal of persistent Atlantic multidecadal variability in Arctic sea ice. *Geophys. Res. Lett.* **2013**, *41*, 463–469. [CrossRef]

65. Gray, S.T.; Graumlich, L.J.; Betancourt, J.L.; Pederson, G.T. A tree-ring based reconstruction of the Atlantic Multidecadal Oscillation since 1567 AD. *Geophys Res. Lett.* **2004**, 31. [CrossRef]

66. Mann, M.E.; Park, J.; Bradley, R.S. Global interdecadal and century-scale climate oscillations during the past five centuries. *Nature* **1995**, *378*, 266–268. [CrossRef]

67. Berger, W.H. Solar modulation of the North Atlantic Oscillation: Assisted by the tides? *Quat. Int.* **2008**, *188*, 24–30. [CrossRef]

68. White, W.B.; McKeon, G.; Syktus, J. Australian drought: The interference of multi-spectral standing modes and travelling waves. *Int. J. Climatol.* **2003**, *23*, 631–662. [CrossRef]

69. McGregor, S.; Timmermann, A.; Stuecker, M.F.; England, M.; Merrifield, M.; Jin, F.-F.; Chikamoto, Y. Recent Walker circulation strengthening and Pacific cooling amplified by Atlantic warming. *Nat. Clim. Chang.* **2014**, *4*, 888–892. [CrossRef]

70. McGregor, S.; Timmermann, A.; Timm, O. A unified proxy for ENSO and PDO variability since 1650. *Clim. Past* **2010**, *6*, 1–17. [CrossRef]

71. Peale, S.J.; Cassen, P.; Reynolds, R.T. Melting of Io by tidal dissipation. *Science* **1979**, *203*, 892–894. [CrossRef] [PubMed]

72. Turcotte, D.L.; Schubert, G. *Geodynamics*, 2nd ed.; Cambridge University Press: Cambridge, UK, 2002.

73. Spencer, R.W.; Braswell, W.D. The Role of ENSO in Global Ocean Temperature Changes during 1955–2011 Simulated with a 1D Climate Model. *Asia-Pac. J. Atmos. Sci.* **2014**, *50*, 229–237. [CrossRef]

74. Abraham, J.P.; Kumarl, S.; Bickmore, B.R.; Fasullo, J.T. Issues Related to the Use of One-dimensional Ocean-diffusion Models for Determining Climate Sensitivity. *J. Earth Sci. Clim. Chang.* **2014**, *5*, 1–7. [CrossRef]

75. Morana, C.; Sbrana, G. *Temperature Anomalies, Radiative Forcing and ENSO*; Management and Statistics Working Paper no. 361; University of Milan-Bicocca Department of Economics: Milan, Italy, 2017. Available online: https://ssrn.com/abstract=2915022 (accessed on 15 October 2017).

76. Wilson, E.O. *Consilience: Unity of Knowledge*; Vintage Books: New York, NY, USA, 1998.

77. Wood, F.J. *Tidal Dynamics: Coastal Flooding, and Cycles of Gravitational Force*; D. Reidel: Norwell, MA, USA, 1996.

78. Pelletier, J.D. Analysis and modeling of the natural variability of climate. *J. Clim.* **1997**, *10*, 1331–1342. [CrossRef]

7

Demonstrating the Effect of Forage Source on the Carbon Footprint of a Canadian Dairy Farm Using Whole-Systems Analysis and the Holos Model: Alfalfa Silage vs. Corn Silage

Shannan M. Little [1,*], Chaouki Benchaar [2], H. Henry Janzen [1], Roland Kröbel [1], Emma J. McGeough [3] and Karen A. Beauchemin [1] (iD)

[1] Agriculture and Agri-Food Canada, Lethbridge Research and Development Centre, Lethbridge, AB T1J 4B1, Canada; Henry.Janzen@agr.gc.ca (H.H.J.); Roland.Kroebel@agr.gc.ca (R.K.); Karen.Beauchemin@agr.gc.ca (K.A.B.)

[2] Agriculture and Agri-Food Canada, Sherbrooke Research and Development Centre, Sherbrooke, QC J1M 0C8, Canada; Chaouki.Benchaar@agr.gc.ca

[3] Department of Animal Science, University of Manitoba, Winnipeg, MB R3T 2N2, Canada; Emma.Mcgeough@umanitoba.ca

* Correspondence: Shannan.Little@agr.gc.ca

Abstract: Before recommending a feeding strategy for greenhouse gas (GHG) mitigation, it is important to conduct a holistic assessment of all related emissions, including from those arising from feed production, digestion of these feeds, managing the resulting manure, and other on-farm production processes and inputs. Using a whole-systems approach, the Holos model, and experimentally measured data, this study compares the effects of alfalfa silage- versus corn silage-based diets on GHG estimates in a simulated Canadian dairy production system. When all emissions and sources are accounted for, the differences between the two forage systems in terms of overall net GHG emissions were minimal. Utilizing the functional units of milk, meat, and total energy in food products generated by the system, the comparison demonstrates very little difference between the two silage production systems. However, the corn silage system generated 8% fewer emissions per kg of protein in food products as compared to the alfalfa silage system. Exploratory analysis of the impact of the two silage systems on soil carbon showed alfalfa silage has greater potential to store carbon in the soil. This study reinforces the need to utilize a whole-systems approach to investigate the interrelated effects of management choices. Reported GHG reduction factors cannot be simply combined additively because the interwoven effects of management choices cascade through the entire system, sometimes with counter-intuitive outcomes. It is necessary to apply this whole-systems approach before implementing changes in management intended to reduce GHG emissions and improve sustainability.

Keywords: agriculture; carbon footprint; carbon sequestration; dairy; enteric methane; greenhouse gas emissions; life cycle assessment; livestock; mitigation; soil organic carbon

1. Introduction

The livestock industry is challenged with reducing greenhouse gas (GHG) emissions to limit the negative impacts of climate change. Using a life cycle approach, it has been estimated that livestock production contributes about 14.5% of global anthropogenic GHG emissions, with dairy production accounting for about 20% of this figure [1]. Dairy production and the demand for milk products continue to rise, thus GHG emissions associated with livestock production will increase further unless mitigation strategies and new technologies are adopted [2].

Approximately 70–85% of total GHG emissions associated with milk consumption in industrialized countries can be attributed to activities on the farm (i.e., "cradle-to-farm gate emissions" [3,4]). This highlights the opportunity for dairy producers to make management choices that can significantly reduce the overall carbon footprint (e.g., sum of GHG emissions and removals in a product system) or GHG intensity (e.g., sum of GHG emissions and removals expressed relative to a kilogram of fat and protein corrected milk (FPCM)) of dairy products.

While overall milk production in Canada has increased slightly since 1990, the dairy cattle population has been decreasing steadily due to increased productivity per cow [5]. GHG emissions attributed to the Canadian dairy sector have remained stable during the last decade, largely due to greater feed consumption of cows, leading to greater enteric methane (CH_4) emissions [6]. Farm-based life cycle assessments of Canadian dairy production have identified methane (CH_4), mostly from enteric fermentation, as the largest source of GHG from dairy farms [7,8].

A number of comprehensive reviews document potential strategies for reducing enteric CH_4 emissions from dairy cows [9–12]. Strategies for reducing enteric CH_4 include supplementing diets with feed additives and dietary fats, offering diets high in starch content, improving forage quality, and changing forage type (e.g., corn silage and legume forages). For example, the high starch content of corn silage, compared to perennial grass or legume silage, tends to shift ruminal fermentation patterns towards the formation of more propionate and less acetate, which lowers hydrogen availability for methanogens that use hydrogen and carbon dioxide (CO_2) to produce CH_4. Legumes tend to decrease CH_4 synthesis compared with grasses because of their shorter residence time in the rumen, although comparisons can be confounded by differences in physiological maturity and growing conditions. In addition, the source of forage in the diet can increase animal productivity, thereby lowering enteric CH_4 emissions per unit of product. While the use of alfalfa and corn silage rather than grasses may reduce enteric CH_4 emissions, the availability of forage crops depends on local growing conditions.

Some dietary mitigation strategies can be implemented without large-scale changes to dairy infrastructure. However, a change in forage feeding affects the feed production system as well as milk production and composition due to differences in intake and nutrient availability (i.e., digestibility). For example, Hassanat et al. [13] reported that, compared to lactating dairy cows fed alfalfa silage, cows fed corn silage produced 10% less enteric CH_4 and 6% more milk, but the milk contained less fat and more protein. Similarly, Benchaar et al. [14] reported that dairy cows fed corn silage produced 14% less CH_4 and 16% more milk, with less fat and more protein, compared with cows fed barley silage diets. While the forage source can affect CH_4 emissions and milk production from the dairy system, these effects must be considered in tandem with effects on emissions from affiliated changes in the agronomic system. The overall effect of changing the forage source on the total GHG emissions from dairy systems remains unclear; for example, planting perennial forages such as alfalfa may also reduce agricultural GHG emissions by sequestering CO_2 in agricultural soils [15–17]. While changes in soil carbon can affect the carbon footprint of agricultural systems, they are rarely included in GHG analysis of dairy products because of the complexity and lack of consensus on methodology [4,7,18,19].

Before recommending a change in forage management as a strategy for enteric CH_4 mitigation, it is important to conduct a holistic—or whole-farm—assessment of all contributing emissions, including those from feed production and digestion, managing the resulting manure, and other on-farm production processes and inputs. While agriculture can help reduce global GHG emissions, finding the best approach is challenging; for example, a dietary change to reduce enteric CH_4 production may inadvertently increase emissions elsewhere in the system [20,21].

The cumulating and cascading effects of management practices on emissions from agricultural systems can rarely be measured directly, so whole-farm emissions are normally estimated using mathematical models [22]. An example of such a model is Holos, a whole-farm model and software tool developed by Agriculture and Agri-Food Canada to estimate GHG emissions from Canadian farm systems (www.agr.gc.ca/holos-ghg, [23]). Holos was designed as an exploratory tool to test possible ways of reducing GHG emissions prior to implementation. The objective of this study was to compare GHG emissions from corn silage- and alfalfa silage-based dairy systems, using whole-farm analyses, including effects on soil carbon.

2. Materials and Methods

This analysis followed the baseline scenario described by McGeough et al. [8], spanning the entire lifespan of an average dairy cow rather than simulating a single, point-in-time condition, which allowed us to account for the dynamics of the herd over time. The simulated farm was located in Napierville in the Montérégie region of Québec, Canada, within Ecodistrict 541 in the Mixed Woods Plains Ecozone. Precipitation and evapotranspiration during the growing season (May–October) averaged 559 and 529 mm, respectively [24]. The soil was a Humic Gleysol with a fine texture [25]. Input data for lactating cows were based on the study of Hassanat et al. [13], conducted at a nearby research facility (Ecodistrict 483), which examined the effects of forage source on enteric CH_4 emissions, nitrogen (N) excretion, and milk performance. Utilizing these input data, we compared two scenarios, using (1) alfalfa or (2) corn as the silage component of the system diet (for lactating cows, dry cows, and heifers).

2.1. Animal Management

2.1.1. Herd Dynamics and Diet

The scenario represents a dairy farm with 60 lactating Holstein cows. The herd dynamics considered the rearing of heifers, subsequent birth of female and male calves, production of veal calves, and average number of lactations per cow (Figure 1). The analysis begins with the birth of calves and ends with the cull of the final group of lactating cows during the sixth year of the production cycle.

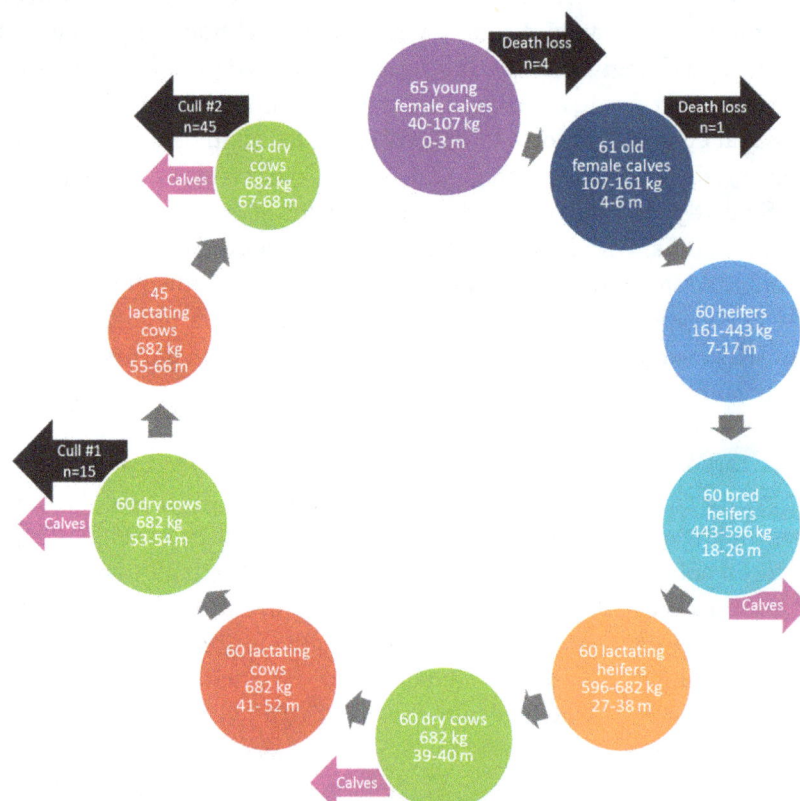

Figure 1. Dairy production timeline including dairy cows and offspring (m = months). Diets for each animal category are given in Table 1.

To achieve a herd size of 60 lactating cows while factoring in death loss, 65 female calves were required. Calves were fed milk replacer and concentrate for the first three months [8] (Table 1). At four months of age, 61 calves transitioned to a mixed hay (legume and grass) and concentrate diet. At seven months, 60 calves transitioned to a mixed hay and silage diet (alfalfa or corn, depending on scenario). Death losses of 7.8 and 1.8% for each of these time periods, respectively, were assumed [26].

Table 1. Feed ingredients and characteristics of diets for each animal category.

Animal Group	Lactating Cows		Dry Cows, Bred Heifers, Heifers		Calves	Calves	Veal Calves	Veal Calves
Scenario/Age Group	Alfalfa Silage	Corn Silage	Alfalfa Silage	Corn Silage	4–6 Months	0–3 Months	4–6.5 Months	0–3 Months
Feed ingredient (% dry matter (DM))								
Alfalfa silage	56.4	-	27.0	-	-	-	-	-
Corn silage	-	56.4	-	27.0	-	-	-	-
Corn grain, ground	25.5	12.4	-	-	1.75	10.5	100.0	40.0
Barley grain	-	-	-	-	1.75	10.5	-	-
Soybean meal	2.2	16.2	-	-	1.25	7.5	-	-
Soybean hulls	5.9	5.8	-	-	-	-	-	-
Grass/legume hay	3.2	3.2	73.0	73.0	95.0	-	-	-
Corn gluten feed	2.1	3.0	-	-	-	-	-	-
Rumen inert fat	2.0	-	-	-	-	-	-	-
Urea	-	0.2	-	-	-	-	-	-
Calcium carbonate	0.5	0.6	-	-	-	-	-	-
Potassium carbonate	-	0.4	-	-	-	-	-	-
Mineral/vitamin supplement	2.3	1.8	-	-	0.25	1.5	-	-
Milk replacer	-	-	-	-	-	70.0	-	60.0
Diet characteristics								
Total digestible nutrient content (% DM)	67.7	70.5	58.5	60.2	59.5	89.4	88.0	89.2
Crude protein (% DM)	16.8	15.6	18.9	15.3	20.4	19.4	9.0	15.6
Y_m (% gross energy intake)	5.85	5.27	6.80	6.50	6.50	0.90	3.00	1.20

Heifers were artificially inseminated at 18 months and continued on the mixed hay and silage diet until calving at 27 months, when the initial lactation cycle began. Primiparous cows (1st lactation), weighing 596 kg, were fed the lactation diet for 12 months prior to entering into a two-month non-lactating period (dry cow) where they were fed the dry cow diet. Cows reached maturity at the beginning of this dry period at 682 kg. The lactation diet, including either alfalfa silage or corn silage at 56.4% of dietary dry matter (DM), was described by Hassanat et al. [13]. Additional ingredients were added to the diets to ensure adequate crude protein, mineral, and vitamin levels. The dry cow diet consisted mainly of mixed hay and was supplemented with the corresponding silage depending on the scenario (Table 1).

Dry matter intake for mature cows was 21.7 kg head^{-1} day^{-1} (alfalfa-silage diet) or 24.6 kg head^{-1} day^{-1} (corn-silage diet), based on measurements by Hassanat et al. [13]. At the end of the second lactation and dry period, 15 cows were culled from the herd, leaving 45, simulating a cow replacement rate of 31%, with cows retained for 2.75 lactations, representing a typical Québec dairy [27]. After the third and final lactation, dry period, and calving, at 69 months, the remaining 45 cows were culled and sent to slaughter. These scenarios represent average practices, recognizing that animals would be culled throughout the cycle in common practice.

The 135 calves not required for replacement of lactating cows entered into the veal system and were fed milk replacer and corn grain from birth to three months. Factoring in death loss, at four months, 127 veal calves were fed corn grain until their slaughter at 6.5 months weighing 273 kg.

2.1.2. Milk Production and Housing

Assumed daily average milk production and composition were from Hassanat et al. [13] (Table 2). Milk production for first lactation heifers was 93% of mature cow production [27]. During lactation, cows were housed in individual tie-stalls and bedded with barley straw grown on the farm. Manure was removed regularly and stockpiled (solid storage). Dry cows, heifers, and all calves were group housed in deep-bedded pens with straw bedding added to absorb moisture. All manure was spread on the farm's lands once per year.

Table 2. Milk production and composition.

Item	Alfalfa Silage Scenario	Corn Silage Scenario
Milk production (kg head^{-1} day^{-1})	32.3	34.3
Fat and protein corrected milk (FPCM) (kg head^{-1} day^{-1}) [a]	30.7	30.5
Milk fat (%)	3.88	3.26
Milk protein (%)	3.04	3.22

[a] Standardized to 4% fat, 3.3% protein.

2.2. Crop Production and Imported Feed

Alfalfa silage, corn silage, mixed hay, corn grain, barley grain, and soybean for meal and hulls were grown on the simulated farm, with area of each based on feed requirements, yield of each crop, and losses from harvest, storage, and feed wastage. Yield, fertilizer rates, and lime application rates were representative of those found in the region (Table 3), except for the alfalfa silage and corn silage, where we used data from Hassanat et al. [13]. Crops were grown using reduced tillage and were not irrigated. Alfalfa silage and mixed hay crops were assumed to be a four-year perennial stand.

Table 3. Characteristics of crops grown on farm.

Characteristic	Alfalfa Silage	Corn Silage	Mixed Hay	Corn Grain	Barley Grain	Soybean	Barley Silage [f]
Agronomic characteristics							
Yield (kg ha^{-1}) [a]	6000	10,000	5020	8300	3100	2700	7000
Nitrogen fertilizer rate (kg N ha^{-1}) [b]	0	150	0	160	80	20	n.a.
Phosphorus fertilizer rate (kg P$_2$O$_5$ ha^{-1}) [b]	0	50	0	50	35	40	n.a.
Herbicide use [b]	No	Yes	No	Yes	Yes	Yes	n.a.
Harvest/ storage loss (%) [c]	12	12	12	3	3	3	n.a.
Feed wastage (%) [c]	5	5	20	0	0	0	n.a.
Moisture content (%) [d]	0	0	13	15	12	14	0
Lime application (kg CaCO$_3$ ha^{-1}) [e]	0	500	0	500	300	200	n.a.
Relative dry matter allocation							
Yield ratio [d]	0.40	0.72	0.40	0.47	0.38	0.30	0.72
Above ground residue ratio [d]	0.10	0.08	0.10	0.38	0.47	0.45	0.13
Below ground residue ratio [d]	0.50	0.20	0.50	0.15	0.15	0.25	0.15
Residue nitrogen content (kg N kg^{-1})							
Above ground [d]	0.015	0.013	0.015	0.005	0.007	0.006	n.a.
Below ground [d]	0.015	0.007	0.015	0.007	0.010	0.010	n.a.
Yearly crop area (ha)							
Alfalfa silage scenario	29.05	0	18.62	9.83	0.08	4.33	n.a.
Corn silage scenario	0	18.17	17.75	5.79	0.08	15.49	n.a.

n.a. = not applicable; [a] Yield is expressed as fresh weight except for alfalfa, corn, and barley silage which are expressed on the basis of dry matter. Alfalfa silage and corn silage yields from study data. Barley silage yield from Guyader et al. [28]. Other yields from CRAAQ [29]; [b] Silage rates from study data. Other rates from CRAAQ [29]; [c] Rotz and Muck [30]; [d] Janzen et al. [31]; [e] CRAAQ [29]; [f] Barley silage was used in ICBM simulation only. Therefore, not all characteristics were required.

Ingredients imported onto the farm (corn gluten feed, rumen inert fat, urea, calcium carbonate, potassium carbonate, mineral/vitamin supplement, and milk replacer) were included in the overall analysis, using published emission factor values, recognizing their contributions were small (Table 4). We assumed no losses or wastage from imported ingredients.

Table 4. Greenhouse gas emission factors associated with crop production and processing and imported feed.

Source	Emission Factor	Unit	Source
Crop production			
Cropping corn silage	161.0	kg ha^{-1}	Little et al. [23]
Cropping alfalfa silage	56.7	kg ha^{-1}	Little et al. [23]
Cropping mixed hay	56.7	kg ha^{-1}	Little et al. [23]
Cropping corn grain	161.0	kg ha^{-1}	Little et al. [23]
Cropping barley grain	126.0	kg ha^{-1}	Little et al. [23]
Cropping soybean	149.1	kg ha^{-1}	Little et al. [23]
Herbicide manufacture for corn silage	0.696	kg ha^{-1}	Little et al. [23]
Herbicide manufacture for corn grain	0.696	kg ha^{-1}	Little et al. [23]
Herbicide manufacture for barley grain	1.392	kg ha^{-1}	Little et al. [23]
Herbicide manufacture for soybean	0.696	kg ha^{-1}	Little et al. [23]
N fertilizer manufacture	3.59	kg CO_2e (kg N)$^{-1}$	Nagy [32]
P fertilizer manufacture	0.5699	kg CO_2e (kg P_2O_5)$^{-1}$	Nagy [32]
Lime manufacture and transport	0.043	kg CO_2e (kg $CaCO_3$)$^{-1}$	O'Brien et al. [33]
Lime degradation	0.44	kg CO_2e (kg $CaCO_3$)$^{-1}$	IPCC [34]
N_2O emissions-direct	0.017	kg N_2O-N (kg N)$^{-1}$	Rochette et al. [35]
Leaching/runoff fraction	0.3	kg N (kg N)$^{-1}$	Rochette et al. [35]
N_2O emissions-indirect due to leaching/runoff	0.0075	kg N_2O-N (kg N)$^{-1}$	IPCC [34]
Volatilization fraction	0.1	kg N (kg N)$^{-1}$	IPCC [34]
N_2O emissions-indirect due to volatilization	0.01	kg N_2O-N (kg N)$^{-1}$	IPCC [34]
Crop processing			
Drying corn	0.014	kg CO_2e kg^{-1}	Vergé et al. [7]
Grinding corn	0.0119	g CO_2e kg^{-1}	Environment Canada [36]; Dabbour et al. [37]
Processing and transport of soy meal	0.0738	kg CO_2e kg^{-1}	Derived from CGB [38]; USDA [39]
Transport of soy hulls	0.0013	kg CO_2e kg^{-1}	Derived from Mc Geough et al. [8]; CGB [38]; USDA [39]
Imported feed			
Corn gluten feed	1.061	kg CO_2e kg^{-1}	O'Brien et al. [33]
Rumen inert fat	0.66	kg CO_2e kg^{-1}	Adom et al. [40]
Urea	3.30	kg CO_2e kg^{-1}	Adom et al. [40]
Calcium carbonate	0.013	kg CO_2e kg^{-1}	Adom et al. [40]
Potassium carbonate [a]	1.59	kg CO_2e kg^{-1}	Adom et al. [40]
Mineral/vitamin supplement	1.59	kg CO_2e kg^{-1}	Adom et al. [40]
Milk replacer	0.00134	kg CO_2e kg^{-1}	O'Brien et al. [33]

[a] Value for other trace minerals.

2.3. Quantification of GHG Emissions

2.3.1. Holos Model

Greenhouse gas emissions from each scenario (alfalfa silage and corn silage) were estimated using the Holos model, Version 2. The system boundary of the analysis was the farm-gate, including the following emissions: CH_4 from enteric fermentation and manure management; nitrous oxide (N_2O), direct and indirect, from soils and cropping and manure management; and CO_2 from energy use. Also included were emissions associated with production of inputs (fertilizers, herbicide, and imported feeds), application of lime, and the production and processing of some diet ingredients (Figure 2). Emissions associated with the production of capital goods were not included. Emissions were expressed as net GHGs or as CO_2-equivalents (CO_2e) utilizing the 100-year global warming potential (GWP) values of CO_2 = 1, CH_4 = 28, N_2O = 265 [41].

2.3.2. GHG Emissions from Livestock Management

Enteric CH_4 emissions were estimated as a function of DM intake and CH_4 conversion factor (Y_m) for the diets (Table 1); Y_m values for diets fed to lactating cows were from Hassanat et al. [13], while values for the other diets are from Mc Geough et al. [8], except for the alfalfa-based diet fed to the dry cows and heifers, where the Y_m value was adjusted to maintain the percentage difference between the alfalfa silage and corn silage reported by Hassanat et al. [13]. The DM intake of each animal group was estimated based on the net energy requirements of the animal, factoring in maintenance, activity, growth, pregnancy, lactation, and the estimated net energy content of the diet [42], except for lactating cows, where measured intake values were used [13].

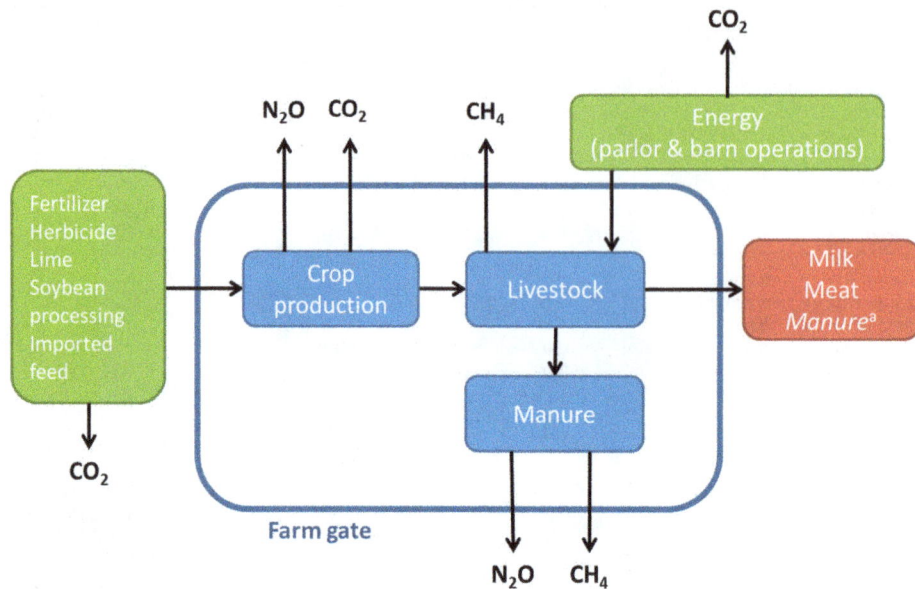

Figure 2. Diagram of dairy system including on- and off-farm emissions and sources and co-products.
[a] Manure is not included as a co-product in core scenarios.

Methane emissions from manure management were estimated based on DM intake and volatile solids production. Volatile solids production was multiplied by the CH_4 producing capacity of the animals ($B_o = 0.24$ for dairy animals) and the CH_4 conversion factor (MCF) of the manure handling system. For solid storage, MCF = 0.02; for deep bedding, MCF = 0.17 [34].

Manure N was estimated based on DM intake, crude protein content of the diet, and N retention of cattle [34], except for mature, lactating dairy cows, where we used measured data from Hassanat et al. [13] to estimate N excretion as a percentage of crude protein intake (alfalfa silage diet = 68.7%; corn silage diet = 61.5%). Direct N_2O emissions were estimated by multiplying manure N by an emission factor specific to the manure handling system (solid storage = 0.005; deep bedding = 0.01; [34]). Nitrogen contributions from straw bedding, along with associated emissions, were assumed to be the same as if residue was left on fields after harvest, so these contributions were accounted for as cropping emissions.

Indirect N_2O emissions from leaching and runoff and volatilization were also estimated. We assumed that leaching and runoff losses during storage were negligible, but that 30% of manure N was volatilized. Indirect N_2O emissions from volatilized N was estimated as 0.01 kg N_2O-N (kg manure N)$^{-1}$. Remaining manure N, adjusted for storage losses, was assumed to be applied to cropland once per year.

Emissions of CO_2 due to energy use were calculated using estimated electricity usage of 968 kWh per dairy cow per year (derived from [7]) and the Québec electricity production CO_2 emission rate of 3.4 g CO_2e kWh^{-1} [36]. Emissions of CO_2 associated with diesel fuel use for spreading manure were estimated based on the volume and concentration of manure produced and an emission factor of 0.347 kg CO_2e (kg manure N)$^{-1}$ [23].

2.3.3. GHG Emissions from Crop and Feed Production and Imported Feed

Direct N_2O emissions from soils and cropping were based on N inputs from synthetic N fertilizer, above- and below-ground crop residue decomposition, land applied manure, and mineralized N (assumed to be zero in these simulations). Using coefficients modified from Janzen et al. [31], crop residue N input was calculated based on crop yields [23] (Table 3). To calculate below-ground residue input for perennial crops, it was assumed that 39% of root biomass turns over annually, with a complete return-to-soil in the final year [43]. Soil texture, climate, and tillage modified the impact of these N inputs [35].

Nitrous oxide emission factors were derived using Canada-specific methodology developed for GHG inventory reporting [35] (Table 4).We used a location-specific direct N_2O emission factor, based on Ecodistrict growing season precipitation and potential evapotranspiration. Indirect N_2O emissions due to leaching and runoff and volatilization were calculated based on loss fractions, adjusted for growing season precipitation and potential evapotranspiration, and indirect emission factors.

Emissions of CO_2 from energy used to produce crops were estimated for each crop (Table 4), including those associated with N and phosphorus fertilizer production, herbicide production, and the production, transport, and dissociation of lime. Also considered were CO_2 emissions from processing of feed (drying and grinding of corn, processing of soybean into meal and hulls) and transport of soy ingredients to and from the farm.

2.3.4. Functional Unit and Co-Product Allocation

Whole farm GHG emissions were estimated for each scenario. As dairy systems produce multiple products, or co-products, more than one functional unit (a quantified performance of a product system; [44]) was required as a reference for assessment. Functional units used in the comparison were:

- kg of FPCM (standardized to 4% fat, 3.3% protein);
- kg of meat, live weight;
- kg of meat, carcass weight;
- kg of protein;
- MJ of energy;
- ha of farm land; and
- kg of manure N.

FPCM was calculated as recommended by the International Dairy Federation (IDF) [45]:

$$\text{FPCM (kg yr}^{-1}) = \text{milk (kg yr}^{-1}) \times [0.1226 \times \text{fat (\%)} + 0.0776 \times \text{true protein (\%)} + 0.2534]. \qquad (1)$$

Carcass weight was calculated as 60% of live weight [8]. We assumed protein content of meat was 17.32% of carcass weight and its energy value was 12.18 MJ $(\text{kg carcass})^{-1}$ [46]. The energy value of milk [42] was calculated as:

$$\text{Milk energy (MJ kg}^{-1}) = 4.184 \times [0.0929 \times \text{fat (\%)} + 0.0563 \times \text{true protein (\%)} + 0.192]. \qquad (2)$$

The generation of multiple co-products requires whole-farm GHG emissions to be allocated between these products. Therefore, four co-product allocation methodologies were utilized to assign emissions to the food products of milk and meat. The first method assigned all GHG emissions to milk. The second method assigned emissions to the products of milk and meat based on income from each (i.e., economic allocation), using the following prices: milk (FPCM): $0.80/L (5-y average price in Québec, 2011–2015 [47]); veal: $5.07/kg carcass; culled cows: $2.29/kg carcass (meat prices = 5-y average in Canada, 2010–2015 [48]). The third and fourth methods partitioned GHG emissions between co-products in a way that reflects underlying physical relationships [44]. The third allocation method followed the physical allocation approach of IDF [45] and the allocation factor (AF) for assigning GHG emissions to milk was calculated as:

$$\text{AF for milk} = 1 - 6.04 \times [\text{total meat (kg live weight)}/\text{total FPCM (kg)}] \qquad (3)$$

$$\text{AF for meat} = 1 - \text{AF for milk}. \qquad (4)$$

IDF [45] also provides a default ratio of 0.02 for [total meat/total FPCM], which was used in the fourth allocation method. Functional units of protein, energy, and land area required no allocation.

While manure is commonly considered a waste, manure N can also be considered as a co-product of the dairy system when it is used to replace inorganic N fertilizer. Thus an additional functional unit, 'kg of manure N', was included in the analysis. The price of this third co-product (manure N) was set at \$1.34/kg (5-y average urea fertilizer, 2011–2015 [49]).

Co-product allocation was required for soybean meal and soy hulls production. The environmental burden of, and cropland area for, growing soybeans was distributed between the products of oil, meal, and hulls. We assumed that soybeans were transported 60 km from the farm for processing before the meal and hulls were returned to the farm. Emissions due to processing were also included. The AF was determined to be 0.67 for soy meal and 0.02 for soy hulls using IDF [45] methodology to determine feed product allocation factors, industry published product mass ratios [38], and 2011–2015 USDA soy prices [39].

2.3.5. Soil Carbon

Assessing CO_2 emissions or removals from changes in soil C stocks is complex and subject to many assumptions [19,33,50–52]. We estimated soil C change using the Introductory Carbon Balance Model (ICBM), a two-component model comprised of young and old soil C pools [53], which allowed us to explore, by way of example, changes in crop rotation or residue management [54]. From among various soil organic carbon models considered, we selected ICBM because of its simplicity, its compatibility with the Holos model and computational structure, and its demonstrated applicability to Canadian conditions, based on extensive prior research [55,56].

Soil carbon changes were estimated using the average, yearly land area for the silage crop in each scenario (alfalfa or corn). ICBM runs were initialized, using estimates of steady state values for young and old carbon pools [53], derived by running estimates for a four-year stand of mixed hay long enough to approach steady state. From this point, a typical five-year alfalfa silage rotation was initiated consisting of 4 years of alfalfa silage followed by one year of barley silage. This rotation was simulated for 30 years utilizing average (rotation) annual carbon inputs. To compare the simulated alfalfa silage rotation to a corn silage rotation, a typical eight-year corn silage rotation was also initiated from the mixed hay steady state as before [53]. This rotation consisted of four years of corn silage followed by a four-year perennial hay stand, run for 30 years using average rotation carbon inputs.

ICBM utilizes local factors to assign a climate-dependent soil biological activity parameter that modifies decomposition rates (r_e). For the location and conditions of this simulated farm, $r_e = 0.9823$ [57]. Crop yields, moisture content, and residue ratios were used to calculate above and below ground residue carbon input (Table 3). The humification coefficient was 0.125 for above ground residue and 0.3 for below ground residue [58]. Decomposition constants were set at 0.80 and 0.00605 for the young and old pools, respectively [59]. For perennial crops, we assumed an annual root turnover of 39% with a complete return-to-soil in the final year [43]. Carbon due to manure input was not considered in this analysis of soil carbon.

3. Results

Our life span approach captured emissions from the birth of the dairy cow to the point of entering the dairy herd and accounted for length and number of lactations once in the producing herd [8,28]. Although described in a linear pattern, for simplicity, the approach describes the full cycle, with animals continually entering and leaving, and provides a system perspective of how herd management affects GHG emissions.

3.1. Greenhouse Gas Emissions

Overall farm GHG emissions for the production cycle differed little between the alfalfa- and corn-based systems (Table 5). The corn silage system had lower enteric CH_4 (−6%), manure CH_4 (−6%), manure-direct N_2O (−17%), soils/cropping-direct N_2O (−2%), and indirect N_2O (−5%) emissions, but these were offset by greater emissions for energy CO_2 (91%), liming CO_2 (168%), and imported feed CO_2 (8%).

Table 5. Food production and greenhouse gas (GHG) emissions in dairy production systems using alfalfa silage- or corn silage-based feeding systems.

Item	Alfalfa Silage System	Corn Silage System
Food production per cycle		
Fat and protein corrected milk (FPCM) (kg) [a]	1,802,328	1,789,906
Milk energy (MJ)	5,652,106	5,605,060
Milk protein (kg)	54,005	60,999
Total live weight (kg)	75,591	75,591
Total carcass weight (kg)	45,355	45,355
Meat energy (MJ) [b]	552,419	552,419
Meat protein (kg) [b]	7855	7855
Total energy (MJ) [b]	6,204,525	6,157,479
Total protein (kg) [b]	61,860	68,854
Manure N per cycle (kg)	32,001	27,476
Required land area (ha year^{-1})	61.90	57.28
kg FPCM ha^{-1} year^{-1}	4853	5208
GHG emissions per cycle (kg CO_2e)		
Enteric CH_4	1,036,657	976,453
Manure CH_4	197,380	184,590
Manure N_2O-direct	112,528	92,974
Soils/cropping N_2O-direct	573,340	561,986
Indirect N_2O (all sources)	123,365	117,595
Energy CO_2	84,006	160,408
Liming CO_2 [c]	15,346	41,176
Imported feed (CO_2)	96,787	104,236
Total	2,239,408	2,239,418

[a] Standardized to 4% fat, 3.3% protein; [b] Calculated on a carcass basis; [c] Emissions due to lime dissociation only.

Corn silage, because of its elevated starch content has a lower Y_m value than most other perennial and annual forages [11,12]. In our study, however, the 10% lower Y_m value for corn silage diets only resulted in 6% less enteric CH_4 as compared with the alfalfa silage because the lactating cows fed corn silage had greater feed intake (Table 5).

Manure CH_4 emissions are a function of volatile solids production. Lactating cows fed the alfalfa silage-based diet produced less volatile solids than cows fed corn silage-based diet because of lower measured feed intake. The opposite occurred for the dry cows and heifers due to their greater intake relative to those fed corn silage. These converse results for feed intake occurred because feed intake of the lactating cows was measured, whereas intakes of the dry cows and heifers were estimated based on energy requirements and energy content of feed. In the absence of measured data, estimating feed intake required based on energy requirements is a commonly utilized methodology [60]. Energy requirements of the non-lactating groups were the same for both silage systems, but the greater energy content of corn silage meant that less silage was needed to supply energy requirements. Increased volatile solids production coupled with a greater CH_4 conversion factor for the deep bedding system of the dry cows and heifers compared with the solid manure storage for the lactating cows resulted in greater manure CH_4 emissions for the alfalfa silage based system. Choice of manure handling practices, therefore, may offer potential opportunity for reducing CH_4 emissions [11,61].

The alfalfa silage system produced 14% more manure N than the corn silage system (Table 5) mainly due to greater DM intake for some cattle groups, greater crude protein content of diets, and greater N excretion with alfalfa silage based diets (Table 5). Dietary crude protein level and N excretion from dairy cattle is primarily affected by N intake; about 70–80% of the N consumed by dairy cows is excreted in manure [62]. The lower N excretion of the corn silage-based system resulted in lower manure N_2O emissions and less manure N to spread on fields.

Despite greater fertilizer N demands and yield of corn silage, emissions for soils/cropping direct N_2O were greater for the alfalfa silage-based system (Table 5), largely because of the greater N input from crop residue from the perennial, leguminous alfalfa. This perennial legume not only has greater root yield but also higher root N content than corn [31].

Indirect N_2O emissions, associated with leaching and runoff and volatilization from both manure and from soils and cropping, were 5% less for corn silage than for alfalfa silage (Table 5).

Greater energy CO_2 emissions for corn silage can be attributed to cropping practices, including use of fertilizer, lime, and herbicide. The lower crude protein content of corn versus alfalfa necessitated supplementing corn silage-based diets with protein from soybeans, adding to CO_2 energy emissions associated with feeding corn silage.

3.2. Production

Diet affects not only GHG emissions but also milk production and associated nutritional value. Cows fed corn silage-based diets produced a similar amount of FPCM but 13% more milk protein than cows fed the alfalfa silage-based diet (Table 5). Less than 1% difference existed between the two silage systems when compared on the basis of milk energy production. As herd dynamics and structure were the same with both scenarios, there was no difference in the mass of meat, meat protein, or meat energy produced.

3.3. Emissions Breakdown

For both silage systems, the majority of emissions were associated with enteric CH_4 (46% and 44% for alfalfa silage and corn silage, respectively; Figure 3). The second largest contributor was soils/cropping N_2O direct emissions (26% for alfalfa silage and 25% for corn silage) followed by manure CH_4. Energy-related CO_2 emissions were 7% of total emissions for corn silage, and 4% of total emissions for alfalfa silage, somewhat lower than in other studies because of extensive hydro-electricity production in Québec [7]. With both systems, emissions associated with imported feed made up 5% of total emissions (Figure 3). The relative contributions of GHG sources were similar to other Canadian studies that included the whole, on-farm production cycle including growing and importing feed [7,8,28,63]. The dominance of enteric CH_4 to total GHG emissions demonstrates the importance of developing methods to reduce these emissions in dairy systems.

3.4. Land Base

Corn silage had greater biomass yield and thus used less land area to produce the required feed: 57.28 ha as compared with 61.90 ha for alfalfa silage (Table 5). Thus, on an annual basis, the corn silage-based system produced 7% more FPCM per hectare of land, and greater GHG emissions per unit of land area, than the alfalfa silage-based system (Table 6). But expressing emissions relative to land area can be misleading if estimates of productivity are not also acknowledged; focusing on GHG emissions on a per unit area basis could lead to decline in food production and may not lead to reduction in GHG emissions for a set amount of product [64]. The complexity of acknowledging land base in carbon footprinting studies of agricultural products requires more consideration.

3.5. Greenhouse Gas Intensity

Our study found no difference between the two silage production systems when emissions were expressed per unit of energy produced (0.36 kg CO_2e (MJ total energy)$^{-1}$). Per kg protein produced, however, the corn silage system generated 10% fewer emissions than the alfalfa silage system (Table 6). Expressing GHG intensity with the functional units of energy and protein circumvents the need to partition GHG emissions between milk and meat products.

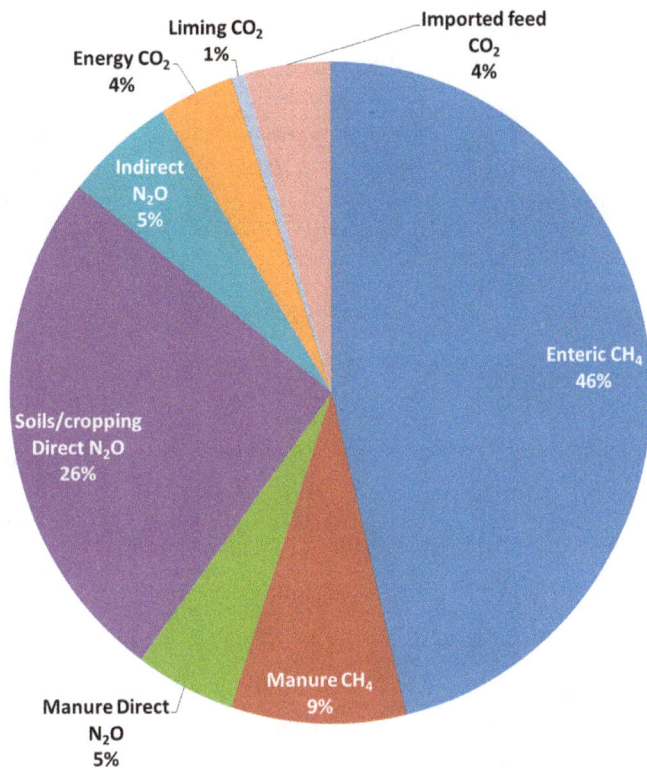

(a)

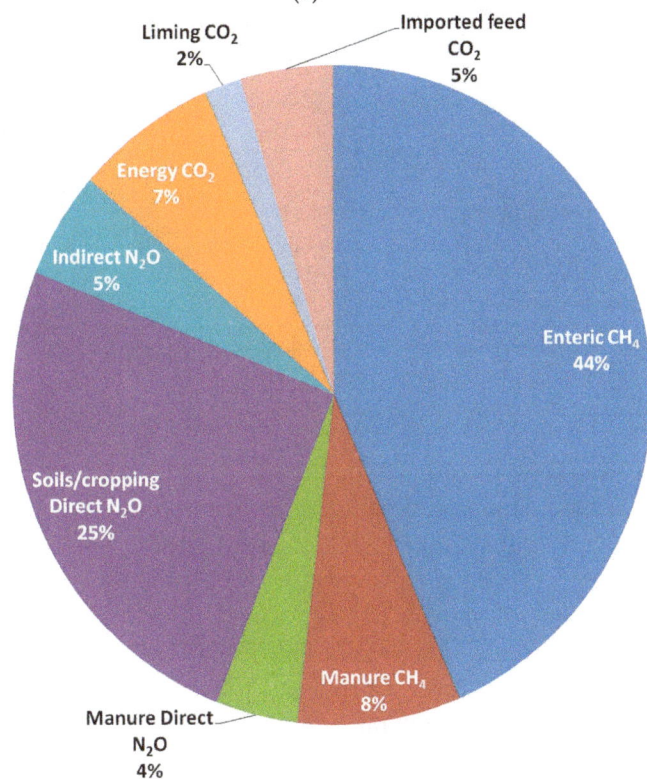

(b)

Figure 3. Greenhouse gas proportions in dairy systems utilizing alfalfa silage (**a**) or corn silage (**b**) as the main forage source in the lactating diets.

Table 6. Greenhouse gas emission intensity of dairy production systems using alfalfa silage- or corn silage-based feeding systems expressed with various functional units and allocation methods.

Item	Alfalfa Silage System	Corn Silage System
kg CO_2e ha^{-1} [a]	36,181	39,098
kg CO_2e (kg total protein)$^{-1}$ [b]	36.20	32.52
kg CO_2e (MJ total energy)$^{-1}$	0.36	0.36
100% allocation to milk		
Emission allocation to milk (%)	100	100
kg CO_2e (kg FPCM)$^{-1}$ [c]	1.24	1.25
kg CO_2e (kg live weight)$^{-1}$	0	0
kg CO_2e (kg carcass weight)$^{-1}$	0	0
Economic allocation		
Emission allocation to milk (%)	89.7	89.6
kg CO_2e (kg FPCM)$^{-1}$ [c]	1.11	1.12
kg CO_2e (kg live weight)$^{-1}$	3.06	3.08
kg CO_2e (kg carcass weight)$^{-1}$	5.10	5.13
Economic allocation—manure as co-product		
Emission allocation to milk (%) [d]	87.3	87.5
kg CO_2e (kg FPCM)$^{-1}$ [c]	1.08	1.10
kg CO_2e (kg live weight)$^{-1}$	2.98	3.01
kg CO_2e (kg carcass weight)$^{-1}$	4.96	5.01
kg CO_2e (kg manure nitrogen)$^{-1}$	1.63	1.65
IDF (2015) allocation—calculated		
Emission allocation to milk (%)	74.7	74.5
kg CO_2e (kg FPCM)$^{-1}$ [c]	0.93	0.93
kg CO_2e (kg live weight)$^{-1}$	7.50	7.56
kg CO_2e (kg carcass weight)$^{-1}$	12.51	12.59
IDF (2015) allocation—default		
Emission allocation to milk (%)	88.0	88.0
kg CO_2e (kg FPCM)$^{-1}$ [c]	1.09	1.10
kg CO_2e (kg live weight)$^{-1}$	3.56	3.56
kg CO_2e (kg carcass weight)$^{-1}$	5.93	5.93

[a] Total cycle emissions/Yearly land area required; [b] Calculated on a carcass basis; [c] Fat and protein corrected milk (FPCM) standardized to 4% fat, 3.3% protein; [d] Emissions allocated to manure N = 2.7% for alfalfa silage and 2.3% for corn silage.

Regardless of how emissions were partitioned between the co-products of milk and meat, the two silage systems showed similar emissions per functional unit (Table 6). When emissions were allocated entirely to milk, the average GHG emission intensity of FPCM was 1.25 kg CO_2e (kg FPCM)$^{-1}$. Economic allocation assigned an average of 89.7% of emissions to milk, which resulted in an approximate GHG emission intensity of 1.12 kg CO_2e (kg FPCM)$^{-1}$. Using the IDF [45] default milk-to-meat ratio, 88.0% of farm emissions were assigned to milk, but using the calculated IDF [45] method to calculate the milk-to-meat ratio led to a much reduced allocation ratio with 74.6% of emissions assigned to milk overall. The default ratio may be more appropriate than the calculated ratio for Canadian dairy production as it is more similar to the other allocation ratios. Using the IDF [45] calculated ratio, the GHG emission intensity per kg of FPCM was lower (0.93 CO_2e, for both alfalfa silage and corn silage) than other calculated intensities while GHG intensity per kg of meat was greater, thereby shifting the emissions towards meat from milk. While the method of allocation affected the GHG intensity values, the differences between the alfalfa silage and corn silage systems remained minimal regardless of allocation method.

The GHG intensity results fall into the range of values from previous studies despite differences in assumptions, inputs, conditions, and GWP factors utilized [7,28,63]. Our values for carbon footprint per functional unit were slightly greater than those of McGeough et al. [8] for similar scenarios, because McGeough et al. [8] used diets with greater digestibility and, hence, lower DM intake which reduced enteric and manure CH_4 emissions. McGeough et al. [8] also used different GWP values and assumed all manure N produced replaced inorganic N fertilizer and calculated an energy CO_2 offset for excess manure N, which may have underestimated associated emissions.

When manure N was considered as a co-product in economic allocation, the overall GHG emission intensity of milk and meat was reduced (Table 6). The GHG emission intensity was 1.63 kg CO_2e per kg manure N for alfalfa silage and 1.65 CO_2e per kg manure N for corn silage. Both systems had excess N, with both synthetic N fertilizer and manure N applied to cropland, implying opportunity to reduce emissions with improved nutrient management by using manure to reduce fertilizer inputs. Considering manure N as a co-product emphasizes that manure has value and is not a waste [45].

The choice of reporting unit affects the ranking and relative merits of proposed practices in carbon footprinting studies. For example, we observed differential results between the two systems when we used 'ha of farm land' or 'kg of protein' as the functional unit. Other functional units, however, elicited differences between systems that were negligible (less than the uncertainty of the estimates). There may be no 'best' functional unit; choice of functional unit, in the end, depends on the underlying questions and aim addressed by the analysis.

3.6. Soil Carbon

A major advantage of ruminant systems is that forage production can recycle carbon back to the soil, thereby lowering atmospheric CO_2 and enhancing soil health [65]. Increasing soil carbon stocks is an effective opportunity to offset enteric GHG emissions in the short term, but its benefit requires long-term evaluation. The ICBM model describes soil organic carbon dynamics in a decadal time frame (30 to 50 years), and estimates parameters from information that is usually available on the farm [54]. We compared ICBM simulations of alfalfa silage and corn silage rotations from an initial steady state under hay. After a period of 30 years, the model predicted a gain of 87,449 kg soil carbon under the alfalfa system and a loss of 6384 kg carbon under the corn system (Figure 4). If over 30 years the dairy system completed five cycles, producing approximately 11,000,000 kg CO_2e, the potential carbon gain from alfalfa and the potential carbon loss from corn represents less than 1% of emissions, implying little impact on soil carbon of switching from hay to an alfalfa silage or corn silage system.

In the long run, ICBM predicted a steady state value of 79,126 kg C ha^{-1} for the mixed hay stand, 96,148 kg C ha^{-1} for the alfalfa silage rotation, and 77,098 kg C ha^{-1} for the corn silage rotation, largely reflecting differences in residue input. While steady state is rarely if ever attained in ecosystems, because of shifting conditions and practices, these predictions illustrate the overall, long-term potential for these systems to capture and withhold carbon from the atmosphere.

ICBM is very responsive to carbon input from residues, especially those entering the soil below-ground. Improved estimates of the mass and turnover of roots and root-derived carbon from perennial crops would reduce uncertainty in current estimates of net GHG emissions from dairy systems, particularly those based on forage crops. Nevertheless, our findings demonstrate the potentially dominant influence that changes in soil carbon can impose on the overall carbon footprint of dairy production systems. The amount of soil carbon gained or lost depends on previous management practices [66]; consequently, estimates of net CO_2 exchange from soil must be derived for local soils, conditions, and management practices.

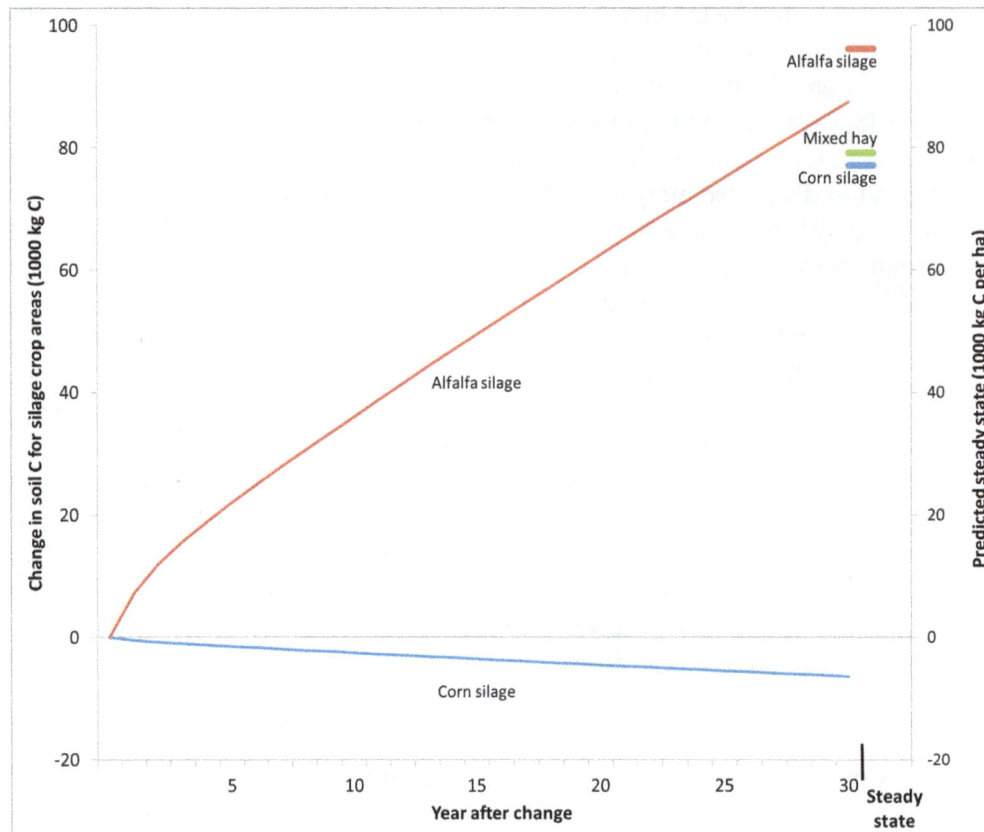

Figure 4. Predicted gain or loss of soil carbon due to rotation change from a mixed hay steady state to alfalfa silage or corn silage for the entire forage cropland (29.05 ha for alfalfa silage and 18.17 ha for corn silage) over 30 years since change (**left side**); and predicted steady state per hectare for each forage rotation (**right side**).

4. Conclusions

Earlier, Hassanat et al. [13] showed that lactating dairy cows fed a corn silage-based diet emitted 10% less enteric CH_4 per unit gross energy intake than cows on an alfalfa silage-based diet. Expanding the analysis to encompass the entire system—the entire dairy herd as well as the production of crops for feed and handling of manure—showed that enteric CH_4 was 6% lower for the corn silage-based system than for the alfalfa silage-based system. However, when all sources and GHGs were considered, the differences between the two forage systems on overall net GHG emissions of the dairy production system were minimal (<0.01%). Over the long-term, however, the alfalfa silage rotation had greater potential to store soil carbon than the corn silage rotation, though not enough to offset GHG emissions from dairy production. Our findings illustrate the importance of soil carbon changes in assessing overall carbon footprint of dairy systems, and the need for improved ways of predicting these changes.

This study reinforces the merits of using a whole-systems or life cycle approach rather than focusing on single elements of a farm system without investigating interrelated effects of management choices. Reported GHG reduction factors cannot be simply combined additively, because interwoven effects of management choices cascade through the entire system, sometimes with counter-intuitive outcomes. A whole-systems approach is most likely to steer policy and management toward reduced GHG emissions and improved sustainability.

Acknowledgments: The study was funded by Emissions Reduction Alberta (Project EOI# B140002) and Agriculture and Agri-Food Canada.

Author Contributions: Shannan M. Little designed and conducted the modeling; Shannan M. Little and Karen A. Beauchemin interpreted the results and wrote the initial and final drafts; Chaouki Benchaar and Emma J. McGeough advised on enteric CH_4 emission results and discussion and contributed to the final draft; H. Henry Janzen and Roland Kröbel advised on soil N_2O emission and soil carbon results and discussion and contributed to the final draft.

References

1. Food and Agriculture Organization of the United Nations (FAO). *Tackling Climate Change through Livestock. A Global Assessment of Emissions and Mitigation Opportunities*; Food and Agriculture Organization of the United Nations—Animal Production and Health Division: Rome, Italy, 2013.

2. O'Mara, F.P. The significance of livestock as a contributor to global greenhouse gas emissions today and in the near future. *Anim. Feed Sci. Technol.* **2011**, *166–167*, 7–15. [CrossRef]

3. Food and Agriculture Organization of the United Nations (FAO). *Greenhouse Gas Emissions from the Dairy Sector: A Life Cycle Assessment*; Food and Agriculture Organization of the United Nations—Animal Production and Health Division: Rome, Italy, 2010.

4. Thoma, G.; Popp, J.; Nutter, D.; Shonnard, D.; Ulrich, R.; Matlock, M.; Kim, D.S.; Neiderman, Z.; Kemper, N.; East, C.; et al. Greenhouse gas emissions from milk production and consumption in the United States: A cradle-to-grave life cycle assessment circa 2008. *Int. Dairy J.* **2013**, *31*, S3–S14. [CrossRef]

5. Canadian Dairy Information Centre. Historical Milk Production. Available online: dairyinfo.gc.ca/index_e.php?s1=dff-fcil&s2=msp-lpl&s3=hmp-phl (accessed on 16 October 2017).

6. Environment Canada. *National Inventory Report 1990–2013: Greenhouse Gas Sources and Sinks in Canada*; Environment Canada: Gatineau, QC, Canada, 2015.

7. Vergé, X.P.C.; Dyer, J.A.; Desjardins, R.L.; Worth, D. Greenhouse gas emissions from the Canadian dairy industry in 2001. *Agric. Syst.* **2007**, *94*, 683–693. [CrossRef]

8. Mc Geough, M.J.; Little, S.M.; Janzen, H.H.; McAllister, T.A.; McGinn, S.M.; Beauchemin, K.A. Life-cycle assessment of greenhouse gas emissions from dairy production in Eastern Canada: A case study. *J. Dairy Sci.* **2012**, *95*, 5164–5175. [CrossRef] [PubMed]

9. Eckard, R.J.; Grainger, C.; de Klein, C.A.M. Options for the abatement of methane and nitrous oxide from ruminant production: A review. *Livest. Sci.* **2010**, *130*, 47–56. [CrossRef]

10. Martin, C.; Morgavi, D.P.; Doreau, M. Methane mitigation in ruminants: From microbe to the farm scale. *Animal* **2010**, *4*, 351–365. [CrossRef] [PubMed]

11. Hristov, A.N.; Oh, J.; Lee, C.; Meinen, R.; Montes, F.; Ott, T.; Firkins, J.; Rotz, A.; Dell, C.; Adesogan, A.; et al. *Mitigation of Greenhouse Gas Emissions in Livestock Production—A Review of Technical Options For Non-CO₂ Emissions*; FAO Animal Production and Health Paper No. 177; Gerber, P.J., Henderson, B., Makkar, H.P.S., Eds.; Food and Agriculture Organization of the United Nations: Rome, Italy, 2013.

12. Knapp, J.R.; Laur, G.L.; Vadas, P.A.; Weiss, W.P.; Tricarico, J.M. Invited review: Enteric methane in dairy cattle production: Quantifying the opportunities and impact of reducing emissions. *J. Dairy Sci.* **2014**, *97*, 3231–3261. [CrossRef] [PubMed]

13. Hassanat, F.; Gervais, R.; Julien, C.; Massé, D.I.; Lettat, A.; Chouinard, P.Y.; Petit, H.V.; Benchaar, C. Replacing alfalfa silage with corn silage in dairy cow diets: Effects on enteric methane production, ruminal fermentation, digestion, N balance, and milk production. *J. Dairy Sci.* **2013**, *96*, 4553–4567. [CrossRef] [PubMed]

14. Benchaar, C.; Hassanat, F.; Gervais, R.; Chouinard, P.Y.; Petit, H.V.; Massé, D.I. Methane production, digestion, ruminal fermentation, nitrogen balance, and milk production of cows fed corn silage- or barley silage-based diets. *J. Dairy Sci.* **2014**, *97*, 961–974. [CrossRef] [PubMed]

15. Paustian, K.; Andrén, O.; Janzen, H.H.; Lal, R.; Smith, P.; Tian, G.; Tiessen, H.; Van Noordwijk, M.; Woomer, P.L. Agricultural soils as a sink to mitigate CO_2 emissions. *Soil Use Manag.* **1997**, *13*, 230–244. [CrossRef]

16. Follett, R.F. Soil management concepts and carbon sequestration in cropland soils. *Soil Till. Res.* **2001**, *61*, 77–92. [CrossRef]

17. Jarecki, M.K.; Lal, R. Crop Management for Soil Carbon Sequestration. *Crit. Rev. Plant Sci.* **2003**, *22*, 471–502. [CrossRef]

18. Rotz, C.A.; Montes, F.; Chianese, D.S. The carbon footprint of dairy production systems through partial life cycle assessment. *J. Dairy Sci.* **2010**, *93*, 1266–1282. [CrossRef] [PubMed]

19. Flysjo, A.; Henriksson, M.; Cederberg, C.; Ledgard, S.; Englund, J.-E. The impact of various parameters on the carbon footprint of milk production in New Zealand and Sweden. *Agric. Syst.* **2011**, *104*, 459–469. [CrossRef]

20. Del Prado, A.; Chadwick, D.; Cardenas, L.; Misselbrook, T.; Scholefield, D.; Merino, P. Exploring systems responses to mitigation of GHG in UK dairy farms. *Agric. Ecosyst. Environ.* **2010**, *136*, 318–332. [CrossRef]

21. Vellinga, T.V.; Hoving, I.E. Maize silage for dairy cows: Mitigation of methane emissions can be offset by land use change. *Nutr. Cycl. Agroecosyst.* **2010**, *89*, 413–426. [CrossRef]

22. Janzen, H.H.; Angers, D.A.; Boehm, M.; Bolinder, M.; Desjardins, R.L.; Dyer, J.A.; Ellert, B.H.; Gibb, D.J.; Gregorich, E.G.; Helgason, B.L.; et al. A proposed approach to estimate and reduce net greenhouse gas emissions from whole farms. *Can. J. Soil Sci.* **2006**, *86*, 401–418. [CrossRef]

23. Little, S.M.; Lindeman, J.; Maclean, K.; Janzen, H.H. *Holos—A Tool to Estimate and Reduce GHGs from Farms*; Methodology and Algorithms for Version 2.0; Agriculture and Agri-Food Canada: Ottawa, ON, Canada, 2013.

24. Marshall, I.B.; Schut, P.H.; Ballard, M. *A National Ecological Framework for Canada: Attribute Data*; Agriculture and Agri-Food Canada: Ottawa, ON, Canada, 1999. Available online: sis.agr.gc.ca/cansis/nsdb/ecostrat/1999report/index.html (accessed on 18 September 2017).

25. Soil Landscapes of Canada Working Group (SLC). Soil Landscapes of Canada Version 3.2. Agriculture and Agri-Food Canada. (Digital Map and Database at 1:1 Million Scale). Available online: sis.agr.gc.ca/cansis/nsdb/slc/v3.2/index.html (accessed on 25 September 2017).

26. United States Department of Agriculture (USDA). *Dairy 2007, Heifer Calf Health and Management on US Dairy Operations. # 550.0110*; United States Department of Agriculture: Animal and Plant Health Inspection Service: Veterinary Services (USDA:APHIS:VS): Centers for Epidemiology and Animal Health (CEAH): Fort Collins, CO, USA, 2010.

27. Valacta Inc. L'évolution de la production laitière québécoise. In *Le Producteur de Lait Québécois*; Valacta Inc.: Sainte-Anne-de-Bellevue, QC, Canada, 2009.

28. Guyader, J.; Little, S.; Kröbel, R.; Benchaar, C.; Beauchemin, K. Comparison of greenhouse gas emissions from corn- and barley-based dairy production systems in Eastern Canada. *Agric. Syst.* **2017**, *152*, 38–46. [CrossRef]

29. CRAAQ. *Référence Economique: Foin AGDEX 120/854; Maïs-Fourrager AGDEX 111/821a; Maïs-Grain AGDEX 111/821b; Orge d'alimentation Animale AGDEX 114/821a; Soya AGDEX 141/821*; Centre de Référence en Agriculture et Agroalimentaire du Québec: Québec, QC, Canada, 2010.

30. Rotz, C.A.; Muck, R.E. Changes in forage quality during harvest and storage. In *Forage Quality, Evaluation, and Utilization*; Fahey, G.C., Collins, M., Mertens, D.R., Moser, L.E., Eds.; American Society of Agronomy, Crop Science Society of America, Soil Science Society of America: Madison, WI, USA, 1994; pp. 828–868, ISBN 9780891181194.

31. Janzen, H.H.; Beauchemin, K.A.; Bruinsma, Y.; Campbell, C.A.; Desjardins, R.L.; Ellert, B.H. The fate of nitrogen in agroecosystems: An illustration using Canadian estimates. *Nutr. Cycl. Agroecosyst.* **2003**, *67*, 85–102. [CrossRef]

32. Nagy, C.N. Energy and Greenhouse Gas Emissions Coefficients for Inputs Used in Agriculture. In *Report to the Prairie Adaptation Research Collaborative*; Centre for Studies in Agriculture, Law and the Environment: Saskatoon, SK, Canada, 2000.

33. O'Brien, D.; Capper, J.L.; Garnsworthy, P.C.; Grainger, C.; Shalloo, L. A case study of the carbon footprint of milk from high-performing confinement and grass-based dairy farms. *J. Dairy Sci.* **2014**, *97*, 1835–1851. [CrossRef] [PubMed]

34. Intergovernmental Panel on Climate Change (IPCC). *2006 IPCC Guidelines for National Greenhouse Gas Inventories*; National Greenhouse Gas Inventories Programme; Eggleston, H.S., Buendia, L., Miwa, K., Ngara, T., Tanabe, K., Eds.; IGES: Prefecture, Japan, 2006.

35. Rochette, P.; Worth, D.E.; Lemke, R.L.; McConkey, B.G.; Pennock, D.J.; Wagner-Riddle, C.; Desjardins, R.L. Estimation of N_2O emissions from agricultural soils in Canada. I. Development of a country-specific methodology. *Can. J. Soil Sci.* **2008**, *88*, 641–654. [CrossRef]

36. Environment Canada. *National Inventory Report 1990–2012—Greenhouse Gas Sources and Sinks in Canada*; Environment Canada: Gatineau, QC, Canada, 2014.

37. Dabbour, M.; Bahnasawy, A.; Ali, S.; El-Haddad, Z. Energy Consumption in Manufacturing of Different Types of Feeds. Available online: www.academia.edu/8373073/ENERGY_CONSUMPTION_IN_MANUFACTURING_OF_DIFFERENT_TYPES_OF_FEEDS (accessed on 20 September 2015).

38. CGB Enterprises, Inc. Soybean Processing. Available online: www.cgb.com/businessunits/soybeanprocessing.aspx (accessed on 15 June 2016).

39. United States Department of Agriculture (USDA). Agricultural Marketing Service. Available online: marketnews.usda.gov/mnp/ls-report-config (accessed on 21 June 2016).

40. Adom, F.; Workman, C.; Thoma, G.; Shonnard, D. Carbon footprint analysis of dairy feed from a mill in Michigan, USA. *Int. Dairy J.* **2013**, *31* (Suppl. 1), S21–S28. [CrossRef]

41. Myhre, G.; Shindell, D.; Bréon, F.-M.; Collins, W.; Fuglestvedt, J.; Huang, J.; Koch, D.; Lamarque, J.-F.; Lee, D.; Mendoza, B.; et al. Anthropogenic and Natural Radiative Forcing. In *Climate Change 2013: The Physical Science Basis. Contribution of Working Group I to the Fifth Assessment Report of the Intergovernmental Panel on Climate Change*; Stocker, T.F., Qin, D., Plattner, G.-K., Tignor, M., Allen, S.K., Boschung, J., Nauels, A., Xia, Y., Bex, V., Midgley, P.M., Eds.; Cambridge University Press: Cambridge, UK; New York, NY, USA, 2013.

42. National Research Council (NRC). *Nutrient Requirements of Dairy Cattle: Seventh Revised Edition*; The National Academies Press: Washington, DC, USA, 2001. [CrossRef]

43. Bolinder, M.A.; Janzen, H.H.; Gregorich, E.G.; Angers, D.A.; VandeBygaart, A.J. An approach for estimating net primary production and annual carbon inputs to soil for common agricultural crops in Canada. *Agric. Ecosyst. Environ.* **2007**, *118*, 29–42. [CrossRef]

44. International Organization for Standardization (ISO). *Greenhouse Gases—Carbon Footprint of Products—Requirements and Guidelines for Quantification and Communication, Technical Specification 14067*; ISO: Geneva, Switzerland, 2013.

45. International Dairy Federation (IDF). *A Common Carbon Footprint Approach for Dairy—The IDF Guide to Standard Lifecycle Assessment Methodology for the Dairy Sector*; International Dairy Federation: Brussels, Belgium, 2015.

46. United States Department of Agriculture (USDA). National Nutrient Database for Standard Reference Release 28. Available online: www.nal.usda.gov/fnic/foodcomp/search/ (accessed on 20 September 2017).

47. Institut de la Statistique du Québec (ISQ). Gouvernement du Québec. Available online: www.stat.gouv.qc.ca/statistiques/agriculture/production-laitiere/statistiques_qc_mrc_cre.html (accessed on 21 September 2017).

48. Agriculture and Agri-food Canada (AAFC). Red Meat Market Information—Price Reports. Available online: www.agr.gc.ca/redmeat/pri_eng.htm (accessed on 21 September 2015).

49. Alberta Agriculture and Rural Development. Alberta Farm Input Survey Prices. Available online: www.agric.gov.ab.ca/app21/farminputprices (accessed on 21 September 2017).

50. Schils, R.L.M.; Verhagen, A.; Aarts, H.R.M.; Šebek, L.B.J. A farm level approach to define successful mitigation strategies for GHG emissions from ruminant livestock systems. *Nutr. Cycl. Agroecosyst.* **2005**, *71*, 163–175. [CrossRef]

51. O'Brien, D.; Shalloo, L.; Patton, J.; Buckley, F.; Grainger, C.; Wallace, M. A life cycle assessment of seasonal grass-based and confinement dairy farms. *Agric. Syst.* **2012**, *107*, 33–46. [CrossRef]

52. Del Prado, A.; Mas, K.; Pardo, G.; Gallejones, P. Modelling the interactions between C and N farm balances and GHG emissions from confinement dairy farms in northern Spain. *Sci. Total Environ.* **2013**, *465*, 156–165. [CrossRef] [PubMed]

53. Andrén, O.; Kätterer, T. ICBM: The introductory carbon balance model for exploration of soil carbon balances. *Ecol. Appl.* **1997**, *7*, 1226–1236. [CrossRef]

54. Kröbel, R.; Bolinder, M.A.; Janzen, H.H.; Little, S.M.; Vandenbygaart, A.J.; Kätterer, T. Canadian farm-level soil carbon change assessment by merging the greenhouse gas model Holos with the Introductory Carbon Balance Model (ICBM). *Agric. Syst.* **2016**, *143*, 76–85. [CrossRef]

55. Bolinder, M.A.; VandenBygaart, A.J.; Gregorich, E.G.; Angers, D.A.; Janzen, H.H. Modeling soil organic carbon stock change for estimating whole-farm greenhouse gas emissions. *Can. J. Soil Sci.* **2006**, *86*, 419–429. [CrossRef]

56. VandenBygaart, A.J.; Gregorich, E.G.; Angers, D.A.; Bolinder, M.A.; Janzen, H.H.; Campbell, C.A. Modeling soil organic carbon change in Canadian agroecosystems: Testing the Introductory Carbon Balance Model. In *Soil Carbon Sequestration and the Greenhouse Effect*; SSSA Spec. Publ. 57; Lal, R., Follett, R.F., Eds.; Soil Science Society of America: Madison, WI, USA, 2009; pp. 13–28, ISBN 978-0-89118-859-9.

57. Bolinder, M.A.; Andrén, O.; Kätterer, T.; Parent, L.-E. Soil organic carbon sequestration potential for Canadian agricultural ecoregions calculated using the introductory carbon balance model. *Can. J. Soil Sci.* **2008**, *88*, 451–460. [CrossRef]

58. Kätterer, T.; Andersson, L.; Andrén, O.; Persson, J. Long-term impact of chronosequential land use change on soil carbon stocks on a Swedish farm. *Nutr. Cycl. Agroecosyst.* **2008**, *81*, 145–155. [CrossRef]

59. Andrén, O.; Kätterer, T.; Karlsson, T. ICBM regional model for estimations of dynamics of agricultural soil carbon pools. *Nutr. Cycl. Agroecosyst.* **2004**, *70*, 231–239. [CrossRef]

60. National Academies of Sciences, Engineering, and Medicine (NASEM). *Nutrient Requirements of Beef Cattle: Eighth Revised Edition*; The National Academies Press: Washington, DC, USA, 2016. [CrossRef]

61. Chadwick, D.; Sommer, S.; Thorman, R.; Fangueiro, D.; Cardenas, L.; Amon, B.; Misselbrook, T. Manure management: Implications for greenhouse gas emissions. *Anim. Feed Sci. Technol.* **2011**, *166–167*, 514–531. [CrossRef]

62. Dong, R.L.; Zhao, G.Y.; Chai, L.L.; Beauchemin, K.A. Prediction of urinary and fecal nitrogen excretion by beef cattle. *J. Anim. Sci.* **2014**, *92*, 4669–4681. [CrossRef] [PubMed]

63. Jayasundara, S.; Wagner-Riddle, C. Greenhouse gas emissions intensity of Ontario milk production in 2011 compared with 1991. *Can. J. Anim. Sci.* **2013**, *94*, 155–173. [CrossRef]

64. O'Brien, D.; Shalloo, L.; Buckley, F.; Horan, B.; Grainger, C.; Wallace, M. The effect of methodology on estimates of greenhouse gas emissions from grass-based dairy systems. *Agric. Ecosyst. Environ.* **2011**, *141*, 39–48. [CrossRef]

65. Guyader, J.; Janzen, H.H.; Kroebel, R.; Beauchemin, K.A. Invited Review: Forage utilization to improve environmental sustainability of ruminant production. *J. Anim. Sci.* **2016**, *94*, 3147–3158. [CrossRef] [PubMed]

66. Smith, P. Do grasslands act as a perpetual sink for carbon? *Glob. Chang. Biol.* **2014**, *20*, 2708–2711. [CrossRef] [PubMed]

8

Assessment of Urban Heat Islands in Small- and Mid-Sized Cities in Brazil

Renata dos Santos Cardoso *, Larissa Piffer Dorigon *, Danielle Cardozo Frasca Teixeira * and Margarete Cristiane de Costa Trindade Amorim *

Department of Geography, São Paulo State University (UNESP), 305 Roberto Simonsen Street, Presidente Prudente, São Paulo 19060-900, Brazil

* Correspondence: renatacardoso16@gmail.com (R.d.S.C.); laridorigon@hotmail.com (L.P.D.); danielle.frasca@hotmail.com (D.C.F.T.); mccta@fct.unesp.br (M.C.d.C.T.A.)

Academic Editors: Valdir Adilson Steinke and Charlei Aparecido da Silva

Abstract: Urban heat islands (UHIs) in large cities and different climatic regions have been thoroughly studied; however, their effects are becoming a common concern in smaller cities as well. We assessed UHIs in three tropical cities, analyzing how synoptic conditions, urban morphology, and land cover affect the heat island magnitude. Data gathering involved mobile surveys across Paranavaí (Paraná), Rancharia (São Paulo), and Presidente Prudente (São Paulo), Brazil, during summer evenings (December 2013–January 2014). Temperature data collected over five days in each city point to heat islands with magnitudes up to 6 °C, under calm synoptic conditions, whereas summer average UHI magnitudes peak at 3.7 °C. In addition, UHI magnitudes were higher in areas with closely spaced buildings and few or no trees and building materials that are not appropriate for the region's climate and thermal comfort.

Keywords: urban climate; urban heat island; mobile traverses; tropical cities

1. Introduction

The transformation of natural surfaces by built urban forms affects several climatological variables, especially air temperature, and consequently humans and the environment. Given the increasing number of people living in cities and the atmospheric changes caused by urbanization, urban heat islands (UHIs) have been detected within very different sized cities and differentiated by their intensities or magnitudes [1–5].

Early studies of UHIs focused almost exclusively on large cities or metropolitan regions [6–12], owing to the greater importance placed on air circulation and pollution [13]. However, a number of studies for small- and medium-sized cities have described problems in the air and temperature of these cities that are unique [13].

Different thermal patterns in the atmosphere over smaller urban areas have been reported [13–16]. Besides the specific characteristics of the studied areas, these studies showed that nocturnal heat islands were evident in small and nonindustrial cities with low population densities. Apparently, heat islands can develop under ideal conditions (i.e., light winds and clear skies), and surface relief tends to modify UHIs regardless of the city size.

Heat islands are strongly related to the size and morphology of urban areas [17], however, even in small cities that are neither polluted nor have excessive verticalization, UHIs are associated with thermal and hygrometric discomfort because of the rise in temperature [18]. Gartland ([19], p.11) points out that not only do heat islands cause minor additional discomfort, but also the higher temperature and lack of shading and the consequent increase in air pollution increase the mortality rates and affect the population health.

The importance of urban climate studies is justified by the fact that small- and mid-sized cities, which are generally not well studied, show increases in population and socioeconomic, political, and environmental significance. In addition, urban planning in these cities, owing to their stages of development, is more effective than in large and metropolitan ones [20].

The use of urban climate studies, however, as a support to urban planning is still incipient in Brazilian cities. Urban space in tropical cities prioritizes the economic rather than the social and environmental aspects, subdividing the land into small lots, removing the vegetation cover, and using building elements that increase the ambient room temperatures. These factors lead to an urban thermal environment that impacts the residents' health, especially the low-income population, who are less likely to adapt to climate changes owing to their limited access to cooling facilities.

In light of these concerns, we attempted to analyze the heat islands detected in three tropical cities to unravel the main factors that are responsible for heat island development in small- and mid-sized cities. By using information about land use, land cover, and atmospheric systems, we identified the maximum nocturnal UHI magnitudes during stable weather conditions, distributed among densely built areas with reduced vegetation cover.

2. Materials and Methods

2.1. Study Areas

Paranavaí is a medium-sized city in the Northwest region of Paraná State (Figure 1). The city is approximately 530 m above sea level, and according to the Brazilian Institute of Geography and Statistics (IBGE) [21], Paranavaí has an estimated population at 87,316, with nearly 90% of them residing in urban areas, which correspond to 41 km^2.

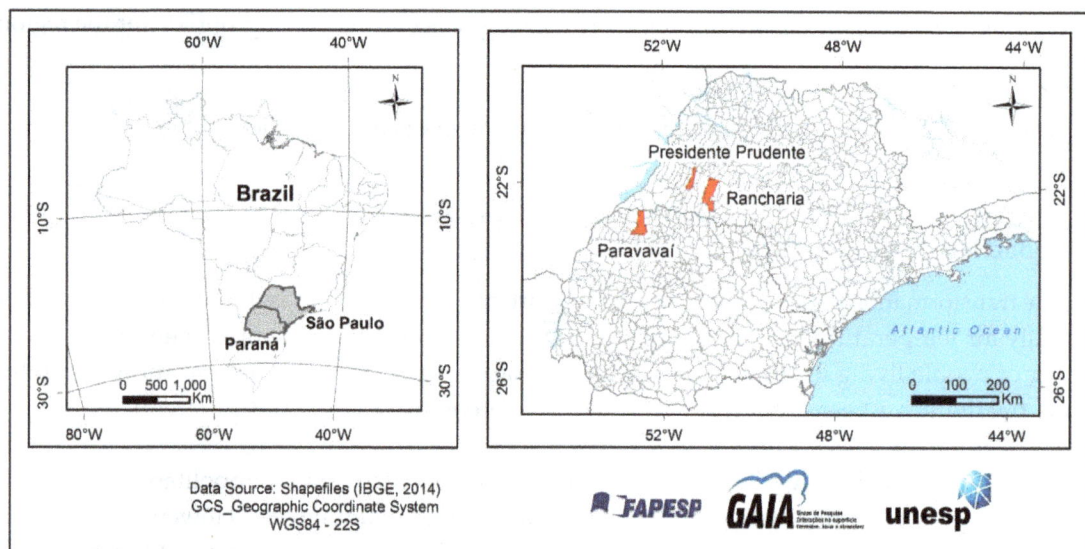

Figure 1. Location map of the study areas.

The history of the city's occupation and colonization began in the 1930s. The colonization in the Northwest of Paraná was based on the construction of roads and small urban centers, such as the medium-sized cities of Maringá, Paranavaí, Cianorte, and Umuarama [22]. Even though the occupation of Paranavaí was initially planned, it subsequently followed housing market demand without adhering to the initial urban plan.

Both Rancharia and Presidente Prudente are in the Western region of São Paulo State (Figure 1). Rancharia is a small city at approximately 520 m above sea level, and its establishment was associated with the construction of the Sorocabana Railroad. Urban expansion occurred along the west and south.

In the southeast, areas of high slopes were urbanized and occupied by low-income housing. Currently, almost 90% of the population (25,828) lives in the urban area [21].

Presidente Prudente is a medium-sized city at 470 m above mean sea level. Its population is estimated at 223,749 people [21], and most of them are living within the urban area (~60 km^2) as well. The oldest neighborhoods were built between 1950 and 1970, and they correspond to densely built areas with trees scattered across streets and in residential yards. Residential areas built in the 1980s and 1990s have some scattered buildings, low plants, and trees. However, most of them are characterized by low-income housing with building materials that are not appropriate for the region's climate and thermal comfort (i.e., thin walls and fiber cement roofs) [23,24].

The main types of air masses that affect these three cities are the Atlantic Tropical mass (aTm), the Continental Tropical mass (cTm), the Continental Equatorial mass (cEm), and the Atlantic Polar mass (aPm), among other atmospheric systems. The tropical air masses are associated with high temperatures during spring and summer, whereas extratropical systems, such as frontal systems (FS) and aPm in autumn and winter cause temperatures to drop. Generally, the climate of the cities is warm and wet from October to March, but milder and drier from April to September, when temperatures decrease with the advance of polar masses throughout the region [25].

2.2. Measuring UHIs in Small- and Mid-Sized Cities

Heat islands are defined by easily measurable parameters, and various studies on urban climate have reported UHIs in Brazil [18,23,24,26–29]. Typically, air temperatures are measured using either thermometers via mobile surveys or a stationary network of sensors. However, because of the financial difficulties in maintaining a network of meteorological stations in and around medium- and small-sized cities, using fixed-point stations for short periods, and mobile measurements across a city has been found to be efficient for assessing urban heat islands in Brazil [30].

Mobile surveys have been extensively used in detecting UHIs in urban and rural areas [6,13,31]. The routes may be linear or circuitous and provide a dense sample of temperatures in a relatively short period of time. When using automobiles during this type of surveys, the sensors can be placed on top of the cars or attached to the side of the vehicle, ~1.8 m above the ground, to minimize the effect of the engine heat on the measured temperatures. The recommended average vehicle speed is 30 km h^{-1}, and traverses are conducted a few hours after sunset, on evenings with light winds and clear sky because, then, micro and local climate differences are maximized [18,32–35].

To assess UHIs in Paranaví, Rancharia, and Presidente Prudente, mobile surveys were conducted after sunset during summer (December–January) evenings to gather temperatures from different traverse routes (Figure 2). The measurements were performed using digital thermohygrometers (model 7664.01.0.00 in Paranaví and Presidente Prudente, and model TH-03B in Rancharia, with precision of ±0.1 °C), fixed on a 1.5-meter-long wooden rod attached to the side of the vehicle.

Data gathering involved two traverse modes. Only one car was used to cover the routes in approximately 1 h in both Paranaví and Rancharia. The route in Paranaví was 8 km long from east to west and 8.3 km long from north to south. In Rancharia, the west–east traverse was 4.6 km and the north–south traverse was 2.2 km. Two vehicles were simultaneously used in Presidente Prudente to complete the west–east (18.3 km) and south–north (14.8 km) traverses, which took approximately 1 h.

The automobile surveys measured air temperatures at 118 sites in Paranaví, 65 in Rancharia, and 275 in Presidente Prudente (average distance of ~100 m between points). The cities' UHI magnitudes were calculated for each day of the traverses to facilitate analysis (Section 3.1).

Average temperatures were then computed for each city individually. The average minimum temperature was subtracted from the other values registered through the traverses to obtain the summer average UHI magnitudes. The average UHI magnitudes were plotted as a function of the elevation profile of each city to depict the temperature differences along the traverses (Section 3.2).

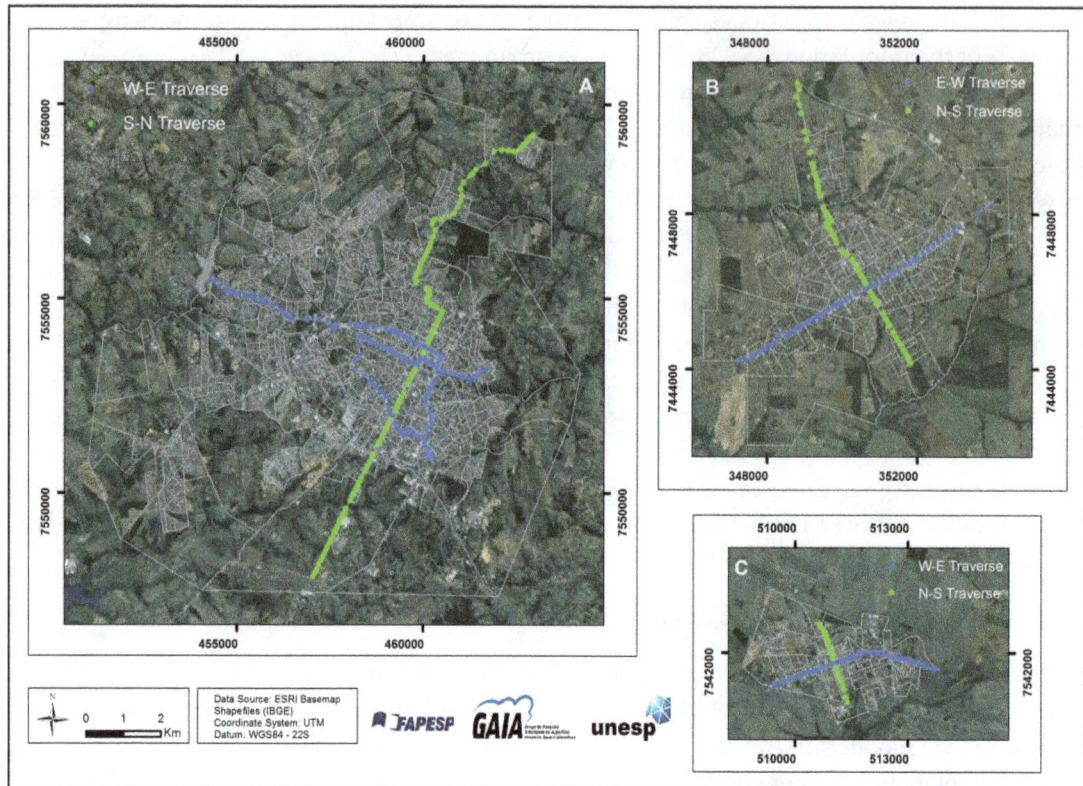

Figure 2. Traverses across (**a**) Presidente Prudente (São Paulo), (**b**) Paranavaí (Paraná), and (**c**) Rancharia (São Paulo).

3. Results and Discussion

3.1. UHIs and Synoptic Conditions

Urban heat islands (UHIs) have been documented for over a century and their formation is associated with changes in the Earth's surface and atmosphere caused by urbanization [1,36–42]. A combination of factors, such as the removal of vegetation, thermal and physical properties of construction materials, building morphology, surface roughness, and anthropogenic heat sources, modifies the local energy and water balances, leading to increases in atmospheric temperature in cities compared to their surroundings [1,3,5,42,43].

Urban heat islands are subsurface, surface, and atmospheric types, with the third being the most typical. Considering heat islands in the atmosphere, two types are distinguished [1,44]. The *canopy-layer* urban heat island (CLHI) occurs close to the surface in cities and extends to approximately the mean building height. Microscale processes govern the UHI climate, and this concerns geographers and climatologists because it affects the spaces where human activities take place every day [3,45]. Above the CLHI lies the *boundary-layer* urban heat island, which is governed by local or mesoscale processes and refers to that portion of the planetary boundary layer whose characteristics are affected by an urban area.

The magnitude or intensity of the UHI is traditionally defined as the temperature difference between urban and rural areas *($\Delta Tu\text{-}r$)* [43]. According to previous studies [9,17,39,42,45], synoptic conditions strongly exert control on the heat island magnitude, which is pronounced at night when the sky is cloudless and the winds are weak. In addition, the prevailing weather conditions during heat island measurements, as well as antecedent conditions leading up to a heat island event, strongly affect the observed temperature differences between urban and rural sites [45].

Since synoptic weather conditions greatly affect the UHI magnitude, we selected data from three cities in the countryside of Brazil where mobile traverses were performed during summer evenings. Tables 1–3 list the nocturnal UHI magnitudes in Paranavaí, Rancharia, and Presidente Prudente as a function of wind speed and direction, and atmospheric systems.

Table 1. Urban heat island magnitude and atmospheric systems during nighttime traverses in Paranavaí, January 2014.

Date	ΔTu-r (°C)	Wind Speed (m s^{-1})	Wind Direction	Atmospheric Systems
11 January 2014	4.7	1.9	NE	modified Atlantic Tropical mass (aTm)
12 January 2014	3.5	1.9	N	modified Atlantic Tropical mass (aTm)
28 January 2014	2.3	1.5	NE	modified Atlantic Tropical mass (aTm)
29 January 2014	5.4	0.7	E	modified Atlantic Tropical mass (aTm)
30 January 2014	5.5	0.6	S	modified Atlantic Tropical mass (aTm)

Data source: Dorigon [46].

Table 2. Urban heat island magnitude and atmospheric systems during nighttime traverses in Rancharia, January 2014.

Date	ΔTu-r (°C)	Wind Speed (m s^{-1})	Wind Direction	Atmospheric Systems
5 January 2014	2.2	2.8	E	Atlantic Tropical mass (aTm)
9 January 2014	2.7	0.6	E	Frontal System (FS)
19 January 2014	3.9	0.2	SE	modified Atlantic Polar mass (aPm)
20 January 2014	2.9	1.7	E	modified Atlantic Polar mass (aPm)
29 January 2014	4.7	0.6	E	Atlantic Tropical mass (aTm)

Data source: Teixeira [47].

Table 3. Urban heat island magnitude and atmospheric systems during nighttime traverses in Presidente Prudente, December 2013.

Date	ΔTu-r (°C)	Wind Speed (m s^{-1})	Wind Direction	Atmospheric Systems
11 December 2013	6	1,6	E	Continental Equatorial mass (cEm)
12 December 2013	3.8	2,5	SE	Continental Equatorial mass (cEm)
13 December 2013	5.5	2,4	SE	Continental Equatorial mass (cEm)
15 December 2013	3.5	1,1	E	Continental Tropical mass (cTm)
16 December 2013	5.1	0	-	Continental Tropical mass (cTm)

Data source: Cardoso and Amorim [48].

In Paranavaí (Table 1), the traverses were carried out in January 2014 (five nights), when the city was under the influence of the Atlantic Tropical mass (aTm). With the aging of the aTm owing to its displacement toward the interior of the continent, the air becomes hotter and drier, and the weather conditions stabilize. Under these conditions and without precipitation, the urban heat island magnitudes in the city ranged from 2.3 °C to 5.5 °C, with higher values when the wind speed was below 1 ms^{-1} (January 29 and 30).

The field campaigns in Rancharia also occurred during five nights in January 2014 (Table 2). It rained two days before the measurements (50 mm), but the greatest urban heat island magnitudes were associated with atmospheric stability, which was governed by light winds and no precipitation.

The phenomenon of urban heat island was intensified when Rancharia was affected by both the modified Atlantic Polar mass (aPm) and Atlantic Tropical mass (aTm). As the aPm, a high-pressure system that causes atmospheric stability, moves toward the continental area, it becomes hot owing to the region's low latitude. Even highly modified, the polar mass (January 19), together with the aTm (January 29), provided calm conditions that highlighted the temperature differences.

The measurements in Presidente Prudente were carried out in December 2013, and because of atmospheric instability, we were not able to perform the mobile traverses under completely calm conditions. On the other hand, the minimum UHI magnitude verified during all field campaigns was 3.5 °C and the maximum 6 °C (Table 3). These values are significant for a mid-sized city, considering that even after an episode of atmospheric instability and precipitation (57 mm on December 14), the temperature difference recorded on the next day reached 3.5 °C [48].

Besides gathering temperatures during the summer, the nocturnal UHIs of the cities were maximized under clear skies and light winds, defined as <3 m s^{-1} by Oke and Maxwell [6], with moderate to strong magnitudes [2].

The differences registered in Paranavaí and Presidente Prudente exceeded 5 °C and agree well with previous work for medium-sized cities elsewhere. For instance, Unger [49] found temperature differences of approximately 4 °C during clear and windless nights in Szeged, Hungary, and for winds equal or inferior to 3 m s^{-1}, and Balkestahl et al. [50] verified the occurrence of heat islands in Oporto, Portugal, with temperature differences up to 7 °C.

In smaller cities, a similar correspondence between the UHI magnitudes in Rancharia and those in Oke [31] were found during a survey along the St. Lawrence Lowland region in the Province of Quebec, Canada. Based on four traverses in the summer, the magnitudes varied from 2.3 °C to 5.2 °C in Saint-Hubert, and from 3.8 °C to 6 °C in Saint-Hyacinthe, whereas in Rancharia they varied from 2.2 °C to 4.7 °C.

3.2. Urban Morphology, Land Cover, and UHI Magnitude

It has already been shown that cities can significantly alter the natural surface conditions and atmospheric properties, resulting in different heating patterns within urban areas [40,43,51]. Moreover, there is a strong relation between land cover and temperature, especially regarding vegetation and built density [1,35,52,53], which allows inferring that different types of land cover within each city can affect the spatial pattern of its UHI [3].

Based on land cover materials and built densities, Oke [1] organized the classic heat island profile. According to the physical properties of the materials, thermal conductivity, and heat capacity, the materials with higher thermal conductivity conduct heat to their interior easier, whereas materials with high heat capacity store more heat and, as more heat is retained, the temperature of the material increases.

Normally, the distribution of temperature depends on the urban arrangement. The classic urban heat island described by Oke [1] is characterized by high thermal gradient at the urban–rural boundary, followed by a gradual rise in temperature toward the city's core; that is, the closer to downtown (high built density, commercial areas, less vegetation) an area is, the higher the temperature is, whereas the farther away from downtown (fewer buildings and more vegetation) an area is, the lower the temperature is.

Nevertheless, a different pattern has been identified throughout small- and medium-sized cities in Brazil, where neighborhoods far from the urban core show higher temperatures [18,23,54–56]. Surface heterogeneities are associated with the formation of one or more heat islands that are not necessarily located in the old core. To determine whether a similar pattern occurs in Paranavaí, Rancharia, and Presidente Prudente, summer average UHI magnitudes were analyzed.

As mentioned above, these cities have different population sizes and present different forms of urbanization; nonetheless, the temperature distribution pattern was very similar in all of them. As Figures 3–5 show, the UHIs with the highest magnitudes were clearly related to the density, whereas the lowest ones were in areas with fewer buildings and greater vegetation cover.

In Paranavaí (Figure 3a), the east–west route was characterized by low surface homogeneity, considering that there were no big variations in building morphology, land cover, or surface relief. The main steep temperature gradient was verified at the very beginning of this traverse, where there was a sparsely built area. The predominance of commercial buildings was responsible for the highest UHI magnitude (3.5 °C) and the magnitude just started to decrease when the traverse passed residential and rural sites, with more open areas and vegetation.

The north–south traverse (Figure 3b) showed greater heterogeneity in surface morphology and land cover, with an elevation difference of 55 m between the highest and lowest points along the traverse. These characteristics increased the variability along the profile and lowered the UHI magnitudes compared to the east–west traverse. The lowest magnitude in both traverses was found along this route, near a densely-vegetated area, whereas the peak of the heat island (2.6 °C) was detected in the commercial area of the city.

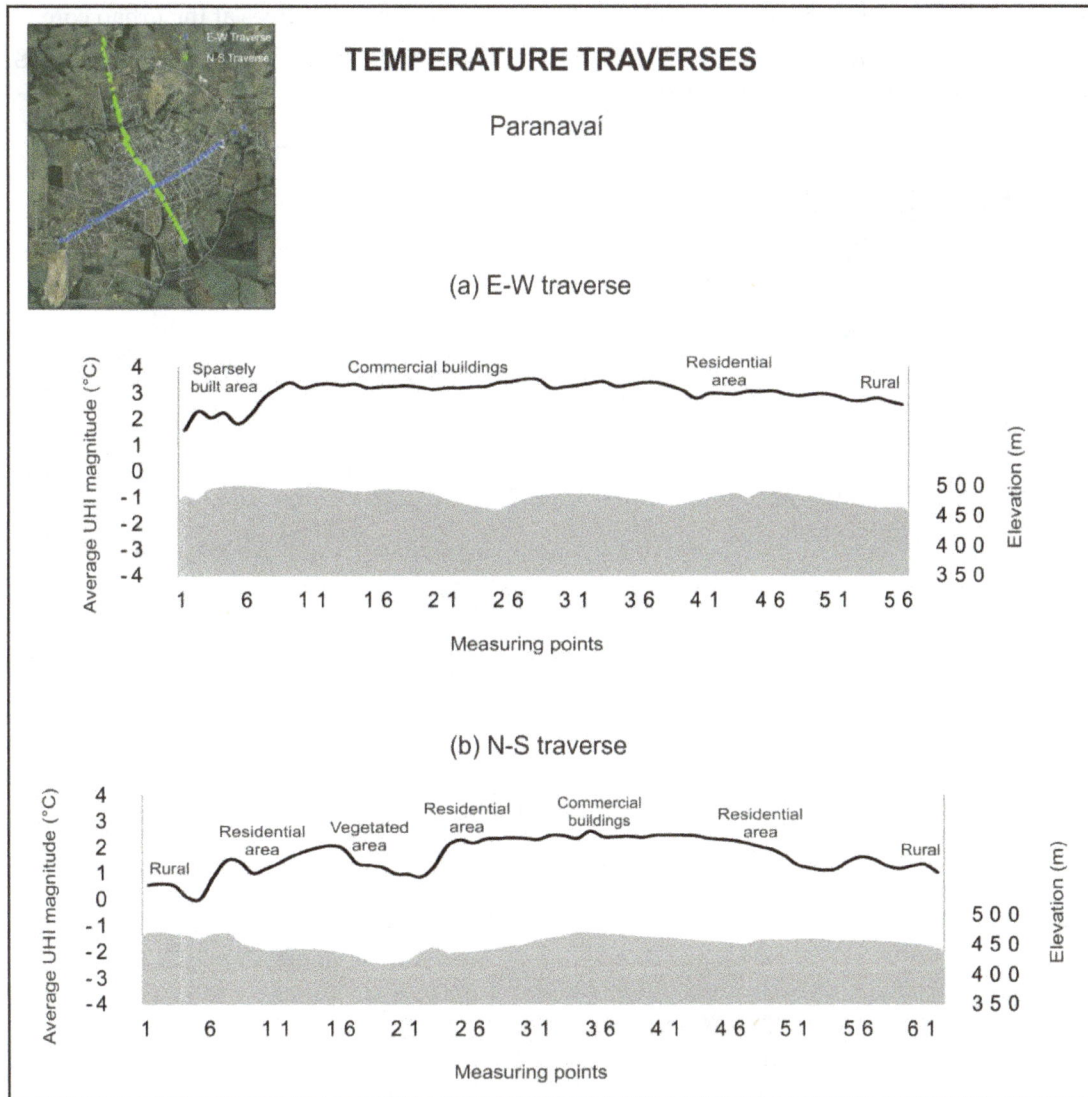

Figure 3. Summer average urban heat island (UHI) magnitudes in Paranavaí during nighttime traverses, January 2014 (5 = five nights): (**a**) east–west traverse; and (**b**) north–south traverse. Horizontal lines indicate the temperature profile. Surface relief is shown the along bottom axes.

In Rancharia (Figure 4), the classic center–periphery heating pattern was identified, with greater values along densely built areas. The urban density and the presence of tree cover were mainly responsible for the UHI magnitude variations.

Along the first route (west–east), the heating from the rural area to commercial buildings downtown was significant, and it can be related to changes in urban morphology (Figure 4a). The UHI reached maximum magnitude (2.7 °C) at the urban core, corresponding with the locations of the commercial buildings. Moving forward, the magnitude decreased as the land cover changed from residential to rural areas.

The north–south traverse (Figure 4b) showed a similar pattern, although the magnitudes were smaller. The highest value was 2.2 °C and was again found among commercial buildings. Residential areas with scattered trees had magnitudes below 2 °C, followed by rural sites at the end of the traverse, where dense trees and lower elevations were associated with decreasing temperatures.

Along the west–east traverse in Presidente Prudente (Figure 5a), the temperature profile frequently varied. At the beginning of the traverse, rural with little vegetation areas and residential areas had a magnitude of 2.7 °C, moving forward the magnitude varied between 3 °C and 3.7 °C. Besides crossing an area with high and midrise commercial and residential buildings, the trees at the urban core were responsible for low magnitudes. Leaving this area, the temperature decreased again when passing by a valley bottom with few trees and low plants.

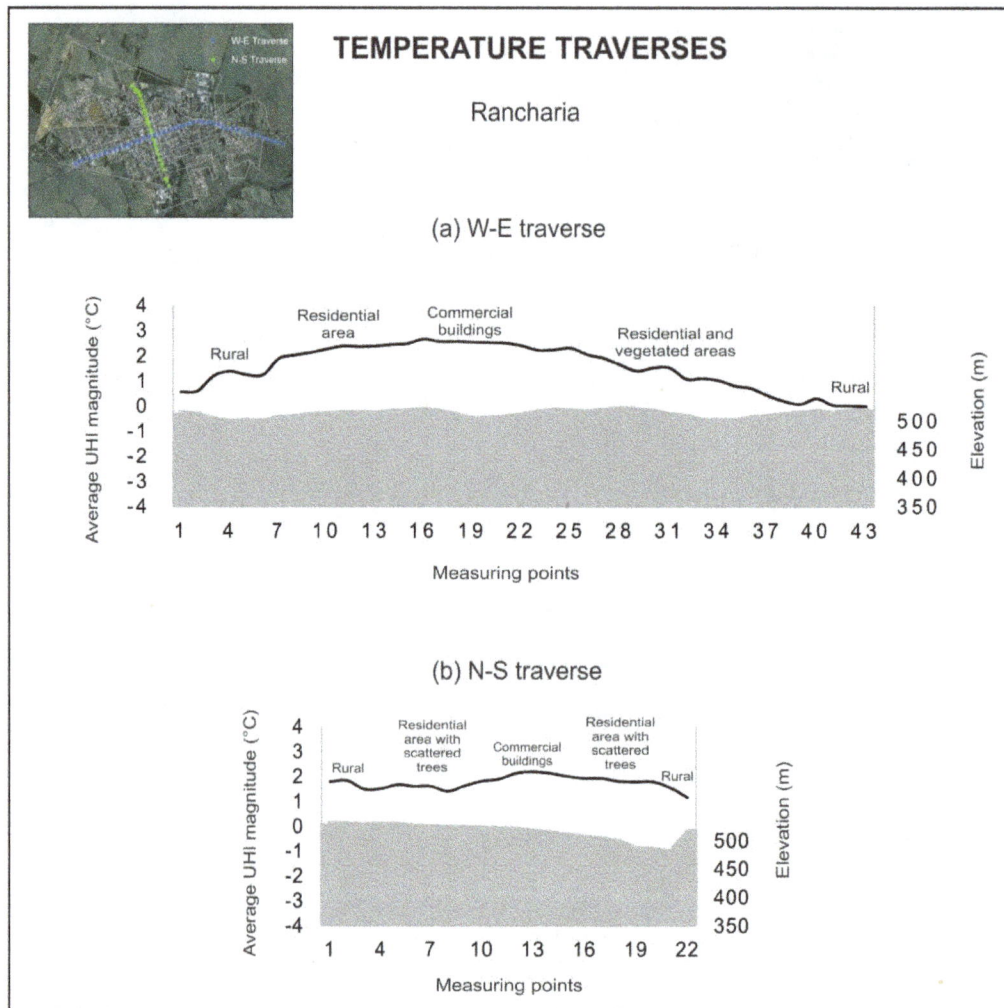

Figure 4. Summer average UHI magnitudes in Rancharia during nighttime traverses, January 2014 (five nights): (**a**) west–east traverse; and (**b**) north–south traverse. Horizontal lines indicate the temperature profile. Surface relief is shown along the bottom axes.

The UHI magnitude increased once more across midrise commercial buildings (up to 2.7 °C). When getting into a neighborhood with low-rise commercial and residential buildings, the temperature decreased from 2.2 °C to 1.9 °C. Reaching the end of the traverse, the magnitude was 1.3 °C at a vegetated valley bottom and rose to 1.8 °C in the last residential area.

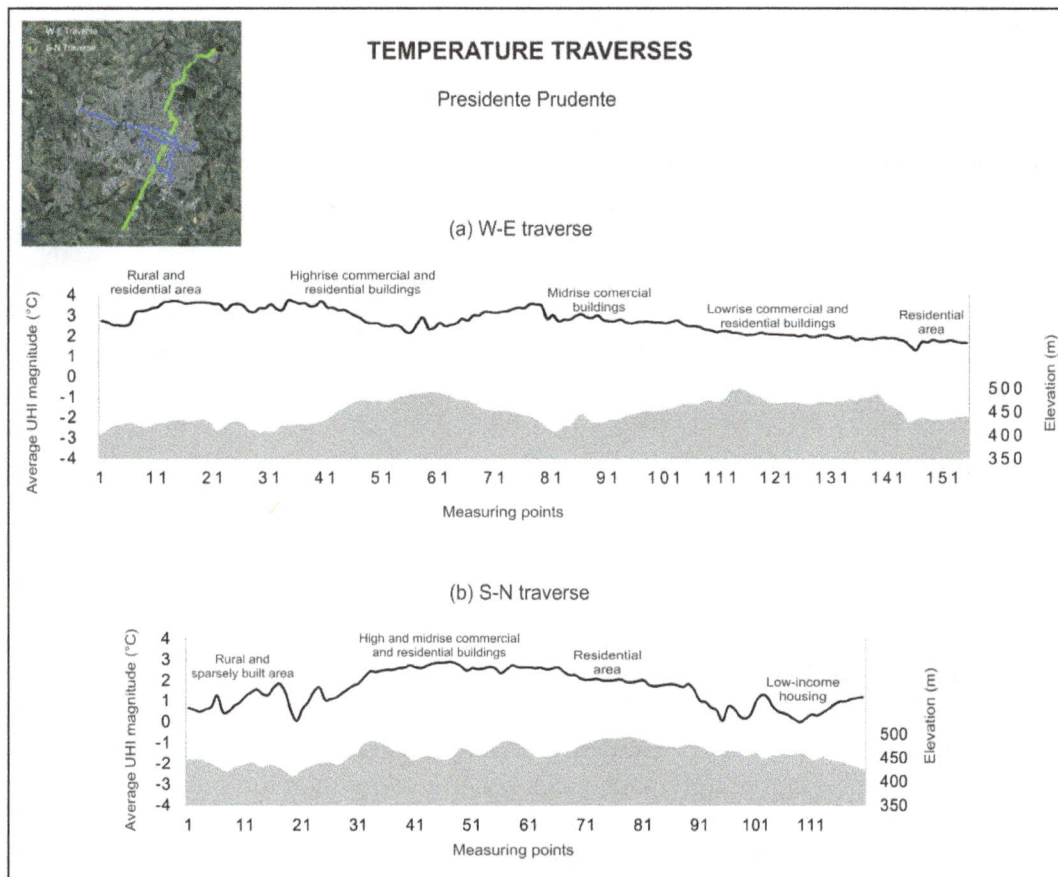

Figure 5. Summer average UHI magnitudes in Presidente Prudente during nighttime traverses, December 2013 (five nights): (**a**) west–east traverse; and (**b**) south–north traverse. Horizontal lines indicate the temperature profile. Surface relief is shown along the bottom axes.

The south–north traverse (Figure 5b) was similar with the classic UHI pattern. High magnitudes up to 2.9 °C were identified in areas with high and midrise commercial and residential buildings, whereas the lowest values were found at the extremes of the route.

At the beginning of this traverse, the temperatures were up to 1.8 °C along rural (predominance of tree and low plants cover) and sparsely built areas and rose above 2 °C when crossing densely built areas. From closely spaced buildings toward a residential area with scattered trees, the UHI magnitudes varied from 2.6 °C to 1.7 °C. Passing low-income housings, where buildings were in compact arrangement but scattered trees and abundant plant cover predominated in the surroundings, the temperatures decreased, and the magnitudes were less than 1.2 °C.

The temperature profiles in this section showed how intraurban temperature patterns and UHI magnitudes vary along different surface characteristics. Besides the classic UHI profile [1], the cities' new arrangements were more heterogeneous, leading to different patterns of increasing temperatures within the urban area, as in Presidente Prudente.

Compared to other studies [57,58], the UHI magnitudes from traverse data and the type of urban morphology and land cover were correlated. For instance, in Nagano, Japan, densely built areas with tall buildings were 3 °C warmer than areas with low plants and open-set trees [57]. At our traverses, a total difference of 3.5 °C (Paranavaí) and 3.7 °C (Presidente Prudente) separated the warmest and the coolest areas.

Lelovics et al. [59] used mobile measurements to examine thermal variations in Szeged, Hungary. The average UHI magnitudes derived here were consistent with their results, especially the ones for sparsely built areas, with magnitudes under 1 °C. Compact areas with midrise buildings were 4.2 °C higher than field sites with low plant cover. Based on our traverse data, the difference between the corresponding areas was, on average, 3.5 °C.

Regarding our findings in Rancharia, these are similar to those reported by Doyle and Hawkins [16] for the small urban area of Shippensburg, Pennsylvania. The average difference between the most urban and most rural areas in Shippensburg was 1.9 °C, whereas the average UHI magnitudes in this study suggest that urban areas were up to 2.7 °C warmer than rural.

4. Conclusions

Although urban areas cover only a small fraction of the Earth's surface, their effects on local climate are large and increasing, especially regarding the development of urban heat islands within cities of different sizes. Furthermore, the inhabitants cannot always counterbalance the warmer conditions in the cities and this makes them increasingly vulnerable to heat waves or other extreme conditions. Thus, it is crucial to investigate the main factors behind the formation of heat islands and insight into mitigation strategies [5].

Assessing UHIs involves several procedures (measurement variables, time and scale definition, instruments, data collection, reporting, etc.); moreover, the magnitude of impact is strongly affected by synoptic conditions, especially precipitation, cloudiness, and wind. In this study, we show the relationship between atmospheric stability and heat islands in tropical cities, where the temperature differences are pronounced under calm conditions.

Urban morphology and land cover are important to the formation of UHIs in Paranavaí, Rancharia, and Presidente Prudente. By analyzing the temperature profiles of the cities, we verified the role of built density in increasing the UHI magnitudes, whereas vegetation is directly related to lower temperatures. Although high magnitudes are identified in densely built areas, the effects of surface relief on lowering the city temperatures are noteworthy, especially when traversing vegetated valleys.

The UHI magnitudes depend on the method of reporting [43]. For example, the average UHI magnitude was up to 2.7 °C in Rancharia, 3.5 °C in Paranavaí, and 3.7 °C in Presidente Prudente for five summer nights (December 2013–January 2014). In contrast, using daily data from the traverses, magnitudes were 4.7 °C, 5.5 °C, and 6 °C, respectively.

Tropical cities are constantly warm and sunny, and the heat islands can intensify the thermal discomfort in urban areas. Therefore, the urban climate of small- and medium-sized cities in the tropics needs to be studied. As urban areas are growing, knowing where UHIs are going to happen and what affects their magnitude will allow for better urban design and planning methods.

Acknowledgments: The authors thank the São Paulo Research Foundation (FAPESP) for the financial support, grants 2013/02057-0, 2013/02056-3, and 2013/02081-8.

Author Contributions: The authors contributed equally to this work.

Conflicts of Interest: The authors declare no conflict of interest. The founding sponsor had no role in the design of the study; in the collection, analyses, or interpretation of data; in the writing of the manuscript, and in the decision to publish the results.

References

1.   Oke, T.R. *Boundary Layer Climates*, 2nd ed.; Methuen and Co.: New York, NY, USA, 1987; pp. 262–303.

2.   García, F.F. *Manual de Climatología Aplicada: Clima, Medio Ambiente y Planificación*, 1st ed.; Editorial Síntesis: Madrid, España, 1995.

3.   Yow, D.M. Urban heat islands: Observations, impacts and adaptation. *Geogr. Compass* **2007**, *1*, 1227–1251. [CrossRef]

4.   Souch, C.; Grimmond, S. Applied climatology: Urban climate. *Prog. Phys. Geogr.* **2006**, *30*, 270–279. [CrossRef]

5.   Grimmond, S. Urbanization and global environmental change: Local effects of urban warming. *Geogr. J.* **2007**, *173*, 83–88. [CrossRef]

6. Oke, T.R.; Maxwell, G.B. Urban heat island dynamics in Montreal and Vancouver. *Atmos. Environ.* **1975**, *9*, 191–200. [CrossRef]

7. Graham, E. The urban heat island of Dublin city during the summer months. *J. Ir. Geogr.* **1993**, *26*, 45–57. [CrossRef]

8. Tso, C.P. A survey of urban heat island studies in two tropical cities. *Atmos. Environ.* **1996**, *30*, 507–519. [CrossRef]

9. Kim, Y.; Baik, J. Maximum urban heat island intensity in Seoul. *J. Appl. Meteor.* **2002**, *41*, 651–659. [CrossRef]

10. Hart, M.; Sailor, D.J. Assessing causes in spatial variability in urban heat island magnitude. In Proceedings of the Seventh Symposium on the Urban Environment, San Diego, CA, USA, 10–13 September 2007.

11. Coutts, A.M.; Beringer, J.; Tapper, N.J. Impact of increasing urban density on local climate: Spatial and temporal variations in the surface energy balance in Melbourne, Australia. *J. Appl. Meteor. Climatol.* **2007**, *46*, 477–493. [CrossRef]

12. Holt, T.; Pullen, J. Urban canopy modeling of the New York City metropolitan area: A comparison and validation of single- and multilayer parameterizations. *Mon. Wea. Rev.* **2007**, *135*, 1906–1930. [CrossRef]

13. Kopec, R. Further observations of the urban heat island in a small city. *Bull. Amer. Meteor. Soc.* **1970**, *51*, 602–606. [CrossRef]

14. Duckworth, F.S.; Sandberg, J.S. The effect of cities upon horizontal and vertical temperature gradients. *Bull. Am. Meteor. Soc.* **1954**, *35*, 198–207.

15. Hutcheon, R.J.; Johnson, R.H.; Lowry, W.P.; Black, C.H.; Hadley, D. Observations of the urban heat island in a small city. *Bull. Am. Meteorol. Soc.* **1967**, *48*, 7–9.

16. Doyle, D.; Hawkins, T. Assessing a small summer urban heat island in rural south central Pennsylvania. *Geogr. Bull.* **2008**, *49*, 65–76.

17. Roth, M. Urban heat islands. In *Handbook of Environmental Fluid Dynamics: Systems, Pollution, Modeling, and Measurements*, 1st ed.; Fernando, H.J.S., Ed.; CRC Press/Taylor & Francis Group: Boca Raton, FL, USA, 2012; Volume 2, pp. 143–159.

18. Amorim, M.C.C.T. Intensidade e forma da ilha de calor urbana em Presidente Prudente/SP. *Geosul* **2005**, *20*, 65–82.

19. Gartland, L. *Ilhas de Calor: Como Mitigar Zonas de Calor em Áreas Urbanas*; 1st ed.; Oficina de Textos: São Paulo, Brazil, 2010.

20. Mendonça, F. Clima e Planejamento urbano em Londrina—Proposição metodológica e de intervenção urbana a partir do estudo do campo térmico. In *Clima Urbano*, 1st ed.; Monteiro, C.A.F., Mendonça, F., Eds.; Editora Contexto: São Paulo, Brasil, 2003; pp. 93–120.

21. Cidades. IBGE. Available online: http://cidades.ibge.gov.br/v3/cidades/home-cidades (accessed on 23 November 2016).

22. Passos, M.M. *A Raia Divisória—Geosistema, Paisagem e Eco-História*, 1st ed.; EDUEM: Maringá, Brasil, 2006.

23. Amorim, M.C.C.T. O clima urbano de Presidente Prudente/SP. Ph.D. Thesis, Tese (Doutorado em Geografia) - Faculdade de Filosofia, Letras e Ciências Humanas, Universidade de São Paulo, São Paulo, Brazil, 2000.

24. Amorim, M.C.C.T.; Dubreuil, V.; Quenol, H.; Sant'Ana Neto, J.L. Características das ilhas de calor em cidades de porte médio: Exemplos de Presidente Prudente (Brasil) e Rennes (França). *Confins* **2009**, *7*, 1–16. [CrossRef]

25. Barrios, N.A.Z.; Sant'anna Neto, J.L. A circulação atmosférica no extremo oeste paulista. *Bol. Climatol.* **1996**, *1*, 8–9.

26. Tarifa, J.R. Análise comparativa da temperatura e umidade na área urbana e rural de São José dos Campos (SP). *Geografia* **1977**, *2*, 59–80.

27. Mendonça, F.A. O clima e o planejamento urbano de cidades de porte médio e pequeno. Proposição metodológica para estudo e sua aplicação à cidade de Londrina/PR. Ph.D. Thesis, Tese (Doutorado em Geografia Física)-Faculdade de Filosofia, Letras e Ciências Humanas, Universidade de São Paulo, São Paulo, Brizil, 1994.

28. Sant'Anna Neto, J.L. *Os Climas das Cidades Brasileiras*, 1st ed.; Programa de Pós-Graduação em Geografia da FCT/UNESP: Presidente Prudente, Brasil, 2002.

29. Dumke, E.M.S. Clima urbano/conforto térmico e condições de vida na cidade—Uma perspectiva a partir do Aglomerado Urbano da Região Metropolitana de Curitiba (AU-RMC). Ph.D. Thesis, Tese de Doutorado-Programa de Doutorado em Meio Ambiente e Desenvolvimento, Universidade Federal do Paraná, Curitiba, 2007.

30. Dubreuil, V.; Amorim, M.C.C.T.; Froissard, X.; Quenol, H. Métodos e monitoramento da variabilidade espaçotemporal da ilha de calor em cidades de porte médio—Rennes/França e Presidente Prudente/Brasil. In *Experimentos em Climatologia Geográfica*; 1st ed.; Silva, C.A., Fialho, E.S., Steinke, E.T., Eds.; Editora da UFGD: Dourados, Brasil, 2014; Volume 1, p. 391.

31. Oke, T.R. City size and the urban heat island. *Atmos. Environ.* **1973**, *7*, 769–779. [CrossRef]

32. Oke, T.R. *Initial Guidance to Obtain Representative Meteorological Observations at Urban Sites*; IOM Report 81; World Meteorological Organization: Geneva, Switzerland, 2004.

33. Gomez, A.L.; Garcia, F.F. La isla de calor en Madrid: Avance de un estudio de clima urbano. *Estud. Geogr.* **1984**, *174*, 5–34.

34. Johnson, D.B. Urban modification of diurnal temperature cycles in Birmingham, U.K. *J. Climatol.* **1985**, *5*, 221–225. [CrossRef]

35. Eliasson, I. Intra-urban noctural temperature differences: Multivariate approach. *Clim. Res.* **1996**, *7*, 21–30. [CrossRef]

36. Howard, L. *Climate of London Deduced from Meteorological Observations*, 3rd ed.; Harvey and Darton: London, UK, 1833.

37. Myrup, L. A Numerical model of the urban heat island. *J. Appl. Meteor.* **1969**, *8*, 908–918. [CrossRef]

38. Landsberg, H.E. *The Urban Climate*, 1st ed.; Academic Press: New York, NY, USA, 1981.

39. Yague, C.; Zurita, E.; Martinez, A. Statistical analysis of the Madrid urban heat island. *Atmos. Environ.* **1991**, *25*, 327–332. [CrossRef]

40. Morris, C.; Simmonds, I.; Plummer, N. Quantification of the influences of wind and cloud on the nocturnal urban heat island of a large city. *J. Appl. Meteor.* **2001**, *40*, 169–182. [CrossRef]

41. Voogt, J.A. Urban Heat Islands: Hotter Cities. Available online: http://www.actionbioscience.org/environment/voogt.html (accessed on 15 July 2016).

42. Montavez, J.P.; Rodriguez, A.; Jimenez, J.I. A study of the urban heat island of Granada. *Int. J. Climatol.* **2000**, *20*, 899–911. [CrossRef]

43. Hinkel, K.M.; Nelson, F.E.; Klene, A.E.; Bell, J.H. The urban heat island in winter at Barrow, Alaska. *Int. J. Climatol.* **2003**, *23*, 1889–1905. [CrossRef]

44. Oke, T.R. The distinction between canopy and boundary-layer heat islands. *Atmosphere* **1976**, *14*, 268–277.

45. Stewart, I.D. Redefining the urban heat island. Ph.D. Thesis, The University of British Columbia, Vancouver, BC, Canada, 2011.

46. Dorigon, L.P. Clima urbano em Paranavaí: Análise do espaço intraurbano. Master's Thesis, Dissertação (Mestrado em Geografia)-Faculdade de Ciências e Tecnologia de Presidente Prudente, Universidade Estadual Paulista, São Paulo, Brazil, 2015.

47. Teixeira, D.C.F. O clima urbano de Rancharia (SP). Ph.D. Thesis, Dissertação (Mestrado em Geografia)-Faculdade de Ciências e Tecnologia de Presidente Prudente, Universidade Estadual Paulista, Presidente Prudente, 2015.

48. Cardoso, R.S.; Amorim, M.C.C.T. Variações espaciais das temperaturas noturnas de Presidente Prudente-SP em episódios de verão. *Ra'e Ga* **2017**. under review.

49. Unger, J. Heat island intensity with different meteorological conditions in a medium-sized town: Szeged, Hungary. *Theor. Appl. Climatol.* **1996**, *54*, 147–151. [CrossRef]

50. Balkestahl, L.; Monteiro, A.; Góis, J.; Taesler, R.; Quenol, H. The influence of weather types on the urban heat island's magnitude and patterns at Paranhos (Porto)—A case study from November 2003 to January 2005. In Proceedings of the Sixth International Conference on urban climate, Gothenburg, Sweden, 12–16 June 2006.

51. Arnfield, A.J. Two decades of urban climate research: A review of turbulence, exchanges of energy and water, and the urban heat island. *Int. J. Climatol.* **2003**, *23*, 1–26. [CrossRef]

52. Unger, J.; Sümeghy, Z.; Gulyás, Á.; Bottyán, Z.; Mucsi, L. Land-use and meteorological aspects of the urban heat island. *Met. Apps.* **2001**, *8*, 189–194. [CrossRef]

53. Duarte, D.H.S.; Serra, G.G. Padrões de ocupação do solo e microclimas urbanos na região de clima continental tropical brasileira: Correlaçõe e proposta de um indicador. *Ambient. Constr.* **2003**, *3*, 7–20.

54. Ortiz, G.F.; Amorim, M.C.C.T. Ilhas de Calor em Cândido Mota/SP: Algumas Considerações. *Formação (Presidente Prudente)* **2011**, *18*, 238–257.

55. Ugeda Júnior, J.C.; Amorim, M.C.C.T. Estudo do clima urbano de Jales-SP através do transecto móvel. *Rev. GeoNorte* **2012**, *9*, 365–377.

56. Moreira, J.L. O Clima Urbano em Penápolis/SP: AnáLise da Temperatura e Umidade Intraurbano. Ph.D. Thesis, Dissertação (Mestrado em Geografia)-Faculdade de Ciências e Tecnologia de Presidente Prudente, Universidade Estadual Paulista, São Paulo, Brazil, 2016.

57. Stewart, I.D.; Oke, T.R. Thermal differentiation of local climate zones using temperature observations from urban and rural field sites. In Proceedings of the Ninth Symposium on Urban Environment, Boulder, CO, USA, 1–6 August 2010.

58. Stewart, I.D.; Oke, T.R.; Krayenhoff, E.S. Evaluation of the "local climate zone" scheme using temperature observations and model simulations. *Int. J. Climatol.* **2014**, *34*, 1062–1080. [CrossRef]

59. Lelovics, E.; Gál, T.; Unger, J. Mapping local climate zones with a vector-based GIS method. In Proceedings of the Air and Water Components of the Environment: Conference Dedicated to World Meteorological Day and World Water Day, Cluj-Napoca, Romania, 22–23 March 2013.

9

Climatic Variability and Land Use Change in Kamala Watershed, Sindhuli District, Nepal

Muna Neupane [1] and Subodh Dhakal [2,*]

[1] Central Department of Environmental Science, Tribhuvan University, Kathmandu 44600, Nepal; nmuna125@gmail.com

[2] Department of Geology, Tri-Chandra Campus, Tribhuvan University, Kathmandu 44600, Nepal

* Correspondence: dhakalsubodh@gmail.com

Academic Editor: Maoyi Huang

Abstract: This study focuses on the land use change and climatic variability assessment around Kamala watershed, Sindhuli district, Nepal. The study area covers two municipalities and eight Village Development Committees (VDCs). In this paper, land use change and the climatic variability are examined. The study was focused on analyzing the changes in land use area within the period of 1995 to 2014 and how the climatic data have evolved in different meteorological stations around the watershed. The topographic maps, Google Earth images and ArcGIS 10.1 for four successive years, 1995, 2005, 2010, and 2014 were used to prepare the land use map. The trend analysis of temperature and precipitation data was conducted using Mann Kendall trend analysis and Sen's slope method using R (3.1.2 version) software. It was found that from 1995 to 2014, the forest area, river terrace, pond, and landslide area decreased while the cropland, settlement, and orchard area increased. The temperature and precipitation trend analysis shows variability in annual, maximum, and seasonal rainfall at different stations. The maximum and minimum temperature increased in all the respective stations, but the changes are statistically insignificant. The Sen's slope for annual rainfall at ten different stations varied between −38.9 to 4.8 mm per year. Land use change and climatic variability have been analyzed; however, further study is required to establish any relation between climatic variability and land use change.

Keywords: Nepal; Kamala watershed; land use; climatic variability, Google earth; ArcGIS

1. Introduction

Human activities like population growth, agricultural intensification, and expansion of settlement and industrial areas have created much more land use and land cover changes in the last few decades. Land use refers to the different tasks carried out by humans on the land or the different types of land that provide services and benefits and/or products and resources. Examples of land use include recreation, wildlife reserves, agriculture, roads, and so on, while land cover refers to the biological and physical structures found on the land surfaces [1]. Land cover describes the surface cover on the earth as water bodies, sand, forest, manmade infrastructures, etc. Land cover and land use changes are both often carry-overs from pre-historic periods that are direct and indirect consequences of human activities to obtain essential resources [2]. Thus, the information from both historic and modern time on land use and human activities along with the stability factors may vary rapidly in response to environmental changes or economical needs [3]. Land use change and climate change are interrelated, as these two processes affect each other. Global ecological changes, generated by both land use changes and climate change, are predicted for the future. The interconnectivity between land use change and the atmospheric flux of carbon-dioxide and its consequent impacts on climate are the result of the effects of land use on climate change along with the alteration of climate impact through land management [4].

Land use and land cover changes have important consequences on natural resources through their impacts on soil and water quality, biodiversity, and global climatic systems [5]. Land cover change effects are regional and tend to be offset with respect to global temperatures; however, modifications on regional climate are also associated with global warming [6]. Land use is also an important factor influencing the occurrence of rainfall-triggered landslides throughout the world [7]. The land surface characteristics influence surface temperatures and latent heat flux and the contrasting characteristics of adjacent land cover types could induce convection enhancing cloud and precipitation formation [4]. Within two recent decades (the 1980s and 1990s), the temperature anomalies increased all around the globe [8]. Heavy precipitation events have intensified [9] and the reoccurrence of cold nights, cold days, and frost have become lesser while hot days and hot nights are reoccurring more often over the past 50 years [10]. Driving forces of land degradation processes on hill slopes in the tropics are often directly related to changes in land cover or to extreme climatic conditions. Conversion of natural forest to agricultural land to support growing populations have resulted in major changes in soil physical properties such as increase in bulk density, destruction of soil structure, and decrease of organic carbon content [11]. A previous study dealing with the relationship between land use change and climate change clearly showed that land use change had contributed to greater effects on ecological variables in comparison to the effects of climate change [4]. Human activities change land through land management to mitigate climate change and these adaptation actions will have some ecological effects. Likewise, the climate change simulations executed in the U.S. Department of Energy Parallel Climate Model (DOE-PCM) using different scenarios of landscape change during the current century revealed that future land use change decisions could alter IPCC (Intergovernmental Panel on Climate Change) climate change simulations for those based exclusively on atmospheric composition change [6]. The study and analysis of current climatic trends indicates a significant warming trend in recent decades with the effect beings more prominent at higher altitudes. An increase in average mean temperature of 1.2–3 °C in-between the year 2050 and 2100 is projected. The Himalayan region, particularly Nepal, has also faced increased warming trends, and the significant impacts are prominent [12]. Mann Kendall trend analysis is used for the analysis of rainfall and temperature data. It is a widely used method for trend analysis. The trend analysis helps in identifying the climatic status occurring and climatic changes/variability occurring in a particular area.

In recent years, land use maps have been prepared using satellite images. Remote sensing and Geographic Information System (GIS) are widely used in land use mapping. Different studies around the world incorporate satellite images and image processing in GIS through supervised and unsupervised classifications for land use mapping [13,14]. However, the Landsat images have to be downloaded from Global Land Cover Facility and U. S. Geological Survey, which has a lower spatial resolution of about 30 m. In some cases, a Landsat image of the required year cannot be obtained for some areas. Thus, another way of obtaining satellite images is through the use of Google Earth images, which has a spatial resolution of nearly 1 m. The images are recognizable and the features required can be easily traced. Thus, land use mapping of Kamala watershed was prepared using Google Earth images. However, the images of Google Earth along the south-west region were not clear and were of lower resolution for 2005.

Various developmental activities like migration, deforestation, urbanization, settlement, and infrastructural growth during the last five decades represent the underlying reasons for land cover/ land use change in the Chure (Churiya/ Siwaliks) range of Nepal [14]. The Chure hills are the most recent mountain system of the Himalayan orogeny [15], and the slopes are dry with poor slope development [16]. The fragile geology makes the Chure more susceptible to any form of human disturbances. The maximum temperature data from 49 stations for 1971–1994 showed a warming trend that was increasing by less than 0.03 °C per year in Siwaliks [17]. The Chure range receives a greater amount and higher intensities of rainfall [14]. As a result, rainfall induced landslides, slope degradation, and floods are major hazards in Chure. The land use change and the climatic changes occurring in the Chure region have had adverse implications on both the environment and the livelihood of populations residing in the region. Since 1950s, a huge population has inhabited the Chure range. Settlements and agricultural activities are predominantly located on the flatlands and the river valleys. The slopes are susceptible to debris flow, landslides, bank cutting, and soil erosion, along with flood problems.

The first objective of the study is to identify major land use types along Kamala Watershed and their change pattern between 1995 and 2014. Secondly, the study is conducted to determine the changes in climatic variability within the study area. The detection of land use change provides a brief of the land use practices done and the changes that had been occurred with the pace of industrialization and population centralization to the city areas since past. The land use study covers the changes in different land types; agricultural, settlement, forest and so on. The knowledge on past and present land use practices will ensure the changes that will occur in future and are useful for policy makers and developers. The climatic variability is also the major obstacle for the living in mountainous countries. The information on the changes on climatic parameter will be helpful for the decision makers to get through knowledge, ideas and solutions to cope better with the changing climate and stress. Thus, this research will be helpful in identifying the land use change and climatic variability in Kamala watershed, which is a hazard prone area.

2. Materials and Methods

2.1. Study Area

Sindhuli district is situated north of Mahabharat range and south of Chure range. The total area of Sindhuli is 2,49,100 ha out of which 1,43,496.5 ha is covered by Chure range. It is located between latitude 26°55′ N to 27°22′ N and longitude 85°15′ E to 86°25′ E. The area ranges between 168 m and 2797 m with about 64.83% of land covered by forest, 19.36% as agricultural land, 11.55% landslides and riverbank and 4.41% used as others. The district incorporates 149 rivers [18]. The study area comprises of Lower Siwaliks, Middle Siwaliks, Upper Siwaliks, Precambrian meta-sedimentary rocks and Lakharpata groups [19]. The Main Boundary Thrust (MBT) is present on southern boundary of the study area. The rocks are soft, loose and easily erodible mainly consisting sandstone, siltstone, mudstone, and conglomerate [20]. Sindhuli is hazard prone zone, which is at risk of multiple natural hazards. The landslides, flood, and debris flows are pervasive due to geological and climatic conditions. Landslides and debris flows are prevalent on hills while there is risk of flood on plain areas of Sindhuli [21]. The extreme precipitation events are one of reason for origination of floods and landslides in this district.

The study area (Figure 1), Kamala watershed lies in Chure range at Sindhuli district of Nepal. The watershed drains to Triveni river (Northwest–Southeast) through Kamala river and other associated rivers. The study area covers eight Village Development Committees namely, Bhadrakali, Ranichuri, Ranibas, Hatpate, Nipane, Sirthauli, Tandi and Harsai and two municipalities: Kamalamai and Dudhauli respectively of Kamala watershed. The study area lies between latitude 27°15.84′ N to 26°56.69′ N and longitude 85°53.95′ E to 86°16.75′ E. Total area covered by the study area is 535.50 km^2. The elevation of the study area ranges from 82 to 1700 m. The total population of the study area is 61,793 [22]. The Sindhuli-Bardiwas highway passes through the study area. The Kamalamai municipality is the biggest municipality of the country and headquarters of Sindhuli. The district is populated on city areas since few decades. Figure 2 shows elevation map of study area.

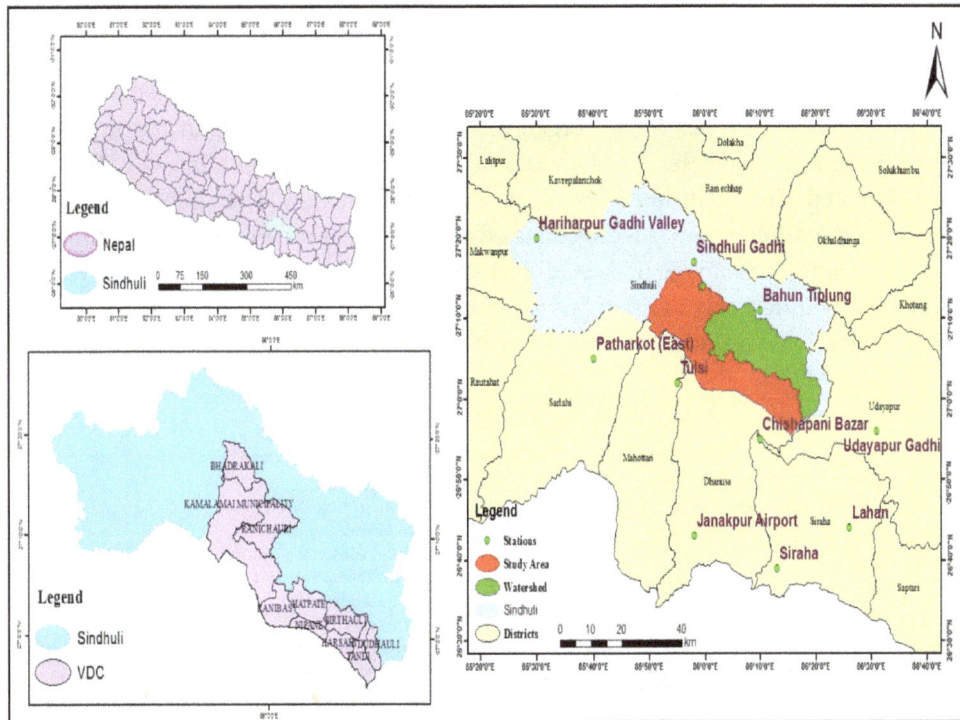

Figure 1. Map of study area.

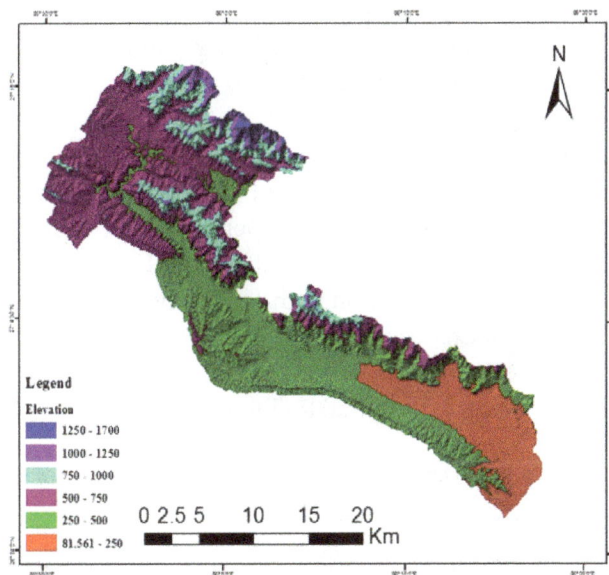

Figure 2. Elevation map of study area.

2.2. Methodology

2.2.1. Land Use Mapping Using Satellite Images and ArcGIS 10.1

The selection of Area of Interest (AOI) was done using clip tool from map of Nepal. Eleven topo-sheets of 1995 of scale 1:25,000 were collected from Department of Survey, Nepal. The maps were scanned and georeferenced using ArcGIS 10.1 software in Universal Transverse Mercator: WGS 1984 45N projection. The Google Earth was downloaded. The Land Use Land Cover (LULC) mapping was carried out in Google Earth images for the year 2005, 2010 and 2014. The AOI was converted to Google Earth compatible format (.KML) from ArcGIS shape file format. Different land use types naming, landslides, forest, cropland, settlement, pond, orchard, river and river terrace were digitized from Google Earth using polygon tool present in it. The digitized LULC types were individually converted to KML file and imported to ArcGIS 10.1 software. The files were then exported to shape file format (.shp). All LULC types were merged using merge tool available in geoprocessing toolbox. The LULC types for all four successive years were also digitized in the similar way. The area for each LULC were calculated. The verification was done through field visit. The Global Positioning System (GPS) locations of the different areas were collected in the field and were overlaid with land use map prepared using Google earth image in ArcGIS 10.1 software as process of validation. The GPS points (150 points in total) of different land use types were collected from each VDC and municipalities within the study area. Table 1 shows land use classification types and their description.

Table 1. Land use classification.

S. No.	LULC Type	Description
1.	Forest	All forest type including shrub land
2.	Cropland	Land used for cultivation
3.	Settlement	Residential, commercial, industrial and minor roads
4.	Orchard	Land used for tree cultivation
5.	Pond	Manmade water body or reservoir
6.	River	Natural water body
7.	River Terrace	River bank including barren land
8.	Landslides	All landslide types including bank cutting, soil erosion and slope failure

2.2.2. Analysis of Climatic Data

The climatic data (temperature and precipitation) from 1985 to 2014 (for 3 decades) were collected from Department of Hydrology and Meteorology, Nepal. The trend analysis was conducted for total annual and seasonal rainfall data while the average, minimum, and maximum temperature trend was analyzed using Mann-Kendall trend test at 95% confidence interval (i.e., alpha = 0.05). The Sen's slope estimator estimates the Sen's slope.

The precipitation data were available from ten nearest stations to the study area namely, Sindhuli Gadhi, Bahun Tiplung, Hariharpur Gadhi, Patharkot (East), Tulsi, Janakpur Airport, Chisapani Bazar, Siraha, Lahan and Udayapur Gadhi. The temperature data were available from only four stations naming Sindhuli Gadhi, Janakpur Airport, Lahan, and Udayapur Gadhi stations (Figure 1). The Sindhuli Gadhi station is the nearest station to the watershed. The missing rainfall values were filled by averaging method [23]. For example, average of total rainfall of 1985 and 1987 was done to fill the missing value of 1986. The trend analysis was conducted using R software (3.1.2 version) and the graph were plotted in Microsoft excel software 2007. Table 2 shows station type and their location.

Table 2. List of precipitation and temperature stations along the Kamala Watershed.

Stations	Station ID	Rainfall (year)	District	Temperature (year)	Average		Altitude(m)	Station Type
					Rainfall (mm)	Temp (°C)		
Sindhuli Gadhi	AS1107	1983–2014	Sindhuli	1989–2014	2380.23	22.23	1463	Climatology
Bahun Tiplung	AS1108	1985–2014	Sindhuli	-	1743	-	1417	Precipitation
Patharkot (East)	AS1109	1985–2014	Sarlahi	-	1713	-	275	Precipitation
Tulsi	AS1110	1985–2014	Dhanusha	-	1703.9	-	457	Precipitation
Janakpur Airport	AS1111	1985–2014	Janakpur	1985–2014	1542.1	25.04	90	Climatology
Chisapani Bazar	AS1112	1985–2014	Dhanusha	-	1535.2	-	165	Precipitation
Hariharpur Gadhi Valley	AS1117	1985–2014	Sindhuli	-	2433.4	-	250	Precipitation
Udayapur Gadhi	AS1213	1984–2014	Udayapur	1989–2014	1641	23.88	1175	Climatology
Lahan	AS1215	1983–2013	Lahan	1984–2014	1234	24.35	138	Agro-meteorology
Siraha	AS1216	1985–2014	Siraha	-	1421.3	-	102	Precipitation

Mann-Kendall Trend Test and Sen's Slope Estimation

Mann-Kendall trend test is a non-parametric statistical trend test that does not require data to be normally distributed. The test is used widely for the analysis of trend in climatologic time series [24] and hydrologic time series [25]. In this test, a null hypothesis is set meaning there is no trend and alternative hypothesis meaning there is a trend in set. The test is based on the calculation of Kendall's tau that is a measure of correlation, which measures the strength of the relationship between two variables, which is itself based on the ranks within the samples. The Kendall's tau's values ranges between −1 and +1. The positive correlation indicates that the ranks of both variables increase together whilst a negative correlation indicates that as the rank of one variable increases, the other decreases. The computational procedure of Mann-Kendall trend test is explained in detail by different authors [24,26].

Sen's Slope is an index to quantify the trend using a non-parametric procedure and developed by Sen [27] and is computed as,

$$Qi = \frac{x_j - x_k}{j - k} \text{ for } i = 1, 2, 3 \ldots N \tag{1}$$

where x_j and x_k are data values at time j and k (j > k), respectively.

3. Results

3.1. Land Use Mapping

The land use types include forest area, agricultural land, settlement, orchard, pond, river, river terrace, and landslide. The forest area is the dominant land cover/land use type on the study area from 1995 to 2014. The agricultural land covers second highest area after forest. The forest area decreased from 1995 to 2014. Figure 3 shows the land use classification for the year 1995, 2005, 2010, and 2014 respectively. Similarly, the area of pond, river, river terrace, and landslide decreased. The agricultural land, settlement, orchard area however increased. Table 3 shows the land use area on four different years from 1995 to 2014.

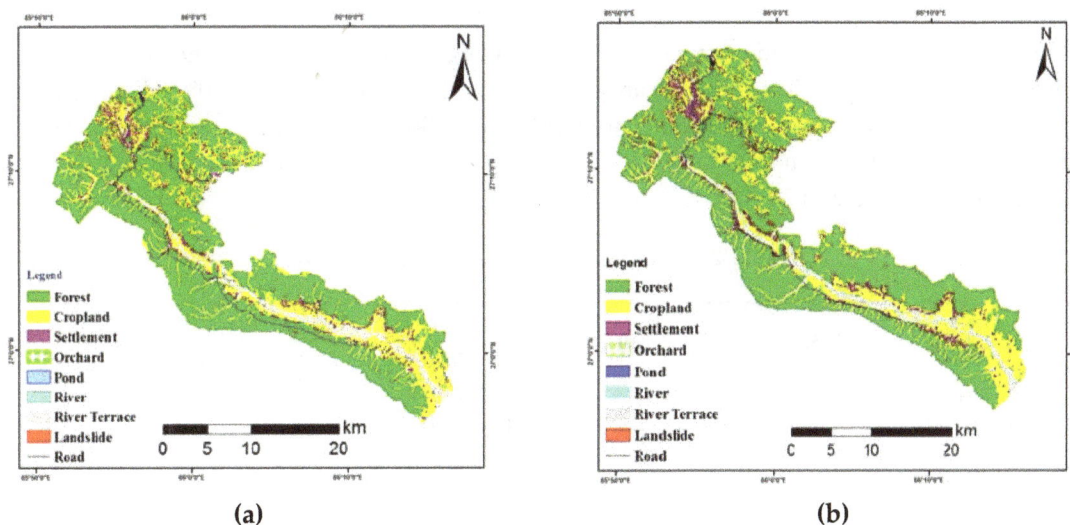

(a)   (b)

Figure 3. *Cont.*

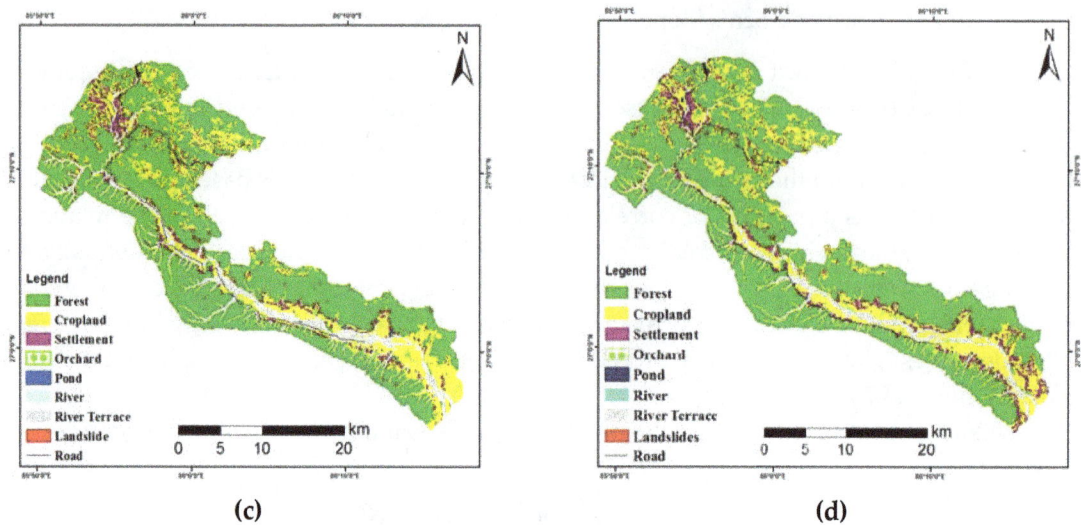

Figure 3. Land use maps of (**a**) 1995; (**b**) 2005; (**c**) 2010; and (**d**) 2014.

Table 3. Land use area in hectare and percentage coverage (1995–2014).

Land Use	1995		2014	
	Area (ha)	% Coverage	Area (ha)	% Coverage
Forest	36,262	67.72	35,410	66.13
Cropland	11,663	21.78	12,994	24.27
Settlement	541	1.01	1398	2.61
Orchard	15	0.03	99.6	0.186
Pond	3	0.01	2.4	0.004
River	875	1.63	554	1.03
River Terrace	4139	7.73	3042	5.68
Landslide	52	0.09	50	0.09

From 1995 to 2014, the land use types went some changes with decrease in pond area by 20%, river by 36.69% forest area by 2.35%, landslide area by 3.85% and river terrace by 26.50% while agricultural land, orchard area and settlement area increased by 11.41%, 546% and 158.41% respectively. However, landslide area coverage from 1995 to 2014 decreased by 2 ha. Table 4 exemplifies the land use change occurred between 1995 and 2014.

Table 4. Land use change dynamics (1995–2014).

Land Use	1995	2005	2010	2014	Change in Land Use from 1995—2014 (%)	
					Increase	Decrease
Forest	36,262	35,785	35,583	35,410	-	2.35
Cropland	11,663	12,581	12,823	12,994	11.41	-
Settlement	541	809	929	1398	158.41	-
Orchard	15	24	26	99.6	564	-
Pond	3	2	1	2.4	-	20
River	875	662	590	554	-	36.69
River Terrace	4139	3617	3539	3042	-	26.50
Landslide	52	70	59	50		3.85

The majority of land have been converted to agricultural and settlement areas. Mostly, the forest and river terraces are used for agricultural and settlement purposes. The agricultural land on the plain are also used for settlement purposes. However, the orchard area is converted from either cropland or forest. Figure 4 shows conversion of agricultural land to settlement areas at Kamalamai municipality from 2005 to 2014 respectively. The GPS points collected from the field on different land use classes were overlaid on the prepared land use map. It was found that 88.66% of the points were correctly placed onthe land use maps indicating that the accuracy of the land use map is 88.66%.

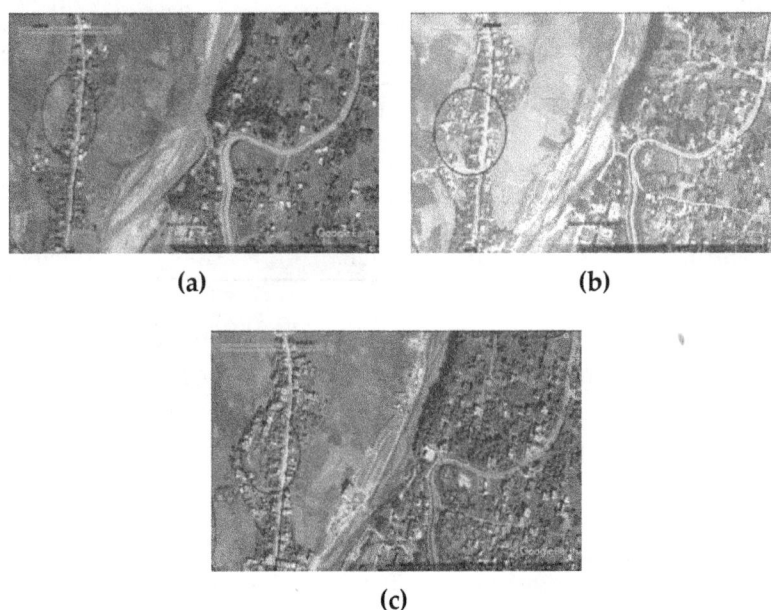

(a)

(b)

(c)

Figure 4. Land use change observed through google earth along Kamalamai municipality in (**a**) 2005, (**b**) 2010 and (**c**) 2014.

3.2. Climatic Data Analysis

3.2.1. Rainfall Trend

The statistical Mann-Kendall trend analysis done using hydro-meteorological data along the study area reveals the variation in annual rainfall. Table 5 demonstrates total annual rainfall trend for 10 stations. The rainfall rate decreased in all stations except Chisapani Bazar and Hariharpur Gadhi Valley. Sindhuli Gadhi, Patharkot (East), Udayapur Gadhi and Lahan obtained significant trend whilst rest other stations have insignificant trend.

Table 5. Total annual rainfall trend for 10 stations.

Stations/Total	p-Value	Tau	Sen's Slope	Significance
Sindhuli Gadhi	0.0037	−0.355	−38.882	Significant
Bahun Tiplung	0.29244	−0.138	−12.34	Insignificant
Patharkot (East)	0.00343	−0.38	−38.74	Significant
Tulsi	0.0586	−0.246	−15.09	Insignificant
Janakpur Airport	0.31775	−0.131	−7.431	Insignificant
Chisapani Bazar	0.4978	0.0897	4.8	Insignificant
Hariharpur Gadhi Valley	0.56806	0.0759	6.391	Insignificant
Udayapur Gadhi	0.0208	−0.295	−14.5	Significant
Lahan	0.00407	−0.366	−18.76	Significant
Siraha	0.56806	0.0759	−6.218	Insignificant

Sindhuli Gadhi, Bahun Tiplung, Patharkot (East), Tulsi, Chisapani Bazar, Lahan noticed negative rainfall trend. Similarly, among these stations, the significant negative trend is observed only on Patharkot (East). Siraha observed positive significant rainfall trend in 30 years. Table 6 shows pre−monsoon rainfall trend for ten stations observed in 3 decades.

Table 6. Pre−monsoon Rainfall trend.

Stations	p-Value	Tau	Sen's Slope	Significance
Sindhuli Gadhi	0.3724	−0.936	−2.158	Insignificant
Bahun Tiplung	0.29244	−0.138	−12.34	Insignificant
Patharkot (East)	0.00343	−0.38	−38.74	Significant
Tulsi	0.4978	−0.0897	−1.2	Insignificant
Janakpur Airport	0.21171	0.163	1.85	Insignificant
Chisapani Bazar	0.78896	−0.0368	−0.2565	Insignificant
Hariharpur Gadhi Valley	0.43245	0.103	2.345	Insignificant
Udayapur Gadhi	0.83048	0.0299	0.6278	Insignificant
Lahan	0.67086	−0.056	−0.8417	Insignificant
Siraha	0.040165	0.267	3.8	Significant

Bahun Tiplung, Chisapani Bazar and Hariharpur Gadhi Valley shows positive but insignificant rainfall trend. The rainfall trend in Sindhuli Gadhi decreased by 30.422 mm per year, which is highest among the ten stations and the trend, is significant. The Patharkot (East) and Lahan have significant negative rainfall trend. Table 7 depicts monsoon rainfall trend received on ten stations.

Table 7. Monsoon Rainfall trend.

Stations	p-Value	Tau	Sen's Slope	Significance
Sindhuli Gadhi	0.00746	−0.345	−30.422	Significant
Bahun Tiplung	0.54826	−0.081	3.319	Insignificant
Patharkot (East)	0.001907	−0.402	−33.81	Significant
Tulsi	0.060985	−0.244	−12.44	Insignificant
Janakpur Airport	0.17513	−0.177	−9.941	Insignificant
Chisapani Bazar	0.30077	0.136	8.912	Insignificant
Hariharpur Gadhi Valley	0.33534	0.126	4.363	Insignificant
Udayapur Gadhi	0.054001	−0.251	−7.39	Insignificant
Lahan	0.011887	−0.32	−17.03	Significant
Siraha	0.11226	−0.207	−9.275	Insignificant

The post−monsoon rainfall reveals negative trend in all stations. The Sindhuli Gadhi, Tulsi and Udayapur Gadhi received significant trend. Table 8 shows post-monsoon rainfall trend noticed in ten stations.

Table 8. Post−monsoon Rainfall trend.

Stations	p-Value	Tau	Sen's Slope	Significance
Sindhuli Gadhi	0.0025	−0.249	−3.723	Significant
Bahun Tiplung	0.91	0.0173	−0.987	Insignificant
Patharkot (East)	0.080391	−0.228	−1.87	Insignificant
Tulsi	0.028178	−0.285	−1.525	Significant
Janakpur Airport	0.45246	−0.1	−0.7	Insignificant
Chisapani Bazar	0.32303	−0.132	−1.56	Insignificant
Hariharpur Gadhi Valley	0.61739	−0.0667	−1.115	Insignificant
Udayapur Gadhi	0.033719	−0.276	−3.011	Significant
Lahan	0.67074	−0.0561	−0.2778	Insignificant
Siraha	0.69459	−0.053	−0.3333	Insignificant

The winter rainfall trend for three decade shows negative decreasing rainfall trend in all ten stations. The trend was however significant for Sindhuli Gadhi, Tulsi, Janakpur Airport, Chisapani Bazar and Udayapur Gadhi. Table 9 presents winter rainfall trend and their respective p−value and Sen's slope obtained during 3 decades.

Table 9. Winter Rainfall trend.

Stations	*p*-Value	Tau	Sen's Slope	Significance
Sindhuli Gadhi	0.00528	−0.333	−1.28	Significant
Bahun Tiplung	0.83639	−0.03	−0.482	Insignificant
Patharkot (East)	0.083379	−0.226	−0.8667	Insignificant
Tulsi	0.028178	−0.285	−1.071	Significant
Janakpur Airport	0.021288	−0.3	−1.346	Significant
Chisapani Bazar	0.01623	−0.323	−0.5733	Significant
Hariharpur Gadhi Valley	0.38193	−0.115	−0.5571	Insignificant
Udayapur Gadhi	0.028111	−0.286	−1.073	Significant
Lahan	0.093608	−0.218	−0.7667	Insignificant
Siraha	0.26079	−0.148	−0.6043	Insignificant

3.2.2. Temperature Trend

The maximum temperature increased with positive trend in all stations. The minimum and average temperature trend also increased in all four stations. The trends are statistically insignificant for all the stations.

The total annual, pre−monsoon, monsoon, post−monsoon, and winter trend for Sindhuli Gadhi station (1983–2014) and Udayapur Gadhi (1984–2014) along with maximum, minimum and average temperature trend for Sindhuli Gadhi (1989–2014) and Udayapur Gadhi (1989–2014) stations are presented in Figures 5–8 respectively.

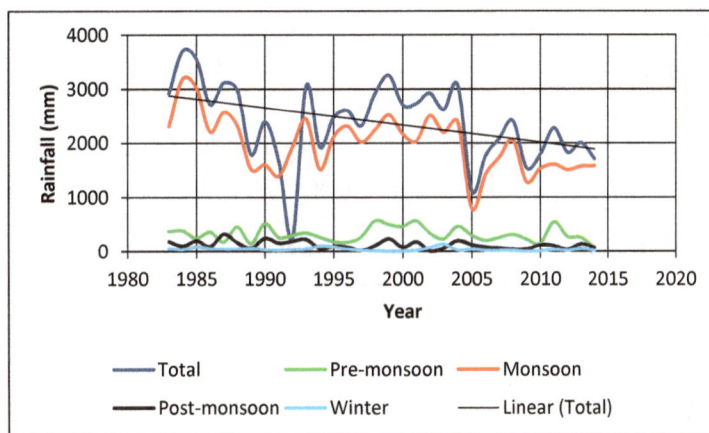

Figure 5. Rainfall trend of Sindhuli Gadhi station.

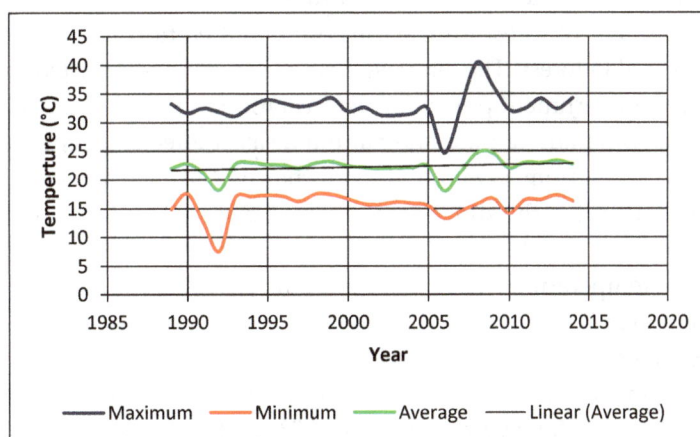

Figure 6. Temperature trend of Sindhuli Gadhi station.

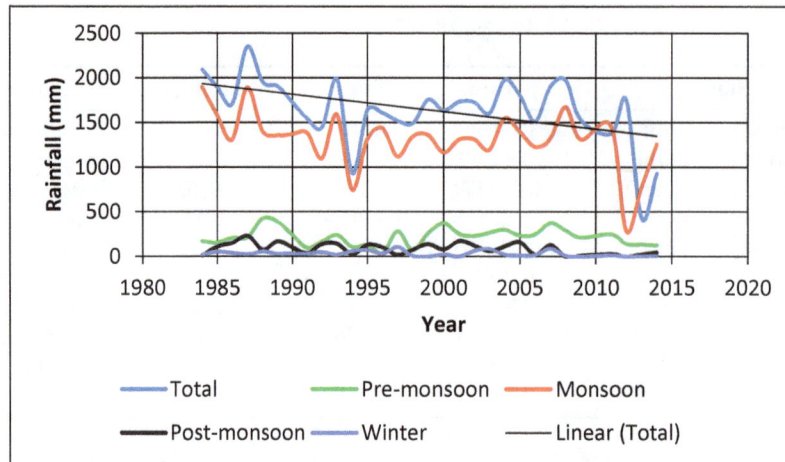

Figure 7. Rainfall trend of Udayapur Gadhi station.

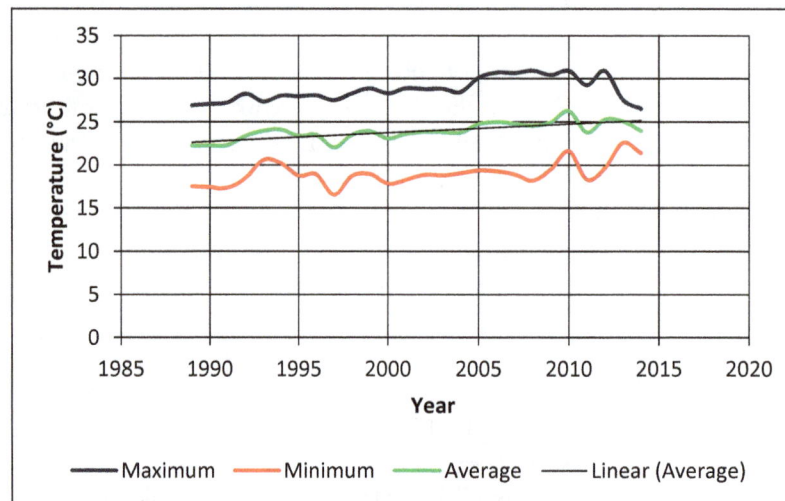

Figure 8. Temperature trend of Udayapur Gadhi station.

4. Discussion and Conclusions

The land use map from 1995 to 2014 showed the decrease in forested area and riverbank whereas agricultural land and settlement area increased. The orchard area increased while the pond area decreased. The increase in settlement and agricultural land might be due to population growth in the study area. The clearance of forest for the purpose of cultivation and settlement are the major reason for changes occurred in forest. The land use change in Chure hills started in 1950s after malaria eradication in Terai with government policy for deforestation to raise land revenue [14] along with launching of resettlement program for poor and natural disaster affected people [28]. The agricultural land had increased by 11.41% from 1995 to 2014 and the river terrace had decreased by 26.50% by modifying to agricultural land in the study area. The pond area also decreased by 20% during the same period in the study area. The decrease in pond area might be due to increase in temperature and decrease in rainfall. The temperature trends along the study area on different stations are increasing and shrinkage of pond area is observed. The land cover and land use changes in Chure and the area under cultivation increased by 0.46% while the forest decreased by 0.40% between 1990 and 2010. In the same study, in Sindhuli between 1995 and 2010, agricultural land was 19.3% and 19.7%, forest was 7.1% and 4.9% while for riverbed it was 64.35% and 66.9% respectively [14]. The land use change in Nepal is due to two causes; one is natural which includes geological structure relief feature, drainage, climate and another is cultural factor that includes population growth, migration of people and infrastructural development [29]. The comparison of landslide area from 1995 to 2005 showed the decrease in landslide area by 3.85%.

The study also includes analysis of temperature and rainfall trend from ten different stations. The temperature trend analysis incorporate average maximum and minimum temperature from four stations whereas the rainfall trend analysis comprises of total, pre—monsoon, monsoon, post—monsoon and winter rainfall trend. The Sen's slope shows the annual rainfall, and seasonal rainfall decreased in Sindhuli Gadhi station. The pre—monsoon rainfall increased along Janakpur Airport, Hariharpur Gadhi, Udayapur Gadhi and Siraha while decreased on other stations. The total annual rainfall increased in Hariharpur Gadhi valley and Chisapani Bazar stations and. The result is in accordance with the result of MoPE (2004) in which the average annual precipitation trend is decreasing in central Nepal [17] and more specifically, all over Nepal is decreasing at the rate of 9.8 mm/decade [30]. The result is similar to a research conducted in Makwanpur district, which also lies in the Chure with the increase in number of warm days and decrease in cooler days [31]. The variations in rainfall trend obtained in the study area declined in pre—monsoon rainfall at Sindhuli Gadhi, Bahun Tiplung, Patharkot (East), Tulsi, Chisapani Bazar and Lahan out of 10 stations and decrease in winter rainfall in all stations is in accordance with different researches conducted along different places of Nepal [17,31]. The significant station wise annual rainfall trend is obtained in Sindhuli Gadhi, Patharkot (East), Udayapur Gadhi and Lahan. Likewise, the monsoon rainfall trend was found significant for Sindhuli Gadhi, Patharkot (East) and Lahan stations only. This result is similar to annual and monsoon precipitation trend in Nepal in which significant change was attained [17]. The pre—monsoon rainfall trend was significant for Patharkot (East) and Siraha stations only. The temperature trend was positive and statistically insignificant for all four stations. The study of temperature conducted on other parts of the country also shows the insignificant trend [17,32]. This signifies the fluctuation of rainfall events and increase in temperature along the study area and in rest parts of the country. The precipitation and temperature changes are extremely changing and are driving to drier and hotter days. The analysis of rainfall data shows the amounts of rainfall in some years are above the average and for some years below the averages. According to the local inhabitants drying of wells, decreasing of water level in different water sources and drought are some phenomenon occurring in the study area. This erratic rainfall induces the landslides and flood events in the study area. The local people in the study area illustrated the flood occurrence is two to three times a year. A report mentioned the incidence of devastating flood along with high intensity precipitation with consequent landslide and debris flow activities in Kulekhani—Sindhuli area on 19 July to 20 July 1993 [30]. The rainfall initiated number of floods and landslides in the study area each year. The land use pattern, the changes and its uses consequently lead to climate change. The consequence of deforestation causes reduced transpiration that results in less cloud formation, less rainfall and increased drying. Also, the climate change effects are determined by the land use patterns and practices [4]. The decrease in forest area and increased settlement might be a cause for increased temperature in the study area.

The study incorporates mapping of land use change (1995–2014) using topographic maps and Google earth images and climatic variability analysis (3 decades or more) using Mann Kendall trend analysis in Kamala watershed of Sindhuli district. The study area lies in Chure that is hazard prone. The study reveals loss in forest, river terrace and pond area while the settlement, agricultural and orchard area increased during 1995 to 2014. The settlement increased mostly on the plain areas and the river terraces are converted to agricultural lands. However, the changes in forest area between 1995 and 2014 are not much protrusive. 88.67% of different features of land use were correctly placed. The climatic variability analysis conducted through Mann—Kendall trend test showed the variability in rainfall and temperature among different stations. The variability in the rainfall pattern among the ten stations over 30 years is estimated. The annual total rainfall increased in Hariharpur Gadhi Valley (6.39 mm/year) and Chisapani Bazar (4.8 mm/year) and decreased on rest of other stations. Similarly, the pre—monsoon rainfall increased on Janakpur Airport, Hariharpur Gadhi Valley, Udayapur Gadhi and Siraha. The monsoon rainfall also increased on Bahun Tiplung, Chisapani Bazar and Hariharpur Gadhi Valley only. The post—monsoon and winter rainfall however, decreased on all stations. The temperature trend increased on all four stations. The total annual rainfall trend was significant for Sindhuli Gadhi, Patharkot (East), Udayapur Gadhi and Lahan while for rest of stations, insignificant trend was obtained. The pre—monsoon rainfall trend was significant for Patharkot (East) and Lahan stations only. Likewise, the monsoon rainfall trend was found significant for Sindhuli Gadhi, Patharkot (East) and Lahan stations. The significant post—monsoon rainfall trend was obtained for Sindhuli Gadhi, Tulsi, and Udayapur Gadhi stations. The winter rainfall trend was attained significant for Sindhuli Gadhi, Tulsi, Janakpur Airport, Chisapani Bazar and Udayapur Gadhi stations while it was insignificant for the maximum, minimum and average temperature trends. The land use practices and changes since few decades in the study area might have accelerated the rate of changes of local climate in the study area.

The information on land use change and climatic variability of local level could be beneficial to understand the local effects of land use change and climatic variability/change in a particular area and in policymaking. The climatic data of longer period would have provided the better illustrations of climatic variability and their changes. The detailed study incorporating relationships of land use and climate could provide prolific results on tracing the effects of climate on land use on a particular area and vice versa.

Acknowledgments: The research was financially supported by President Chure—Tarai Madhesh Conservation Development Board and is the outcome of the dissertation conducted at Central Department of Environmental Science, Tribhuvan University. Authors want to thank Kumod Raj Lekhak,. Niraj Bal Tamang and Suman Panday for their support during research. Sanju Ghimire and Muna Khatiwada are acknowledged for their support during fieldwork.

Author Contributions: Muna Neupane and Subodh Dhakal designed the research together. Muna Neupane conducted the literature study, fieldwork and data analysis. Subodh Dhakal supervised the research and guided throughout the study period. Muna Neupane wrote the manuscript and Subodh Dhakal finalized it.

Conflicts of Interest: The authors declare no conflict of interest.

References

1.   Natural Resource Canada.   Available online: http://www.nrcan.gc.ca/earth$-$sciences/geomatics/satellite$-$imagery$-$air$-$photos/satellite$-$imagery$-$products/educational$-$resources/9373 (accessed on 13 May 2016).

2.   Ellis, E. Land—Use and Land—Cover Change. Available online: http://www.eoearth.org/view/article/154143 (accessed on 15 May 2016).

3.   Karsli, F.; Atasoy, M.; Yalcin, A.; Reis, S.; Demir, O.; Gokceoglu, C. Effects of land-use changes on landslides in a landslide-prone area (Ardesen, Rize, NE Turkey). *Environ. Monit. Assess.* **2009**. [CrossRef] [PubMed]

4.   Dale, V.H. The relationship between land-use change and climate change. *Ecol. Appl.* **1997**, *7*, 753–769. [CrossRef]

5.   Awasthi, K.D.; Sitaula, B.K.; Singh, B.R.; Bajacharaya, R.M. Land-use change in two Nepalese watersheds: GIS and geomorphometric analysis. *Land Degrad. Dev.* **2002**, *13*, 495–513. [CrossRef]

6.   Feddema, J.J.; Oleson, K.W.; Bonan, G.B.; Mearns, L.O.; Buja, L.E.; Meehl, G.A.; Washington, W.M. The importance of land—cover change in simulating future climates. *Science* **2005**, *310*, 1674–1678. [CrossRef] [PubMed]

7.   Glade, T. Landslide occurrence as a response to land use change: a review of evidence from New Zealand. *Catena* **2003**, *51*, 297–314. [CrossRef]

8.   Caesar, J.; Alexander, L.; Vose, R. Large-scale changes in observed daily maximum and minimum temperatures: Creation and analysis of a new gridded data set. *J. Geophys. Res. Atmos.* **2006**, *111*, 1–10. [CrossRef]

9.   Gassner, C.; Promper, C.; Beguería, S.; Glade, T. Climate change impact for spatial landslide susceptibility. In *Engineering Geology for Society and Territory—Volume 1*; Springer International Publishing: Basel, Switzerland, 2015.

10. Field, C.B.; Barros, V.R.; Mach, K.J.; Mastrandrea, M.D.; van Aalst, M.; Adger, W.N.; Arent, D.J.; Barnett, J.; Betts, R.; Bilir, T.E.; et al. *Climate Change 2014: Impacts, Adaptation, and Vulnerability. Part A: Global and Sectoral Aspects*; Contribution of working group II to the fifth assessment report of the intergovernmental panel on climate change; Cambridge University Press: Cambridge, UK, 2014.

11. Matson, P.A.; Vitousek, P.M. Cross-system comparisons of soil nitrogen transformations and nitrous oxide flux in tropical forest ecosystems. *Glob. Biogeochem. Cycles* **1987**, *1*, 163–170. [CrossRef]

12. Agrawal, S.; Raksakulthai, V.; Van Aalst, M.; Larsen, P.; Smith, J.; Reynolds, J. *Development and Climate Change in Nepal: Focus on Water Resources and Hydropower*; Organization of Cooperation and Development: Paris, France, 2003.

13. Otieno, V.O.; Anyah, R.O. Effects of land use changes on climate in the Greater Horn of Africa. *Clim. Res.* **2012**, *52*, 77–95. [CrossRef]

14. Ghimire, M. *Land Use and Land Cover Change in the Churia–Tarai Region, Nepal*; Ministry of Forest and Soil Conservation: Kathmandu, Nepal, 2012.

15. Gansser, A. *Geology of the Himalayas*; John Wiley & Sons: London, UK, 1964.

16. Ghimire, M. Landslide occurrence and its relation with terrain factors in the Siwalik Hills, Nepal: Case study of susceptibility assessment in three basins. *Nat. Hazards* **2011**, *56*, 299–320. [CrossRef]

17. Shrestha, A.B.; Wake, C.P.; Mayewski, P.A.; Dibb, J.E. Maximum temperature trends in the Himalayas and its vicinity: An analysis based on temperature records from Nepal for the period 1971–94. *J. Clim.* **1999**, *12*, 2775–2787. [CrossRef]

18. DDMC. *Disaster Preparedness Plan 2068 BS, Sindhuli*; District Disaster Management Committee: Sindhuli, Nepal, 2011.

19. DMG. *Geological Maps of Petroleum Exploration Block–8, Janakpur, Central Nepal*; [map]; Petroleum Promotion Exploration Project, Department of Mines and Geology, Government of Nepal: Kathmandu, Nepal, 2004.

20. Dhakal, S. Evolution of geomorphologic Hazards in Hindu Kush Himalaya. In *Mountain Hazards and Disaster Risk Reduction*; Springer: Tokyo, Japan, 2015; pp. 53–72.

21. Government of Nepal, Ministry of Environment. *National Adaptation Programme of Action (NAPA) to Climate Change*; Ministry of Environment: Kathmandu, Nepal, 2010.

22. CBS. *Statistical Year Book of Nepal*; Central Bureau of Statistics: Kathmandu, Nepal, 2011.

23. Singh, R.B.; Mal, S. Trends and variability of monsoon and other rainfall seasons in Western Himalaya, India. *Atmos. Sci. Lett.* **2014**, *15*, 218–226. [CrossRef]

24. Mavromatis, T.; Stathis, D. Response of the water balance in Greece to temperature and precipitation trends. *Theor. App. Climatol.* **2011**, *104*, 13–24. [CrossRef]

25. Yue, S.; Wang, C. The Mann–Kendall test modified by effective sample size to detect trend in serially correlated hydrological series. *Water Resour. Manag.* **2004**, *18*, 201–218. [CrossRef]

26. Karmeshu, N. Trend Detection in Annual Temperature & Precipitation using the Mann Kendall Test–A Case Study to Assess Climate Change on Select States in the Northeastern United States. Master's Thesis, University of Pennsylvania, Pennsylvania, PA, USA, 2012.

27. Poudel, S.; Shaw, R. The Relationships between Climate Variability and Crop Yield in a Mountainous Environment: A Case Study in Lamjung District, Nepal. *Climate* **2016**, *4*, 13. [CrossRef]

28. Joshi, A.L.; Shrestha, K.; Sigdel, H. Deforestation and participatory forest management policy in Nepal. *Underlying Causes of Deforestation and Forest Degradation in Asia. World Rainforest Movement*. Available online: http://www.wrm.org.uy/deforestation/Asia/Nepal.html (accessed on 5 August 2016).

29. Shrestha, S.H. A review of land use pattern in Nepal. *Himal. Rev.* **1975**, *7*, 33–42.

30. WECS. *Water Resources of Nepal in the Context of Climate Change*; Water and Energy Commission Secretariat: Kathmandu, Nepal, 2011.

31. Paudyal, P.; Kafle, G. Assessment and prioritization of community soil and water conservation measures for adaptation to climatic stresses in Makawanpur district of Nepal. *J. Wetl. Ecol.* **2012**, *6*, 44–51.

32. Dhakal, S.; Sedhain, G.K.; Dhakal, S.C. Climate Change Impact and Adaptation Practices in Agriculture: A Case Study of Rautahat District, Nepal. *Climate* **2016**, *4*, 63. [CrossRef]

10

A Global ETCCDI-Based Precipitation Climatology from Satellite and Rain Gauge Measurements

Felix Dietzsch * , Axel Andersson [†], Markus Ziese, Marc Schröder, Kristin Raykova, Kirstin Schamm and Andreas Becker

Deutscher Wetterdienst, Frankfurter Straße 135, 63067 Offenbach (Main), Germany;
axel.andersson@dwd.de (A.A.); markus.ziese@dwd.de (M.Z.); marc.schroeder@dwd.de (M.S.);
kristin.raykova@dwd.de (K.R.); kirstin@schamm.de (K.S.); andreas.becker@dwd.de (A.B.)
* Correspondence: felix.dietzsch@dwd.de
† Current address: Deutscher Wetterdienst, Bernhard-Nocht-Straße 76, 20359 Hamburg, Germany.

Academic Editor: Yang Zhang

Abstract: Precipitation is still one of the most complex climate variables to observe, to understand, and to handle within climate monitoring and climate analysis as well as to simulate in numerical weather prediction and climate models. Especially over ocean, less is known about precipitation than over land due to the sparsity of in situ observations. Here, we introduce and discuss a global Expert Team on Climate Change and Indices (ETCCDI)-based precipitation climatology. The basis for computation of this climatology is the global precipitation dataset Daily Precipitation Analysis for Climate Prediction (DAPACLIP) which combines in situ observation data over land and satellite-based remote sensing data over ocean in daily temporal resolution, namely data from the Global Precipitation Climatology Centre (GPCC) and the Hamburg Ocean Atmosphere Parameters and Fluxes from Satellite Data (HOAPS) dataset. The DAPACLIP dataset spans the period 1988–2008 and thus the global ETCCDI-based precipitation climatology covers 21 years in total. Regional aspects of the climatology are also discussed with focus on Europe and the monsoon region of south-east Asia. To our knowledge, this is the first presentation and discussion of an ETCCDI-based precipitation climatology on a global scale.

Keywords: precipitation; satellite; climatology; ETCCDI; rain gauge; global

---

1. Introduction

While precipitation is one of the most important climate variables, it is also one of the most complex variables regarding global distribution and variability. This makes it difficult to handle, e.g., in climate models. The new global precipitation dataset Daily Precipitation Analysis for Climate Prediction (DAPACLIP) which combines in situ and satellite-based remote sensing data is introduced. The main purpose of the development of the dataset is the evaluation of decadal climate predictions of the Mittelfristige Klimaprognosen (MiKlip) project framework [1]. In this paper, the global and regional aspects of DAPACLIP are evaluated. A special emphasis is put on the Expert Team on Climate Change and detection indices (ETCCDI) that are used for the evaluation. Climatological features related to precipitation and based on the ETCCDI are discussed and certain features for Europe and the monsoon region of south-east Asia are emphasized. Results from this work will form the basic evaluation of the MiKlip decadal prediction system.

A lot of effort has been put into the development of remote sensing-based regional and global precipitation datasets in the past years and decades. Frequently used datasets are the Global Precipitation Climatology Project (GPCP) and products from the Tropical Rainfall Measurement Mission (TRMM). GPCP provides gridded datasets on a monthly and a daily basis. It is a combined product from various satellite-based sources, such as the Special Sensor Microwave Imager (SSM/I) and Special Sensor Microwave Imager Sounder (SSMIS), infrared (IR) imagers, and Television and Infrared Observation Satellite (TIROS)-based Observation Vertical Sounder (TOVS) precipitation analysis, additional gauge data, and the Outgoing Longwave Radiation (OLR) precipitation index. Details can be found in [2–4]. A large variety of Tropical Rainfall Measurement Mission (TRMM)-based precipitation products are available, which partly additionally include data from SSM/I, Advanced Microwave Scanning Radiometer-Earth Observation System (AMSR-E), Advanced Microwave Sounding Unit-B (AMSU-B) and IR observations from geostationary platforms. TRMM products do not cover the whole globe, but latitudes up to 60°, dependent on the specific dataset. Details about TRMM data are explained in, e.g., [5].

The main difference between these precipitation datasets and the presented DAPACLIP dataset is that the DAPACLIP dataset combines gauge and satellite data by keeping the individual dataset characteristics and in particular by not changing the uncertainty of the individual datasets. Other datasets use satellite retrievals also over land, which are often calibrated by using gauge data.

ETCCDIs play an important role in the assessment of climate extremes. This is well addressed in [6], which describes the progress over the last decades as well as for the Intergovernmental Panel On Climate Change (IPCC) AR5 report. Also, the indices are used for intercomparisons of hindcasts from CMIP3 and CMIP5, reanalysis data and observational data [7] as well as for CMIP climate projections of CMIP3 and CMIP5 [8]. In [9], it is stated that satellite-based climate datasets now become relevant for climate extreme detection and attribution because their temporal coverage starts to reach sufficient length to be of value for climate analysis. This emphasizes one of the values of the DAPACLIP dataset presented in this study. In addition, precipitation extremes and their distribution were analyzed using ETCCDI and gridded rain gauge data from the GPCC [10]. All of these analyses were carried out using ETCCDI indices only over land and not globally.

After a short introduction of the DAPACLIP dataset and the ETCCDI indices, the ETCCDI-based precipitation climatology is presented and discussed, first on a global scale and then for Europe and the Asian monsoon region.

2. Data and Methodology

This section describes the data, the processing and the methods used for the resulting dataset. For the global dataset, a combination from interpolated and gridded station-based rain gauge measurements is used over land, and a rain rate retrieval algorithm and gridding procedure for satellite-based passive microwave imagers is used over ocean. Both separate datasets are then merged to a single precipitation dataset. The dataset includes the years from 1988 to 2008. It is provided in three different spatial resolutions of 0.5°, 1° and 2.5°. The half degree version is only available for the European domain (30° N to 80° N and 30° W to 65° E). Otherwise, the whole global scale is covered. For the DAPACLIP dataset and the derived climatology, the use of ETCCDI indices is elucidated.

2.1. Rain Gauge Measurements

The GPCC at Deutscher Wetterdienst (DWD) provides daily precipitation products created from station-based rain gauge measurement data. Data are mainly received from national meteorological and hydrological services. Other sources are used as well, e.g., from local authorities, global and regional data collections, and near real-time World Meteorological Organisation Global Telecommunicataion System (WMO-GTS) CLIMAT and SYNOP data. At first, the retrieved station data are checked for quality, e.g., common problems are coding errors in SYNOP data, different latin transcriptions of station names, or transmission errors. Figure 1 shows the number of available stations for the dataset split up into the different sources. The total number of used stations lies approximately between 30,000 in 1988 and 25,000 in 2008.

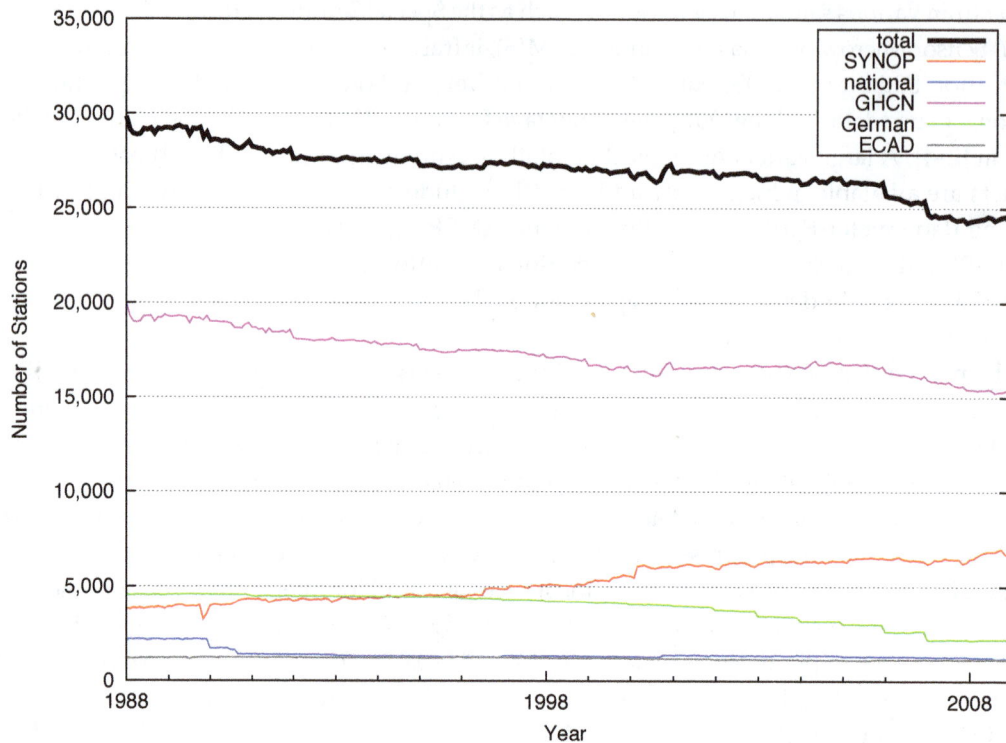

Figure 1. Temporal distribution of used land-surface stations. "SYNOP" are stations distributed via the World Meteorological Organisation Global Telecommunication System (WMO-GTS) , "National" are data provided by national meteorological and/or hydrological services, "GHCN" are data from the Global Historical Climatology Network [11], "German" are stations operated by Deutscher Wetterdienst (DWD) , "ECAD" are data from the European Climate Data and Assessment [12], and "Total" is the set union of all available stations at one date (multiple stations counted only once).

For the DAPACLIP dataset used here, the GPCC developed a "Full Data Daily" product [13]. It consists of daily precipitation sums that are interpolated onto a regular grid. Different interpolation schemes have been investigated in order to find the best performing one: inverse distance weighting, SPHEREMAP (described in [14]), ordinary kriging with different corellogram settings and the mean of the nearest precipitation measurements. Details on kriging are described in [15,16]. To find the optimal interpolation method, a "leave-one-out" approach was applied to all stations. All schemes were tested with interpolation of precipitation totals and relative anomalies. Ordinary block kriging of relative anomalies was chosen as the best performing interpolation method and was used for the gauge data. The background data for the relative anomaly interpolation were taken from the GPCC Full Data Monthly V7 product [17,18]. Verification results on the GPCC climatology compared to other datasets are described in [17]. Uncertainty information is provided with the gauge data. The uncertainty is influenced by station density, precipitation amount, precipitation phase and orography. For the Full Data Daily product, the uncertainty information is split into two components. The kriging interpolation error contains biases caused by the spatial distribution of stations and the grid size. The absolute uncertainty contains the standard deviation of the measured precipitation [13,19]. Figure 2 provides insight into the available features of the GPCC daily precipitation product and gives an impression of data quality and uncertainty.

GPCC Full Data Daily 1° for 29 June 2008

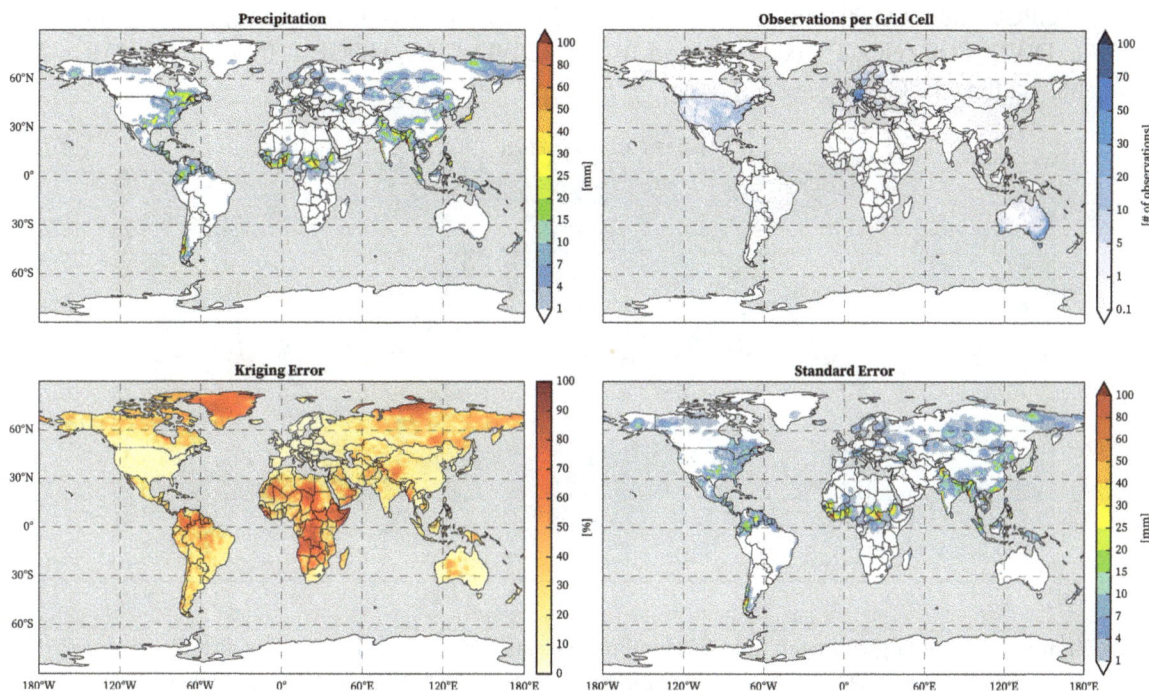

Figure 2. Example of the features of the Global Precipitation Climatology Centre (GPCC) daily precipitation product for 29 June, 2008. (adapted from [13]). Top left: estimated amount of precipitation in mm·d^{-1}. Top right: distribution of precipitation measurement stations. Bottom left: kriging error in %. Bottom right: standard deviation according to the method of [19] in mm·d^{-1}.

2.2. Satellite Remote Sensing

A statistical neuronal network algorithm is used to derive precipitation data from satellite-based passive microwave imagers. The data that have been used here are based on the HOAPS-3.2 dataset [20–23]. This version of HOAPS is provided and distributed by the Satellite Application Facility on Climate Monitoring (CM SAF). The HOAPS dataset itself uses brightness temperatures (BT) from SSM/I. This is explained in detail in [22,24]. The algorithm is applied over ice-free ocean and 50 km off coasts. Figure 3 shows a spatial map of the HOAPS-3.2 precipitation climatology. Verification results of the HOAPS dataset can be found in [22,23], and references therein. For the global precipitation product, HOAPS-3.2 was enhanced by using additional TRMM microwave imager BT data available for 1997 and later. This increases the temporal sampling per region due to the additional instrument overpasses for the respective time range. Thus, the HOAPS rain rate retrieval algorithm was adapted accordingly to allow the processing of TMI data.

After processing the BTs, the rain rates are accumulated to daily precipitation totals and gridded by calculating the Gaussian mean for all pixel values that belong to a certain grid cell. The Gaussian standard error is given as an uncertainty measure.

2.3. Data Merging

After separate processing of precipitation data over land (GPCC) and ocean (HOAPS), both datasets are merged by filling the global grid with values from each dataset. Grid cells where data from both land and ocean are available are filled with ocean values. The maximum fraction of land in grid cells with values from both datasets is 5%. Hence it seems reasonable to neglect the GPCC data here. Remaining grid cells are filled with interpolated values from land and ocean data. In this way, data from both GPCC and HOAPS find entrance into corresponding grid cell values. This leads simultaneously to a smooth gradient at the edge of land and ocean data. Figure 4 shows an exemplary day with global precipitation over land and ocean merged from both datasets. The merged GPCC and HOAPS data record is referenced (DOIs 10.5676/DWD_CDC/HOGP_050/V001, 10.5676/DWD_CDC/HOGP_100/V001, and 10.5676/DWD_CDC/HOGP_250/V001, [25–27]) and is available via the GPCC webpage [28].

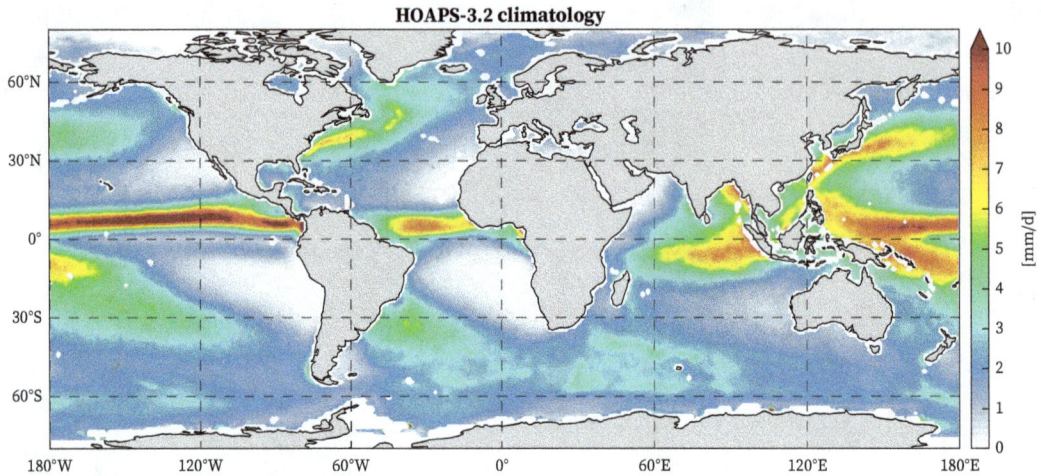

Figure 3. Climatological mean field of Hamburg Ocean Atmosphere Parameters and Fluxes from Satellite data version 3.2 (HOAPS-3.2) precipitation for the years 1988 to 2008 (adapted from [22]).

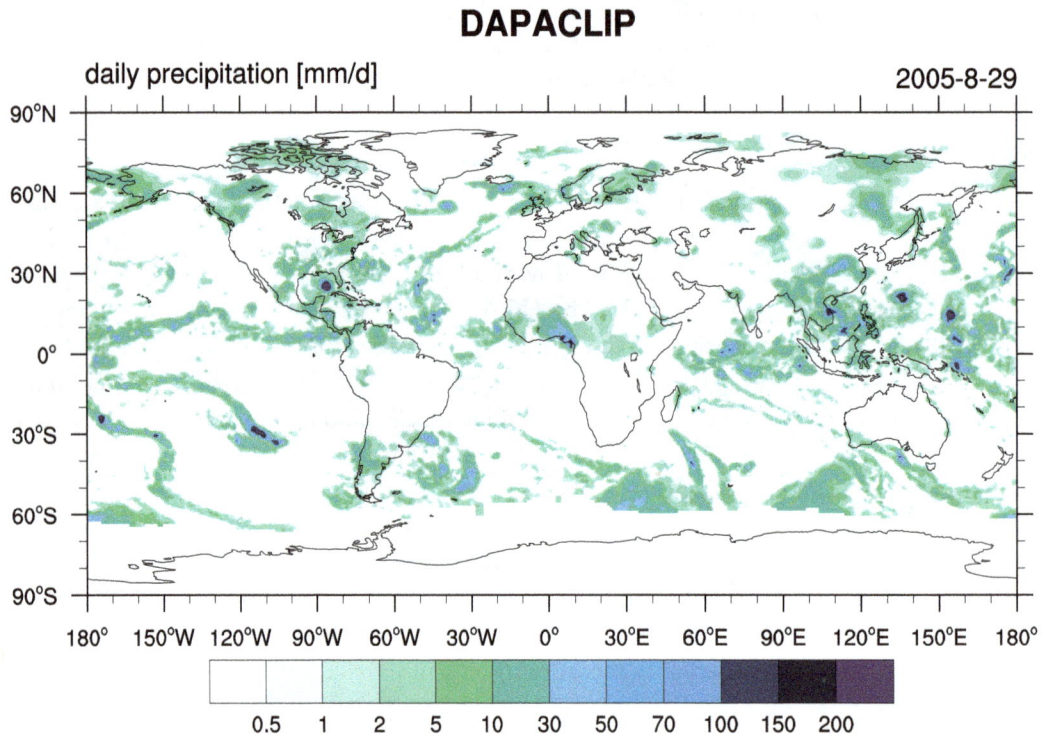

Figure 4. Example of the merged Daily Precipitation Analysis for Climate Prediction dataset (DAPACLIP) from land and ocean data for 29 August, 2005. Remarkable features from distinct events can be seen, for example the precipitation pattern of hurricane Katrina at the Gulf of Mexico and several tropical storms in the western Pacific Ocean.

2.4. ETCCDI

The ETCCDI provides indices and datasets for the assessment of climate variability and change [29]. Here, a subset of these indices is used for the determination of global precipitation characteristics. The selected indices are listed in Table 1.

An alternative definition of the 95th (r95p) and 99th percentile (r99p) indices is used, since the original definition of each of these indices asks for another background climatology, and then gives the precipitation sum of all days that exceed the percentile threshold. This is more useful when it comes to comparisons, e.g., between models and observation data, or trend detection. Here, we use the percentile values themselves, since these indices illustrate the precipitation climatology in a better way. The Simple Daily Intensity Index (sdii) is the mean precipitation amount that is registered on wet days. More detailed descriptions of the indices can be found in [30].

Table 1. List of used Expert Team on Climate Change and Detection precipitation indices (ETCCDI). Wet days are defined by precipitation ≥ 1 mm·day^{-1} [30,31].

Index	Description	Unit
cdd	Maximum number of Consecutive Dry Days	days
cwd	Maximum number of Consecutive Wet Days	days
r10	Number of days with more than 10 mm of precipitation	days
r20	Number of days with more than 20 mm of precipitation	days
r95p	95th percentile of the daily precipitation amount	mm·day^{-1}
r99p	99th percentile of the daily precipitation amount	mm·day^{-1}
rx1	Maximum precipitation sum of one day	mm
rx5	Maximum precipitation sum of five consecutive days	mm
sdii	Simple Daily Intensity Index	mm·day^{-1}

3. Results

3.1. Global Climatology

Figure 5 shows the ETCCDI values for the DAPACLIP dataset on a one degree regular longitude/latitude global grid. The consecutive dry days index (cdd) values are shown in panel (a). Here, we find the longest dry spells in the Saharan desert, the Atacama desert, the Gobi desert and the Californian peninsula. The most extreme value is found in the eastern Saharan desert, where the dry spell duration matches the complete dataset time span from 1988 to 2008; Panel (b) shows the consecutive wet days index (cwd). The largest values appear in the monsoon-influenced and tropical regions. We also find characteristically increased values along the west coast of Europe and North America, where wet spell durations exceed one month. Wet spells of more than three months occur on the tropical and monsoon regions, with maximum values of more than nine months along the Pacific coast of Colombia; Panel (c) shows the r95p index. The patterns here correspond well with the annual mean precipitation, as shown in Figure 2 in [2]. In general, maximum values are found along the Intertropical Convergence Zone (ITCZ), e.g., over the central Pacific Ocean, the central Atlantic Ocean, and the Indian Ocean. Also, the typical extra-tropical cyclone tracks can be seen in the north east Pacific off the coast of Taiwan and Japan, and in the northern Atlantic ocean along with increased percentile values at the coast of north-west America and at the European west coasts especially for Norway. At the U.S. east coast, we find increased percentile values, since it is a typical event for this region, that tropical storm systems and hurricanes turn northwards and convert into extra-tropical low pressure systems on their way over the Atlantic. We also find higher percentiles for the west coast of New Zealand. In the monsoon regions of Southeast Asia, percentile values of 50 mm·day^{-1} are exceeded; Panel (d) shows the r99p index. While the r95p values reach 80 mm·day^{-1}, r99p exceeds 200 mm·day^{-1}. The largest precipitation amounts are found in the western Pacific ocean, the Indian ocean, along the equator, and along the American east coast with more than 60 mm·day^{-1}. Over land, the 99th percentile values are smaller in general, but again we find certain regions among coastal areas with large extreme precipitation values (e.g., Norway, Galicia, Canadian west coast, southern New Zealand); Panel (e) shows the heavy precipitation index r10, which is here defined as the mean of number of days per year, where precipitation amounts of $\geq$10 mm·d^{-1} occur. Here, we find local maxima along the equator with values from 50 to 90 days. The largest values are found over land in tropical and monsoon regions in South America and Southeast Asia with more than 90 days per year. The global maximum is found at the Colombian west coast with a value of 227 days; Panel (f) shows the same index for days with more than 20 mm·d^{-1} of precipitation (r20). The distribution patterns are very similar compared to the preceding index. Here, the global maximum value is 130 days at the Colombian west coast; Panel (g) shows the maximum one day precipitation index (rx1). The values over ocean are consistently larger than over land and can reach 800 mm for a single day. The maximum values are significantly lower over land, but can reach 100 mm. These values are found predominantly in the tropics and subtropics; Panel (h) shows the maximum five day precipitation index (rx5). Similar to the rx1 index, the largest values are found over ocean with local maxima of up to 2000 mm·d^{-1} over the Pacific and Indian ocean. These values are likely to be caused by certain tropical storm events. Over land, five day sums between 100 and 400 mm seem to be characteristic, except for the drier continental climate regions, where the extreme five day precipitation sums are lower (50 mm or less); Panel (i) shows the simple daily intensity index (sdii). Here, we find global maximum values along the ITCZ with values from 12 to 18 mm·d^{-1}. Similar values occur in the already discussed tropical storm regions. The largest values are found again on coastal edges in the tropics and monsoon regions (Western Ghats, Bay of Bengal, Andaman Sea) with values up to 3 mm·d^{-1}. Large values also occur at the Colombian west coast, Japanese south coast, southern Atlantic and Pacific ocean, the West African coast, and the U.S. east coast.

3.2. Europe

Figure 6 shows ETCCDI indices for the European domain from 30° W to 65° E and 30° N to 80° N at a higher spatial resolution of 0.5°. Indices with interesting features were selected. Panel (a) shows the cwd index. The maximum value of more than three months is located at the northwest of the Faroese islands. Values of more than two months are found at the northwest coast of Scotland. Wet spells of more than one month are found over larger areas, such as the Norwegian coast, the south of Iceland, along the European west coasts of Denmark, Germany, Ireland, Wales, but also in central Ukraine; Panel (b) shows the r95p index. We mostly find values of less than 10 mm·d^{-1} especially in continental climate zones. Larger values occur at the southwestern coast of Norway with up to 45 mm·d^{-1}. Large values also occur along the southwestern coast of Turkey with up to 60 mm·d^{-1}. A further maximum is found along the Adriatic coast of Slovenia and Croatia with up to 45 mm·d^{-1}. Over the Atlantic Ocean, single peaks of percentile precipitation amounts are found, and which are distributed heterogeneously. We assume these values to be signatures from extra-tropical cyclones. The distribution and intensity is a matter of further discussion; Panel (c) shows the heavy precipitation index r10. Here, regions with a larger number of significant precipitation events are emphasized well. Along the Norwegian coast, we find up to 120 days per year with more than 10 mm of precipitation on average, which is approximately every third day of the year. A similar value is also shown for a pixel at the Faroese islands. Values of up to 90 days per year occur at southern Iceland, Scotland, southwestern Ireland, and the northeastern Turkish coast. The index has smaller values in the southwestern Mediterranean region indicating that there are fewer events of heavy precipitation, but as shown in panel (b), extreme events on daily scale are present in this region. Local spots over land with an increased number of heavy precipitation events are found in Galicia, the Alps, and the Adriatic coast with up to 70 days per year. Over the Atlantic Ocean, the number of days with heavy precipitation exceeds values of 50 days per year on average, too. All the observed patterns of panels (a–c) are also related to positive and negative North Atlantic Oscillation (NAO) events. The precipitation correlates well with positive NAO events in north and northwest Europe, while it is anti-correlated for negative NAO events in southern Europe, especially for the Adriatic coast and the southwestern Turkish coast. Panel (d) shows the rx1 index. Over land, maximum values between 100 and 200 mm·d^{-1} occur. These values are reached in Norway, the Alps, the French Mediterranean region, the German Ore Mountains, and the Adriatic coast. It is possible to backtrack some of these values to single events. For example, in August 2002, a precipitation amount of 312 mm·d^{-1} was measured at the station of Zinnwald-Georgenfeld in the German Ore mountains due to a Vb-track low [32]. Because of the grid interpolation, that value does not appear in this climatology as such. The percentile values over ocean are generally larger than over land. The maximum values are found in the Atlantic ocean with up to 600 mm·d^{-1} and the Mediterranean Sea with up to 400 mm·d^{-1}.

3.3. Monsoon Asia

Figure 7 shows selected ETCCDI indices for the Monsoon Asia region, which is here defined from 60° E to 180° E and from 15° S to 55° N. Here, cdd, cwd, the number of days of the r10 index, and the sdii are shown. Typical monsoon characteristics can be identified for the regions, where both dry and wet spells occur, as can be seen in panels (a) and (b) for the Indian subcontinent and northern parts of Southeast Asia. The climatology shows a duration of six to nine months for dry spells and mostly two to three months and partly up to four months for maximum wet spell durations. The orography influences the spell duration, for example at the Western Ghat mountains of India, where maximum wet spells can endure longer than in the rest of India. The same monsoon characteristic also appears for northern Australia and the surrounding Pacific ocean. For the southern part of Southeast Asia near the equator, the perennial convective ITCZ zone can be identified by the long wet spell duration of up to four months and shorter dry periods over land and over ocean. The consecutive dry day climatology also shows expected large values for the dry regions of the Gobi desert, southern Pakistan, and Afghanistan. Panel (c) shows the very heavy precipitation index r20. This index emphasizes the orographic induced precipitation distribution. In mountainous regions, the number of days with heavy precipitation reaches values up to 90 days per year on average. The same goes for the west coast of Southeast Asia, while the number of days is smaller for the remaining areas and even smaller on the lee side of the mountains. We also find large numbers of days with heavy precipitation over land in the equatorial zone, where numbers also go up to 90 days per year on average. An increased index value can further be seen on the Indian ocean west of the Sumatra island and along the equatorial pacific. Another feature is the region around Taiwan, Japan and the neighboring Pacific ocean, where the structure of increased index values allows to see paths of tropical storm systems and typhoons that occur there. In dry regions, such as the Gobi desert, the index values drop down to one day per year or less on average. Panel (d) shows the sdii. The distribution of extreme values is similar to that of the heavy precipitation index. We find maximum values at the luff of mountainous regions, and smaller values for the rest. They can reach up to more than 30 mm·d^{-1}. One pixel indicates the Cherrapunjee location with its extraordinary annual precipitation amount of more than 11, 000 mm. The precipitation amount on wet days lies between 27 and 30 mm·d^{-1}. The SDII values in the equatorial region are smaller than in the monsoon region. Values between 8 and 15 mm·d^{-1} are reached there. For the already discussed tropical storm region between Taiwan and Japan along the Ryukyu islands, the SDII shows increased values of up to 24 mm·d^{-1}, corresponding to other indices. In the dry areas, the SDII shows values of less than 2 mm·d^{-1}.

DAPACLIP ETCCDI Climatology

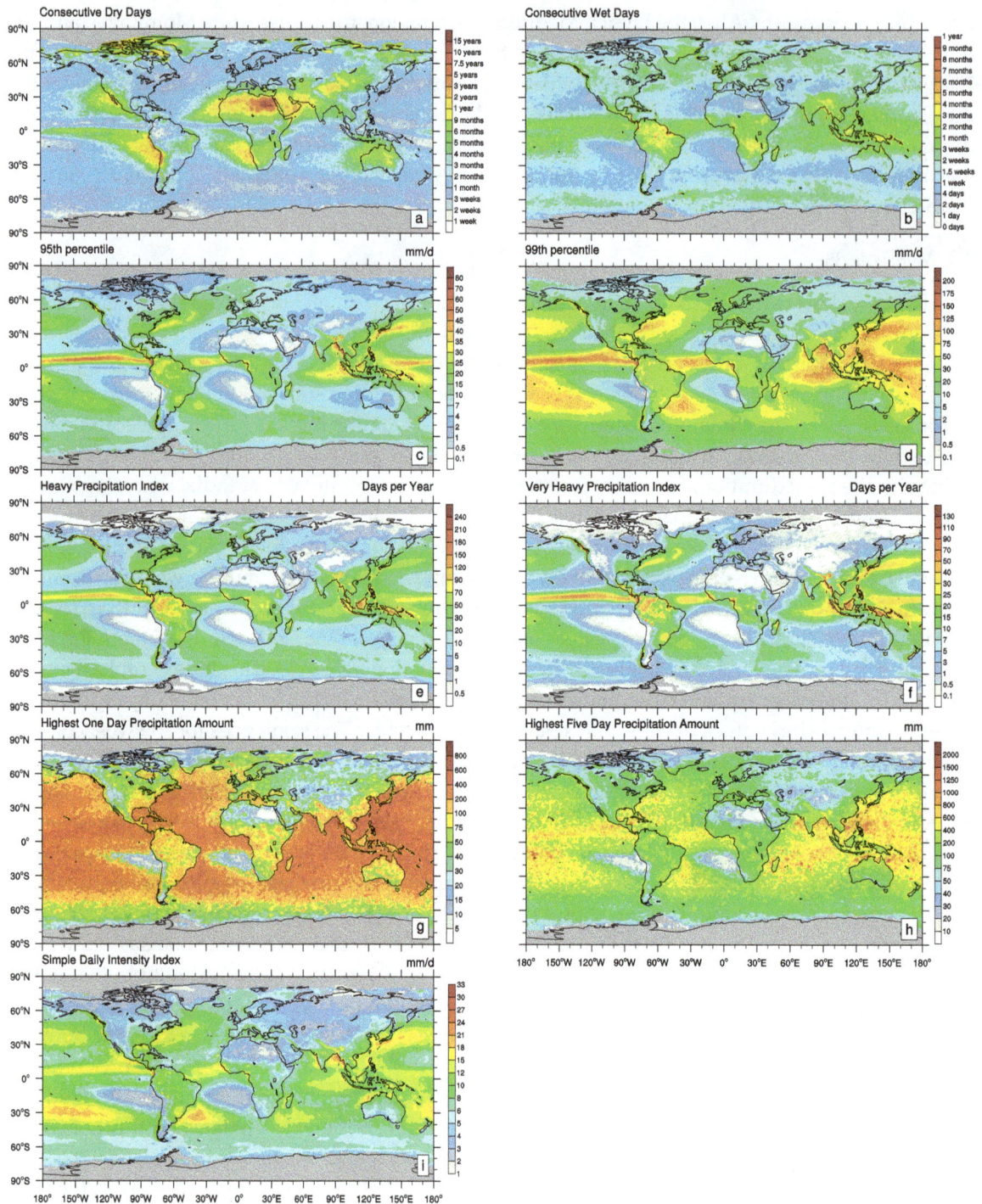

Figure 5. DAPACLIP ETCCDI climatology for the period 1988–2008 for all indices listed in Table 1. Panel (**a**): consecutive dry days (cdd). Panel (**b**): consecutive wet days (cwd). Panel (**c**): 95th percentile of precipitation (r95p). Panel (**d**): 99th percentile of precipitation (r99p). Panel (**e**): Heavy precipitation index (r10). Panel (**f**): Very heavy precipitation index (r20). Panel (**g**): Maximum one day precipitation (rx1). Panel (**h**): Maximum five day precipitation (rx5). Panel (**i**): Simple daily intensity index (sdii).

DAPACLIP ETCCDI Climatology Europe 0.5 degrees

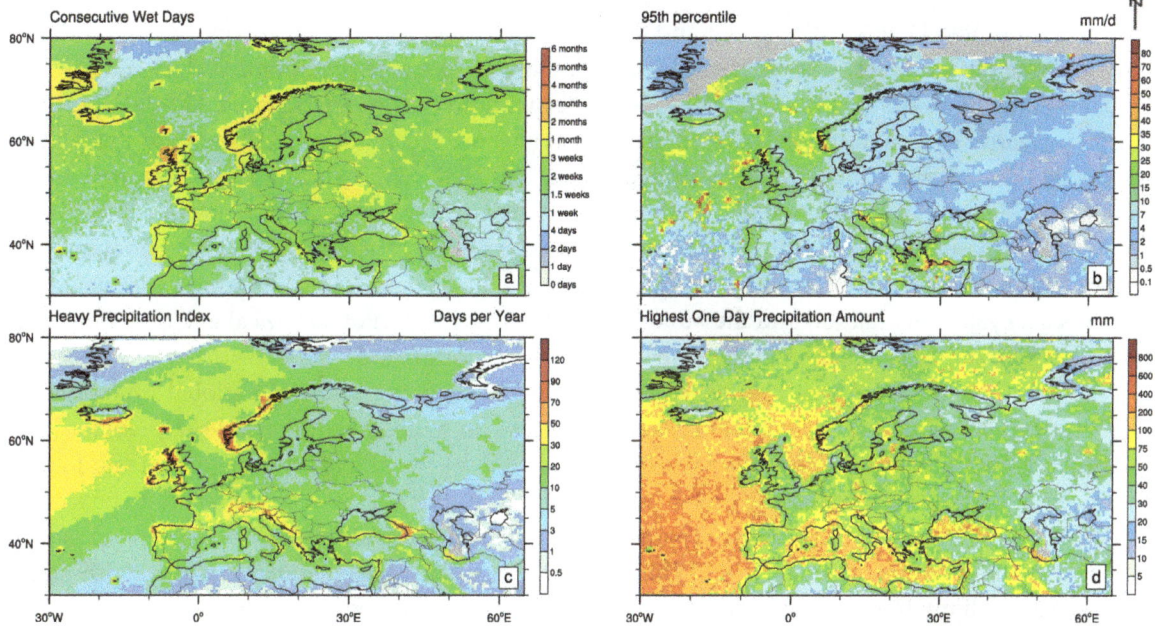

Figure 6. DAPACLIP ETCCDI climatology for the period 1988–2008 for Europe in 0.5° spatial resolution for selected ETCCDI indices. Panel (**a**): cwd index. Panel (**b**): r95p index. Panel (**c**): r10 index. Panel (**d**): rx1 index.

DAPACLIP ETCCDI Climatology Monsoon Asia

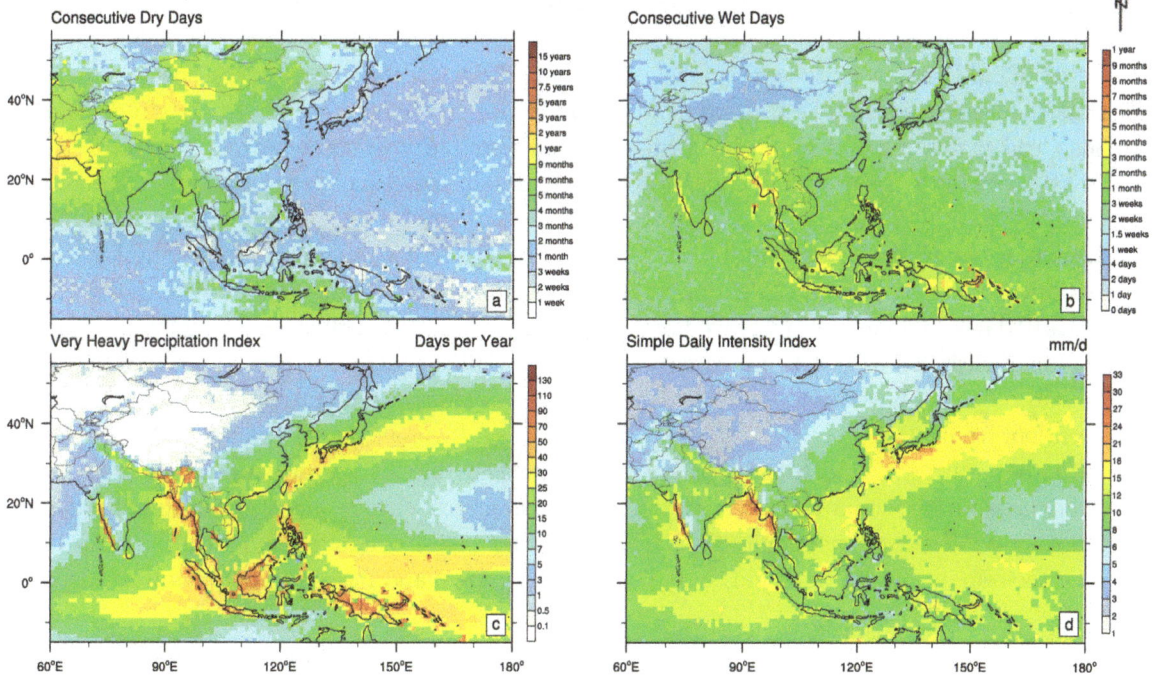

Figure 7. DAPACLIP ETCCDI climatology for the period 1988–2008 for the region of Monsoon Asia in 1.0° spatial resolution for selected ETCCDI indices. Panel (**a**): cdd index. Panel (**b**): cwd index. Panel (**c**): r20 index. Panel (**d**): sdii index.

4. Summary and Discussion

The presented DAPACLIP dataset and the ETCCDI climatology are based on a combination of two separate data sources. Gauge-based GPCC data are used as the precipitation data source over land, and satellite-based data from the HOAPS dataset are used over ocean surfaces. The dataset spans a time range of 21 years from 1988–2008. Three different spatial resolutions are provided: 1° and 2.5° for the global scale and 0.5° for the European domain. Uncertainty measures are also featured within the dataset. The datasets are provided under a digital object identifier (DOI) reference [25–27]. The DAPACLIP dataset is currently used for the hindcast-based evaluation of the MiKlip decadal forecast system and it has been developed for this purpose in the first place. ETCCDI indices were used for the forecast system evaluation and they allowed for the comparison of statistical and climatological features and their representation in the respective decadal model hindcasts.

In addition, ETCCDI was also used as the basis for the generation of a global precipitation climatology. Nine different indices were examined and a climatology for the whole 21 years (i.e., the DAPACLIP dataset temporal coverage) has been presented for the global scale in general and for the European and the Monsoon Asia region for more details. The climatology is based on the 1° version of DAPACLIP and is available on request. It shows different climatological features such as tropical storm paths, orographic effects, and monsoon. While climatological large scale features are well reproduced, there are some limitations for small scale features in some regions. There are various reasons for this that will be discussed further.

The density of gauge stations is occasionally sparse. For example, this is the case in Africa and the polar regions. The lack of available station data leads to large correlation distances for the Kriging interpolation method and corresponding uncertainties of up to 100%. The gauge dataset itself features the Kriging interpolation error as shown in Figure 2, but is not included in the DAPACLIP dataset. The quality of the satellite-based data depends on the available sampling rate. Due to the diurnal distribution of satellite overpasses, not all precipitation events are likely to be detected. In DAPACLIP, the number of available satellites increases with time and so does the sampling rate. The number of available land surface gauges per grid cell varies in time, which influences the spatial sampling essentially. These issues make it difficult to derive trends from this dataset. In certain cases, this can also lead to missing extreme events or the over-representation of some individual strong events. e.g., for these reasons, the rx1 index appears noisy over ocean surfaces. Other indices are influenced by this, too.

Another issue is the interpolation on a regular grid, which leads to weakening of single extreme precipitation peaks, such as the mentioned 2002 flood in eastern Germany. Future versions of the DAPACLIP dataset will include improved uncertainty estimations, also taking into account the Kriging error. Nonetheless, the ETCCDI climatology shows reasonable results for Africa, as one would expect, but it would be useful to evaluate these results in more detail using precipitation products that are independent from GPCC.

Another source of uncertainty for both, satellite and rain gauge measurements are limited detection capabilities for extreme precipitation amounts. Several effects, such as saturation in satellite radiances or wind disturbance at gauges, hamper the exact measurement of extreme events. For these reasons, the rain rate algorithm is currently under development and the next version will be based on a physical retrieval.

5. Outlook

This study presented results based on the current GPCC database and satellite data derived from SSM/I and TMI observations. It is planned to include additional gauge-based data with improved quality control to enhance the spatial coverage in data void regions. The satellite sampling of HOAPS will be improved by including additional satellite data, e.g., from AMSR-E and the Global Precipitation Mission Microwave Imager (GMI).. The rain rate retrieval will be changed towards the use of a 1D-Var algorithm. This allows for the simultaneous determination of precipitation and uncertainty measures.

More uncertainty features such as the Kriging error will be included. The DAPACLIP dataset is supposed to profit from all these enhancements.

The dataset is currently used for the evaluation of decadal climate predictions and first results from global evaluation are briefly summarized here. A core element of this evaluation is the global ETCCDI-based precipitation climatology. Comparing ETCCDI climatologies of decadal MiKlip hindcasts with this climatology exhibits well known features related to the split of the ITCZ in the eastern Pacific, likely related to biases in simulated sea surface temperatures (SST) [33,34]. It also exhibits an apparent misplacement of the ITCZ in the Pacific: the difference in the cdd index shows a strong regional maximum at the ITCZ in the MiKlip decadal hindcasts, while a strong overestimation of the cwd index is observed just north of that region.

Several other datasets are suitable for the generation of global ETCCDI-based climatologies. Comparing ETCCDI-based climatologies from various global precipitation datasets is an obvious next step and is considered here to be valuable such that consistencies and differences can be described and ideally understood.

Acknowledgments: The DAPACLIP dataset was developed within the MiKlip project framework funded by the German Federal Ministry for Education and Research (BMBF). The EUMETSAT Satellite Application Facility on Climate Monitoring is acknowledged for providing the HOAPS v3.2 data. NASA is acknowledged for providing the TMI data.

Author Contributions: Felix Dietzsch developed and analyzed the ETCCDI climatology of the DAPACLIP dataset presented herein, worked on merging the separate datasets to a final product and drafted the manuscript. Axel Andersson was responsible for the development of the satellite-based rain rate retrieval for the HOAPS dataset. Markus Ziese worked on processing the GPCC rain gauge data for the daily global precipitation dataset over land. Kerstin Schamm developed the Kriging interpolation scheme for the GPCC dataset. Kristin Raykova supported the work on the climatology with further analysis of land surface data. Marc Schröder was involved in the development of the satellite-based dataset and the generation of the DAPACLIP dataset as well as the ETCCDI climatology as supervisor and supported the drafting of the manuscript. Andreas Becker managed and supported the development process of the rain gauge dataset.

Conflicts of Interest: The authors declare no conflict of interest.

References

1. Marotzke, J.; Müller, W.A.; Vamborg, F.S.E.; Becker, P.; Cubasch, U.; Feldmann, H.; Kaspar, F.; Kottmeier, C.; Marini, C.; Polkova, I.; et al. MiKlip—A national research project on decadal climate prediction. *Bull. Am. Meteor. Soc.* **2016**, doi:10.1175/BAMS-D-15-00184.1.

2. Adler, R.F.; Huffman, G.J.; Chang, A.; Ferraro, R.; Xie, P.-P.; Janowiak, J.; Rudolf, B.; Schneider, U.; Curtis, S.; Bolvin, D.; et al. The version-2 global precipitation climatology project (GPCP) monthly precipitation analysis (1979–present). *J. Hydrometeorol.* **2003**, *4*, 1147–1167.

3. Huffman, G.J.; Adler, R.F.; Morrissey, M.M.; Bolvin, D.T.; Curtis, S.; Joyce, R.; McGavock, B.; Susskind, J. Global precipitation at one-degree daily resolution from multisatellite observations. *J. Hydrometeorol.* **2001**, *2*, 36–50.

4. Huffman, G.J.; Adler, R.F.; Bolvin, D.T.; Gu, G. Improving the global precipitation record: GPCP Version 2.1. *Geophys. Res. Lett.* **2009**, *36*, L17808, doi:10.1029/2009GL040000.

5. Huffman, G.J.; Adler, R.F.; Bolvin, D.T.; Gu, G.; Nelkin, E.J.; Bowman, K.P.; Hong, Y.; Stocker, E.F.; Wolff, D.B. The TRMM multisatellite precipitation analysis (TMPA): Quasi-global, multiyear, combined-sensor precipitation estimates at fine scales. *J. Hydrometeorol.* **2007**, *8*, 38–55.

6. Alexander, L.V. Global observed long-term changes in temperature and precipitation extremes: A review of progress and limitations in IPCC assessments and beyond. *Weather Clim. Extremes* **2016**, *11*, 4–16.

7. Sillmann, J.; Kharin, V.V.; Zwiers, F.W.; Zhang, X.; Bronaugh, D. Climate extremes indices in the CMIP5 multimodel ensemble: Part 1. Model evaluation in the present climate. *J. Geophys. Res. Atmos.* **2013**, *118*, 1716–1733.

8. Sillmann, J.; Kharin, V.V.; Zwiers, F.W.; Zhang, X.; Bronaugh, D. Climate extremes indices in the CMIP5 multimodel ensemble: Part 2. Future climate projections. *J. Geophys. Res. Atmos.* **2013**, *118*, 2473–2493.

9. Easterling, D.R.; Kunkel, K.E.; Wehner, M.F.; Sun, L. Detection and attribution of climate extremes in the observed record. *Weather Clim. Extremes* **2016**, *11*, 17–27.

10. Raykova, K. Trendanalysen Von Niederschlagsextremen Und Untersuchung Der Extremwertverteilung Basierend Auf Täglichen Stationsmessungen von 1988 bis 2013. Master Thesis, Goethe University Frankfurt, Frankfurt, Germany, 2016.

11. Vose, S.V.; Peterson, T.C.; Schmoyer, R.L.; Eischeid, J.K. The global historical climatology network: A preview of version 2. 75. In Proceedings of the Annual Meeting of the American Meteorological Society, Diamond Anniversary, United States of America, Dallas, TX, USA, 15–20 January 1995.

12. Klein Tank, A.M.G.; Wijngard, J.B.; Können, G.P.; Böhm, R.; Demaree, G.; Gocheva, A.; Mileta, M.; Pashiardis, S.; Hejkrlik, L.; Kern-Hansen, C.; et al. Daily dataset of 20th-century surface air temperature and precipitation series for the European Climate Assessment. *Int. J. Climatol.* **2002**, *22*, 1441–1453.

13. Schamm, K.; Ziese, M.; Becker, A.; Finger, P.; Meyer-Christoffer, A.; Schneider, U.; Schröder, M.; Stender, P. Global gridded precipitation over land: A description of the new GPCC First Guess Daily product. *Earth Syst. Sci. Data* **2014**, *6*, 49–60.

14. Willmott, C.; Rowe, C.; Philpot, W. A sensitivity analysis of some common assumptions associated with grid-point interpolation and contouring. *Am. Cartogr.* **1985**, *12*, 5–16.

15. Krige, D. *Lognormal-de Wijsian Geostatistics for Ore Evolution*; South African Institute of Mining and Metallurgy: Johannesburg, South Africa, 1981; pp. 23–39.

16. Shepard, D. A two-dimensional interpolation function for irregularly-spaced data. In Proceedings of the 1968 23rd ACM National Conference, New York, NY, USA, 1968; pp. 517–524, doi:10.1145/800186.810616.

17. Schneider, U.; Becker, A.; Finger, P.; Meyer-Christoffer, A.; Ziese, M.; Rudolf, B. GPCC's new land surface precipitation climatology based on quality-controlled in-situ data and its role in quantifying the global water cycle. *Theor. Appl. Climatol.* **2014**, *115*, 15–40.

18. Schneider, U.; Becker, A.; Finger, P.; Meyer-Christoffer, A.; Rudolf, B.; Ziese, M. GPCC Full Data Reanalysis Version 7.0 at 1.0°: Monthly land-surface precipitation from rain-gauges built on GTS-based and historic data. *Deutsch. Wetterd.* **2015**, doi:10.5676/DWD_GPCC/FD_M_V7_100.

19. Yamamoto, J. An alternative measure of the reliability of ordinary kriging estimates. *Math. Geol.* **2000**, *32*, 489–509.

20. The Satellite Application Facility on Climate Monitoring. Available online: http://cmsaf.eu (accessed on 21 November 2016).

21. Hamburg Ocean Atmosphere Parameters and Fluxes from Satellite Data. Available online: http://hoaps.org (accessed on 21 November 2016).

22. Andersson, A.; Fennig, K.; Klepp, C.; Bakan, S.; Grassl, H.; Schulz, J. The Hamburg Ocean atmosphere parameters and fluxes from satellite data—HOAPS-3. *Earth Syst. Sci. Data* **2010**, *2*, 215–234.

23. Fennig, K.; Andersson, A.; Bakan, S.; Klepp, C.; Schröder, M. Hamburg Ocean atmosphere parameters and fluxes from satellite data — HOAPS-3.2 — monthly means/6-hourly composites. *Satell. Appl. Facil. Clim. Monit.* **2012**, doi:10.5676/EUM_SAF_CM/HOAPS/V001.

24. Andersson, A.; Fennig, K.; Schröder, M. Algorithm theoretical basis document HOAPS release 3.2. *Satell. Appl. Facil. Clim. Monit.* **2011**, *1*, 31–34.

25. Andersson, A.; Ziese, M.; Dietzsch, F.; Schröder, M.; Becker, A.; Schamm, K. *HOAPS/GPCC European Daily Precipitation Data Record With Uncertainty Estimates Using Satellite and Gauge Based Observations at 0.5°.* Deutscher Wetterdienst: Offenbach, Germany, 2016; doi:10.5676/DWD_CDC/HOGP_050/V001.

26. Andersson, A.; Ziese, M.; Dietzsch, F.; Schröder, M.; Becker, A.; Schamm, K. *HOAPS/GPCC Global Daily Precipitation Data Record With Uncertainty Estimates Using Satellite and Gauge Based Observations at 1.0°*; Deutscher Wetterdienst: Offenbach, Germany, 2016; doi:10.5676/DWD_CDC/HOGP_100/V001.

27. Andersson, A.; Ziese, M.; Dietzsch, F.; Schröder, M.; Becker, A.; Schamm, K. *HOAPS/GPCC Global Daily Precipitation Data Record With Uncertainty Estimates Using Satellite and Gauge Based Observations at 2.5°*; Deutscher Wetterdienst: Offenbach, Germany, 2016; doi:10.5676/DWD_CDC/HOGP_250/V001.

28. HOAPS/GPCC global daily precipitation data record with uncertainty estimates using satellite and gauge based observations Version 1. Available online: ftp://ftp.dwd.de/pub/data/gpcc/html/HOGP_V001.html (accessed on 19 October 2016).

29. CLIMDEX. Datesets for Indices of Climate Extremes. Available online: http://climdex.org (accessed on 26 September 2016).

30. Klein Tank, A.M.G.; Zwiers, F.W.; Zhang, X. *Guidelines on Analysis of Extremes in a Changing Climate in Support of Informed Decisions for Adaptation*; World Meteorological Organisation (WMO): Geneva, Switzerland, 2009.

31. Zhang, X.; Alexander, L.; Hegerl, G.C.; Jones, P.; Klein Tank, A.M.G.; Peterson, T.C.; Trewin, B.; Zwiers, F.C. Indices for monitoring changes in extremes based on daily temperature and precipitation data. *WIREs Clim. Chang.* **2011**, *2*, 851–870.

32. Rudolf, B.; Rapp, J. Das Jahrhunderthochwasser der Elbe: Synoptische Entwicklung und klimatologische Aspekte. In *Klimastatusbericht 2002*; Deutscher Wetterdienst: Offenbach(Main), Germany, 2002; pp. 172–187.

33. Hazeleger, W.; Wang, X.; Severijns, C.; Ştefănescu, S.; Bintanja, R.; Sterl, A.; Wyser, K.; Semmler, T.; Yang, S.; van den Hurk, B.; et al. EC-Earth V2.2: description and validation of a new seamless earth system prediction model. *Clim. Dyn.* **2012**, *39*, 2611–2629.

34. Michael, J.-P.; Misra, V.; Chassignet, E.P. The El Niño and Southern Oscillation in the historical centennial integrations of the new generation of climate models. *Reg. Environ. Chang.* **2013**, *13*, 121–130.

11

Watershed Response to Climate Change and Fire-Burns in the Upper Umatilla River Basin, USA

Kimberly Yazzie [1] and Heejun Chang [2,*

[1] Department of Environmental Science and Management, Portland State University, Portland, OR 97207, USA; kiyazzie@pdx.edu
[2] Department of Geography, Portland State University, Portland, OR 97207, USA
* Correspondence: changh@pdx.edu

Academic Editors: Daniele Bocchiola, Claudio Cassardo and Guglielmina Diolaiuti

Abstract: This study analyzed watershed response to climate change and forest fire impacts in the upper Umatilla River Basin (URB), Oregon, using the precipitation runoff modeling system. Ten global climate models using Coupled Intercomparison Project Phase 5 experiments with Representative Concentration Pathways (RCP) 4.5 and 8.5 were used to simulate the effects of climate and fire-burns on runoff behavior throughout the 21st century. We observed the center timing (CT) of flow, seasonal flows, snow water equivalent (SWE) and basin recharge. In the upper URB, hydrologic regime shifts from a snow-rain-dominated to rain-dominated basin. Ensemble mean CT occurs 27 days earlier in RCP 4.5 and 33 days earlier in RCP 8.5, in comparison to historic conditions (1980s) by the end of the 21st century. After forest cover reduction in the 2080s, CT occurs 35 days earlier in RCP 4.5 and 29 days earlier in RCP 8.5. The difference in mean CT after fire-burns may be due to projected changes in the individual climate model. Winter flow is projected to decline after forest cover reduction in the 2080s by 85% and 72% in RCP 4.5 and RCP 8.5, in comparison to 98% change in ensemble mean winter flows in the 2080s before forest cover reduction. The ratio of ensemble mean snow water equivalent to precipitation substantially decreases by 81% and 91% in the 2050s and 2080s before forest cover reduction and a decrease of 90% in RCP 4.5 and 99% in RCP 8.5 in the 2080s after fire-burns. Mean basin recharge is 10% and 14% lower in the 2080s before fire-burns and after fire-burns, and it decreases by 13% in RCP 4.5 and decreases 22% in RCP 8.5 in the 2080s in comparison to historical conditions. Mixed results for recharge after forest cover reduction suggest that an increase may be due to the size of burned areas, decreased canopy interception and less evaporation occurring at the watershed surface, increasing the potential for infiltration. The effects of fire on the watershed system are strongly indicated by a significant increase in winter seasonal flows and a slight reduction in summer flows. Findings from this study may improve adaptive management of water resources, flood control and the effects of fire on a watershed system.

Keywords: runoff; climate change; fire-burns; water resources; Umatilla River; PRMS

1. Introduction

Anthropogenic influences on climate coupled with natural variability in climate have shifted the spatial and temporal distribution of water resources worldwide [1,2]. Quantifying recharge and streamflow response to climate change is an essential step to developing long-term water resource management plans to increase the understanding of the global energy balance in a hydrologic regime to improve adaptive capacity [3,4]. Identifying changes in basin runoff is important due to the strong effects on water and energy demands [5], which also have important societal and ecological implications.

A marked shift in global mean surface temperature in the 20th century has been widely cited as an indicator of climate change and its direct relationship to changes in the global energy budget [6,7]. In the Pacific Northwest, climate change impacts include shifts in the magnitude and timing of runoff [8–11], reduced proportion of precipitation falling as snow in montane regions [12,13], decreases in snow water equivalent [14,15] and an increase in the frequency and intensity of floods and droughts [11,16,17].

This paper explores the watershed response to climate change in the upper Umatilla River Basin (URB) where runoff behavior is observed before and after fire-burns in the 21st century. Seasonal flows, the ratio of snow water equivalent to precipitation and potential recharge were quantified. Three research questions guided this analysis. (1) How does the hydrologic regime change seasonally and annually in response to climate change in the 21st century in comparison to historical conditions? (2) What are the effects of land cover change after fire-burns on basin runoff? (3) Which water budget components (e.g., seasonal runoff, snow water equivalent) are sensitive to a change in climate and could potentially be considered in water resource management for climate adaptation planning?

While there is extensive research on the Columbia River Basin [18–22], the upper URB is largely understudied. Burns et al. [23] indicated that water levels have declined 3048–9144 cm since 1970 in some of the deeper Columbia River Basalt Group (CRBG) aquifers where the physical characteristics of basalt, depositional environment, folding and faulting impede groundwater flow. Previous studies estimated recharge in the region prior to predevelopment to be 6.90 cm/year and an increase to 10.80 cm/year in the 1980s due to irrigation [24].

Additionally, the effects of wildfire on runoff in the basin have not been thoroughly studied. The effects of wildfire can decrease canopy interception, thus increasing runoff and changing the chemical and physical properties of soil [25,26]. The infiltration rate after a wildfire has been observed to decrease two to seven-fold [25,27,28], with erosion from overland flow [29], an increase in percent area burned and more open landscape [30]. These are reasons for concern for forest resource managers who investigate the effects on the carbon cycle and forest productivity. Changes in peak discharges are more apparent than changes in annual runoff [25], where in some places, peak runoff increases by two orders of magnitude [31,32]. In the Western Northwest, vegetation shifts are projected from conifer to mixed forests by way of increased wildfire occurrence [33].

Surface and groundwater interactions and the effects of climate change on the hydrologic regime in the upper Umatilla River Basin (URB) are largely understudied. The lower URB has four Oregon Water Resources Department Groundwater Restricted Areas as a result of well withdrawals and a heavy dependence on groundwater for agricultural and municipal needs [34]. In the 1920s, the Umatilla Reclamation Project blocked the return of anadromous fish, resulting in a steep decline of salmon return. In 1988, the Umatilla Basin Project Act resulted in a bucket-for-bucket exchange of Umatilla River water for Columbia River water, which improved flows to restore salmonid and steelhead populations [35]. The cultural value of water cannot be understated, making it more important to protect natural resources vital to the Confederated Tribes of the Umatilla Indian Reservation (CTUIR) in the URB. CTUIR implemented the protection of first foods into natural resource management as a form of self-determination for environmental equity and tribal resilience, including water, salmon, deer, cous and huckleberry [36,37]. First foods are the "minimum ecological products necessary to sustain CTUIR culture" [37]. This study thus contributes to the protection of cultural and ecosystem services by providing runoff changes throughout the 21st century for climate adaptation planning.

2. Materials and Methods

2.1. Study Area

The study site is 2365 km^2 in the upper URB in northeastern Oregon on the Columbia Plateau (Figure 1). A significant portion of the Umatilla Indian Reservation, 647 km^2, is within the study site boundaries. The Umatilla River is a 145-km reach that enters into the Columbia River, originating in the Blue Mountains with a gravel-bed channel system and a multi-channel pattern [38]. It is largely groundwater fed; approximately 70%–80% of annual flows are provided in the summer months [39]. The upper basin is approximately 14% in drainage area, but supplies 40%–50% of the average flow to the Umatilla River [40,41]. The URB is mostly semiarid, located east of the Cascades in the rain shadow. It receives 12.7 cm–127 cm in annual precipitation and ranges in elevation from 82 m–1676 m [36]. The study site is 55.1% coniferous tree cover, 0.1% deciduous and mixed tree cover, 21.7% shrub cover, 21.8% grass and 1.3% bare soil [42].

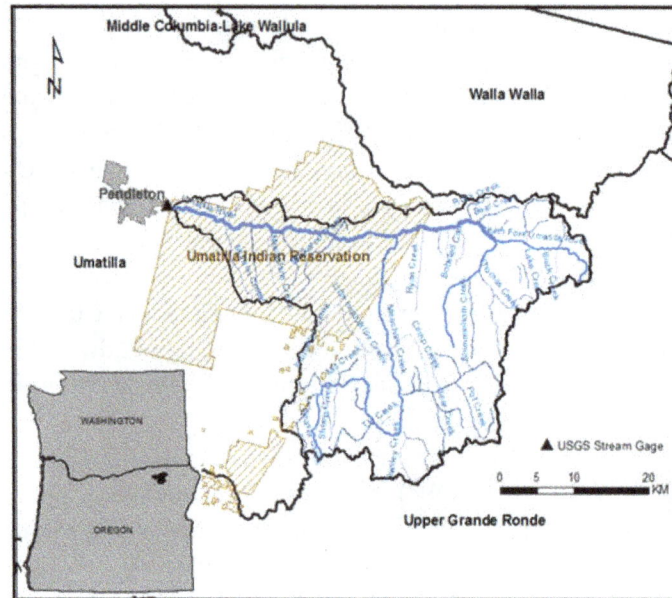

Figure 1. The study site is 2365 km² in the upper Umatilla River Basin in northeastern Oregon.

2.2. Hydrologic Model: Precipitation Runoff Modeling System

The Precipitation Runoff Modeling System (PRMS) (Ver. 3.0.5) is a deterministic, distributed-parameter, physically-based process hydrologic model that estimates water-balance relations, stream flow regimes and soil-water relations [43]. Precipitation, minimum and maximum temperature and solar radiation are the primary inputs into PRMS to compute a water balance and energy balance for individual hydrologic response units (HRUs) with distributed parameters [43]. It has been used extensively in assessing changes in runoff resulting from climate change [8].

2.3. Climate Data

We obtained daily time series of precipitation, minimum and maximum temperature from the University of Idaho Gridded Surface Meteorological data (METDATA) for climate forcings at 4-km (1/24-degree) resolution for model calibration (1995–2010) [44]. Solar radiation data were not obtained, in which case PRMS internally estimates daily shortwave solar radiation [43]. High resolution gridded METDATA were derived from observations and regional reanalysis using a hybrid method, combining spatially-rich data from the Parameter-elevation Regressions on Independent Slopes Model (PRISM) and temporally rich data from the North America Land Data Assimilation System Phase 2 (NLDAS-2) [44–46].

Downscaled simulated historical and future climate data with a resolution of 4 km (1/24-degree) were obtained for the 1980s (1970–1999), 2020s (2010–2039), 2050s (2040–2069) and 2080s (2070–2099). They were derived using the multivariate adaptive constructed analogs statistical downscaling method (MACA), a non-interpolated-based approach [47]. Data from the Coupled Model Inter-Comparison Project 5 (CMIP5), Representative Concentration Pathways 4.5 and 8.5 (RCP 4.5 and RCP 8.5) were used in this analysis [48]. RCP4.5 is a medium stabilization scenario where an additional 4.5 W/m² of radiative forcing energy is trapped in the atmosphere by year 2100 [48]. RCP8.5 has a very high baseline emission scenario at the 90th percentile, where an additional 8.5 W/m^2 is trapped by 2100, with no climate action anticipated [49]. Downscaled CMIP5 climate data were obtained from the University of Idaho for each of ten Global Climate Models used in this study and uploaded to the USGS's Geo Data Portal [50] with a shapefile of the study site, for which the portal provided specific climate data for each individual HRU on a weighted scale.

2.4. Global Climate Models

In a multi-model ensemble approach, ten global climate models (GCMs) were chosen based on model performance in the Pacific Northwest (Table A1), including a low normalized error score [51] and data availability for each GCM. Analysis of model performance for each GCM revealed minor differences in precipitation projections; differences are minimal (<0.02 cm), with greater differences in temperature (<2.64 °C), but not large enough to exclude any GCMs. Further, bias corrections were not made, as studies have found little to no difference in selecting or weighting GCM output [52], and our main purpose was to investigate relative changes of runoff in the future from the historical period.

2.5. Indicators for Detecting Climate Change Impact

We used four climate change impact detection indices to analyze runoff behavior before fire-burns in the 2020s, 2050s and 2080s and the effects of fire-burns in the 2080s in RCP 4.5 and RCP 8.5. The indices are: (1) ensemble mean change in annual and seasonal runoff; (2) center timing of streamflow (CT); (3) the ratio of 1 April snow water equivalent to October–March precipitation (snow water equivalent (SWE)/P); and (4) potential recharge to analyze ground and surface water interactions. These indices were compared between the means of historical and future time periods and deemed statistically significant at the 5% significance level. To compare differences in the aforementioned indices between different periods, we used one-way analysis of variance (ANOVA) and the Kruskal–Wallis test (for small samples). Detection of an overall significant difference required multiple comparisons with Tukey's honest significance test and Kruskal–Wallis' multiple comparisons test.

2.6. Generation of Hydrologic Response Units

HRUs were discretized using watershed boundaries, soils and land cover to derive 107 HRUs (Table A2). A daily water and energy balance is computed for each HRU, and the sum of the responses of all HRUs is weighted on a unit-area basis [43]. HRUs smaller than 4%–5% were avoided for daily-flow computations for basin-wide estimates [53].

2.7. Post-fire Analysis

We modified PRMS model parameters to exemplify post-fire conditions (Table 1). Our method is similar to Konrad [54], who adjusted model parameters pertaining to seasonal vegetation cover, shortwave radiation transmitted through the canopy, upper soil storage capacity and total soil storage capacity. We projected burned areas in the 2080s based on the fire history in the URB obtained from the U.S. Forest Service at Umatilla National Forest [55]. The maximum spatial extent of fires that occurred in history (representing the 1980s) was used for representing post-fire conditions for both the 1980s and the 2080s. Accordingly, the parameters in Table 1 for those HRUs with fire occurrences in the 1980s were changed in the 2080s in both the RCP 4.5 and RCP 8.5 scenarios. The newest version of the Model for Interdisciplinary Research on Climate (MIROC5), an intermediate warm model, was chosen to simulate runoff in RCP 4.5, and the Hadley Global Environment Model 2- Earth System (HadGEM2-ES), a warmer model with acute summer drying, was run in RCP 8.5. These two GCMs were used in a previous study by Turner et al. [30].

Table 1. Initial and assigned model parameter values for forest cover reduction analysis.

	Description	Initial	Assigned	% Δ
COVDEN_SUM	Summer vegetation cover density	0.5	0.1	−80%
COVDEN_WIN	Winter vegetation cover density	0.5	0.1	−80%
RAD_TRNCF	Solar radiation transmission	0.3	0.5	40%
SOIL_RECHR_MAX	Max. storage for soil recharge zone	1.643	0.55	−67%
SOIL_MOIST_MAX	Max. value of water for soil zone	2.14–12.537	1.08	−50%

2.8. Calibration and Verification

Automated and manual calibration was completed with observed (1995–2010) streamflow (Table A3). We used Let Us Calibrate (LUCA, V. 2.0.0), a multiple-objective, stepwise, automated procedure [56], to calibrate water balance, daily flow timing of all flows and of high and low flows. Manual calibration was required to improve manual calibration of simulated peak runoff to observed conditions. Consumptive use is very small to the total volume of streamflow and was determined as negligible and not added to the hydrograph to obtain normal streamflow conditions. Four years of additional data (2010–2014, 26% of the data used for calibration) were used for model verification.

2.9. Model Evaluation

Four statistical analyses were used to analyze model performance (Table 2). The Nash–Sutcliffe efficiency (NSE) indicates accuracy, one being a perfect fit, and a negative coefficient would indicate that the mean value of observed data would be a better predictor than the model [57]. The percent bias (PBIAS) determines under- or over-prediction of simulated data in comparison to observed data with an optimal value of 0; a negative value would indicate underestimation [58]. Kling–Gupta efficiency (KGE) is an alternative to NSE, where 1 is an optimal fit. Different components of the model area are evaluated, such as correlation, bias and variability [59]. Root mean square error (RMSE) is first calculated to indicate the differences, the residuals between simulated and observed values. Normalized RMSE (NRMSE) was also calculated where the lower the value, the lower the variance.

Table 2. Final model goodness of fit statistics based on daily streamflow for Nash-Sutcliffe efficiency (NSE), Percent Bias (PBIAS), Kling–Gupta efficiency (KGE), and Normalized RMSE (NRMSE).

	NSE	% Bias	KGE	NRMSE
Initial Model Results	0.04	4	0.57	97.7
After Calibration (1995–2010)	0.73	3.5	0.81	52.2
Validation (2010–2014)	0.73	3.5	0.83	52.1

$$NSE = 1 - \frac{\sum_{i=1}^{n}\left(y_i^{obs} - y_i^{sim}\right)^2}{\sum_{i=1}^{n}\left(y_i^{obs} - y_i^{mean}\right)^2}$$

y_i^{obs} = the i-th observed flow; y_i^{sim} = the i-th simulated flow value; y^{mean} = mean of observed flow data.

$$PBIAS = \frac{\sum_{i=1}^{n}\left(y_i^{obs} - y_i^{sim}\right) * 100}{\sum_{i=1}^{n}\left(y_i^{obs}\right)}$$

$KGE = 1 - \sqrt{(r-1)^2 + (\alpha - 1)^2 + (\beta - 1)^2}$ r = linear correlation coefficient; α = relative variability in the simulated and observed values; β = ratio of the means of the simulated and observed values.

$$RMSE = \sqrt{\frac{\sum_{i=1}^{n}\left(y_i^{obs} - y_i^{sim}\right)^2}{n}}$$

$$NRMSE = \frac{RMSE}{Y_{obs,max} - Y_{obs,min}}$$

$Y_{obs,max}$ = observed maximum flow; $Y_{obs,min}$ = observed minimum flow.

3. Results

3.1. Change in Mean Annual Temperature

The ensemble mean temperature increases from 8.6 °C in the 2020s to 9.5 °C in the 2050s, to 10.2 °C in the study area in the 2080s in RCP 4.5. In RCP 8.5, an increase from 8.8 °C in the 2020s to 10.2 °C in the 2050s to 12.1 °C in the 2080s is observed (Figure 2). By the end of the 21st century, there is a 3.3 °C increase in mean temperature in RCP 8.5 and an increase of 1.6 °C in the RCP 4.5 scenario.

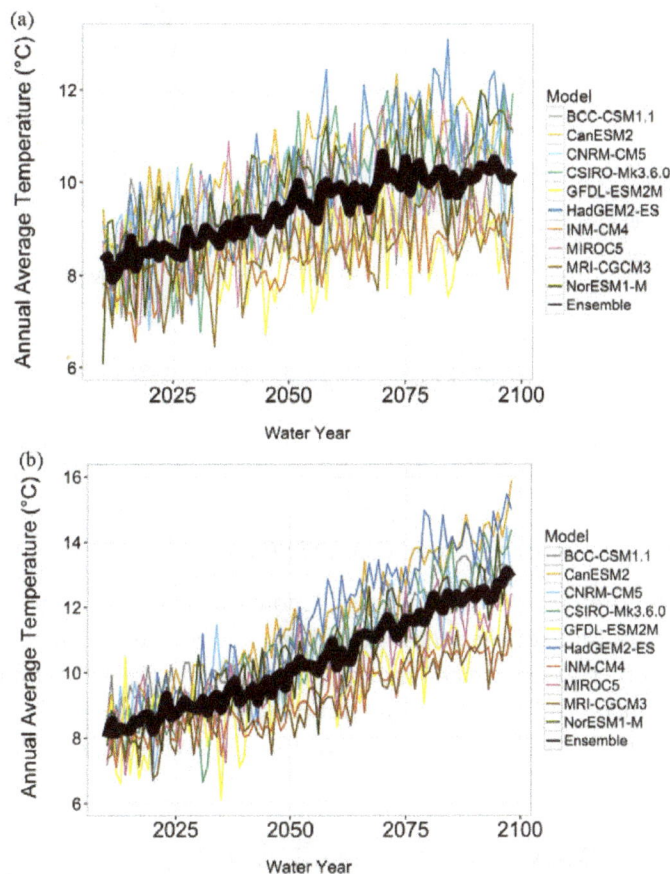

Figure 2. (a) Annual mean temperature for all global climate models (GCMs) for Representative Concentration Pathway (RCP) 4.5 with the ensemble mean in black; (b) Annual mean temperature for all GCMs for RCP 8.5.

BCC-CSM1.1 = Beijing Climate Center – Climate System Model, version 1.1; CanESM2 = Second Generation Canadian Earth System Model; CNRM-CM5 = Centre National de Recherches Météorologiques, Climate Model, version 5; CSIRO-Mk3.6.0 = Commonwealth Scientific and Industrial Research Organisation, Mark 3.6; GFDL-ESM2M = Geophysical Fluid Dynamics Laboratory- Earth System Model 2 Modular; HadGEM2-ES = Hadley Centre Global Environmental Model, version 2 (Earth System); INM-CM4 = Institute for Numerical Mathematics Coupled Model, version 4.0; MIROC5 = Model for Interdisciplinary Research on Climate, version 5; MRI-CGCM3 = Meteorological Research Institute Coupled Global Climate Model, version 3.

3.2. Change of Temperature and Percent Change in Precipitation

The 10 GCMs project a steady increase of temperature with increased model variability toward the end of the 21st century in the study area (Figure 3). In the 2020s, annual temperature is projected to increase ranging from 0.7 °C–2.2 °C in RCP 4.5 and 0.9 °C–2.2 °C in RCP 8.5, and annual precipitation is projected to change from −3.4%–4.5% in RCP 4.5 and −5.7%–6.3% in RCP 8.5. In the 2050s, the change of temperature ranges from 1.2 °C–3.2 °C for RCP 4.5 and 1.7 °C–4.1 °C for RCP 8.5 and a percent change in precipitation from −2.9%–7.3% for RCP 4.5 and −5.3%–12.3% for RCP 8.5. In the 2080s, the change of temperature ranges from 1.6 °C–4.1 °C in RCP 4.5 and 3.1 °C–6.6 °C in RCP 8.5 and a percent change in precipitation from −4.8%–13.6% in RCP 4.5 and −3.6%–11.3% in RCP 8.5.

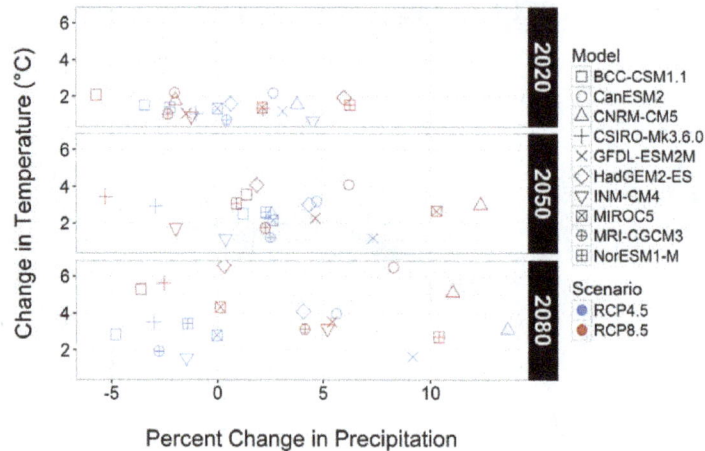

Figure 3. Change of temperature and annual percent change in precipitation in comparison to historical conditions for all GCMs.

3.3. Seasonal Change of Temperature and Percent Change in Precipitation

There is less model uncertainty in the winter with more variation in the summer season throughout the century in both scenarios for both temperature and precipitation. In the summer, there is substantial variability in model predictions in the change of temperature in the 2050s and 2080s. The percent change in precipitation increases slightly throughout the 21st century (Figure 4). In the summer, there is a significant increase in the percent change in precipitation with high model uncertainty in both scenarios in the 2050s and 2080s (Figure 4).

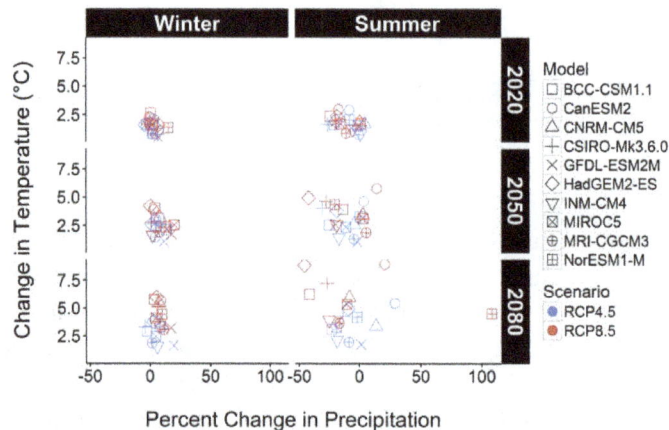

Figure 4. Change in seasonal temperature (T) and percent change in precipitation (P) for all GCMs in comparison to historical records.

3.4. Center Time of Flow after Fire-Burns

Mean center time (CT) in RCP 4.5 in the 2080s before forest cover reduction occurred 37 days earlier than baseline conditions and occurred 40 days earlier after forest cover reduction (Figure 5; Table 3). In RCP 8.5 in the 2080s, mean CT occurred 30 days earlier before forest cover reduction and occurred 32 days earlier than baseline conditions after forest cover reduction (Table 3). In the 1980s after forest cover reduction, Mean CT occurs five days earlier in RCP 4.5 and three days earlier in RCP 8.5. Mean CT is significantly different between historical pre-fire and before and after forest cover reduction, between historical post-fire and before and after forest cover reduction, but not significantly different between before and after forest reduction in both the RCP 4.5 and 8.5 scenarios and before and after fire-burns in historic conditions. For a comparison, ensemble mean basin CT before forest cover reduction occurs 27 days earlier in RCP 4.5 and 32 days earlier in RCP 8.5 in the 2080s, occurring earlier than the model prediction by MIROC5 in RCP 4.5 (Table 3).

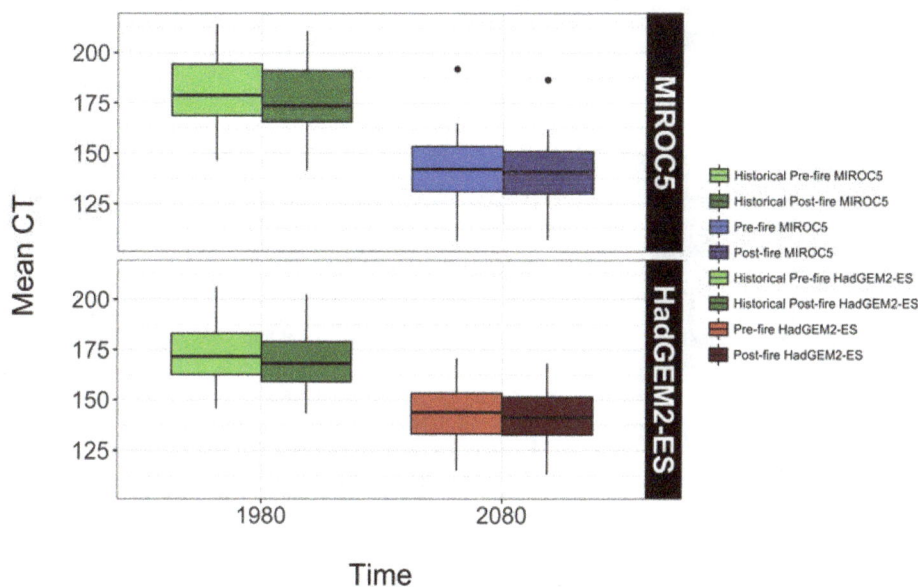

Figure 5. Mean center timing (CT) of flow for MIROC5 in RCP 4.5 and HadGEM2-ES in RCP 8.5 before and after forest cover reduction in comparison to historic conditions.

Table 3. Ensemble mean center timing (CT) of flow before forest cover reduction in historical conditions (1980s), in the 2050s and in the 2080s, in comparison to MOROC5 in RCP 4.5 and HadGEM2-ES in RCP 8.5 in the 2080s before and after forest cover reduction.

		Ensemble Mean					MIROC5				HadGEM2-ES			
	Hist.	RCP 4.5		RCP 8.5			RCP 4.5				RCP 8.5			
	1980s	2050s	2080s	2050s	2080s	Pre 1980s	Post 1980s	Pre 2080s	Post 2080s	Pre 1980s	Post 1980s	Pre 2080s	Post 2080s
Δ in days in μ CT	6/23 175	5/31 152	5/27 148	5/29 150	5/22 143	6/29 181	6/24 176	5/23 144	5/20 141	6/20 172	6/17 169	5/21 142	5/19 140
σ	17.32	14.76	15.83	16.23	15.72	17.26	17.15	14.07	16.26	15.32	13.77	14.23	14.04

3.5. Seasonal Flows after Fire-Burns

Winter runoff in MIROC 5 under RCP 4.5 in the 2080s showed a 92% increase before forest cover reduction and an 85% increase after forest cover reduction compared to the historical period with the same respective land cover conditions (Figure 6; Table 4). In RCP 8.5 in the 2080s, winter runoff increases 79% before forest cover change and increases 72% after forest cover reduction (Table 4). A decrease in summer flows is observed for both before and after forest cover reduction. A 72%

decrease and 67% decrease after land cover change are observed in RCP4.5 and a 72% and 67% decrease before and after forest cover reduction in RCP 8.5 (Figure 6; Table 4). Winter flows are significantly different between historical and both land cover conditions in RCP 4.5 and RCP 8.5 and for summer flows in RCP 4.5. It is not significantly different between summer flows in RCP 8.5 and between both land cover reductions in both scenarios, as well as between before and after forest cover reduction in historical conditions. The ensemble basin mean of winter flows before forest cover reduction substantially increases in the 2080s by 98% in RCP 8.5 in comparison to an increase by 71% in RCP 4.5 (Table 5).

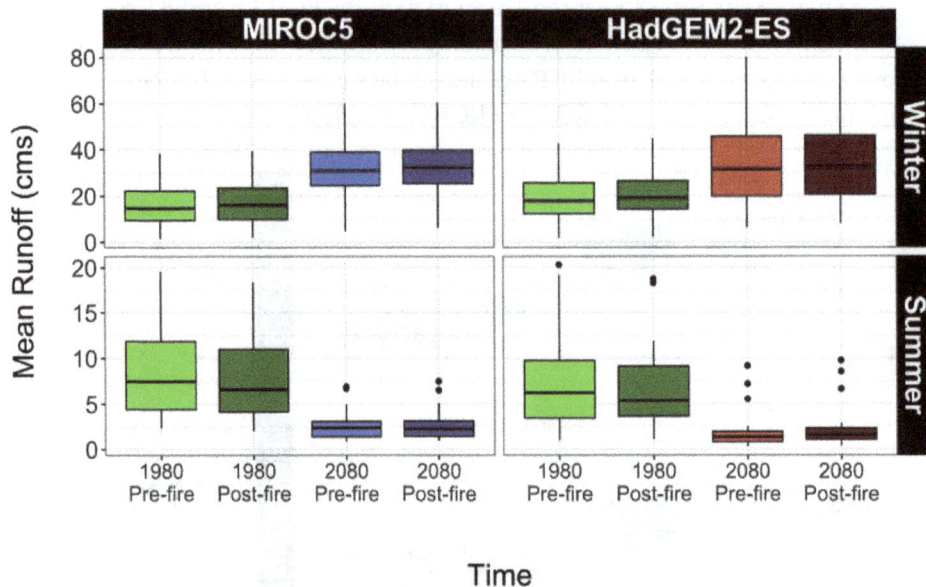

Figure 6. Seasonal flows for MIROC5 in RCP 4.5 and HadGEM2-ES in RCP 8.5 before and after forest cover reduction in comparison to historic conditions.

3.6. Ratio of Snow Water Equivalent to Precipitation before Fire-Burns

The ratio of SWE to P over the whole study area is substantially lower in RCP 4.5 in comparison to the 1980s with a 81% and 90% decrease before and after forest cover reduction (Figure 7; Table 6). The absolute difference of SWE/P is 9.0×10^{-2} and 0.1, before and after forest cover reduction in RCP 4.5 in comparison to the 1980s before canopy reduction (Table 6). Between the two land cover conditions in RCP 4.5, there is an 8% decrease in SWE/P after forest fire (Table 6). In RCP 8.5, the ratio is significantly lower with a 98% decrease and 99% decrease before and after forest cover reduction and an absolute difference of 7.9×10^{-2} and 7.3×10^{-2} in comparison to historical conditions (Table 6). This does not consider expected variability at different elevations and aspect. There is a significant difference between both historical land cover conditions and future land cover conditions in both scenarios. There is no significant difference between both scenarios before and after forest cover reduction in both historical and future periods. The difference between the ensemble basin mean of SWE/P in the 2080s before fire-burn is similar to MIROC5 predictions before forest cover reduction.

Table 4. Seasonal flows before and after forest cover reduction in the 1980s and 2080s for MIROC5 in RCP 4.5 and HadGEM2-ES in RCP 8.5.

	MIROC5								HadGEM2-ES							
	1980s Hist.				2080s RCP 4.5				1980s Hist.				2080s RCP 8.5			
	Pre-fire		Post-fire		Pre-fire		Post-fire		Pre-fire		Post-fire		Pre-fire		Post-fire	
μ Runoff (cm³/s)	WTR	SMR	WTR	SMR	WTR	SMR	WTR	SMR	WTR	SMR	WTR	SMR	WTR	SMR	WTR	SMR
	15.93	8.84	17.22	8.30	30.68	2.47	31.99	2.66	18.95	7.12	20.36	6.77	33.92	1.93	35.03	2.23
Min	1.53	2.34	2.07	2.03	4.77	0.86	6.15	0.95	1.78	1.01	2.32	1.12	6.33	0.32	8.17	0.39
Max	8.36	19.57	39.31	18.38	60.01	6.91	60.38	7.5	42.66	20.3	44.79	18.71	80.29	9.17	80.54	9.78
%Δ					92.59	−72.06	85.77	−67.95					79.00	−72.89	72.05	−67.06
σ	9.71	5.13	9.86	4.72	12.35	1.56	12.16	1.68	8.67	4.67	8.74	4.27	16.43	2.01	16.11	2.25

WTR = winter; SMR = summer.

Table 5. Ensemble mean seasonal flows before forest cover reduction in the 1980s, 2050s, and 2080s.

	Ensemble Mean									
	Hist.		RCP 4.5				RCP 8.5			
	1980s		2050s		2080s		2050s		2080s	
μ Runoff (cm³/s)	WTR	SMR	WTR	SMR	WTR	SMR	WTR	SMR	WTR	SMR
	17.69	6.916	29.26	3.51	30.33	3.27	30.63	3.35	35.36	2.37
Min	10.3	4.746	21.74	1.99	24.30	2.09	22.40	2.20	27.99	1.54
Max	25.26	10.092	36.28	4.69	39.02	5.13	44.50	5.13	44.74	3.48
%Δ			65.40	−49.24	71.45	−52.71	73.14	−51.56	98.89	−65.73
σ	3.97	1.32	3.55	0.61	3.38	0.72	4.82	0.98	4.19	0.50

WTR = winter; SMR = summer.

Table 6. Ensemble mean of snow water equivalent to precipitation (SWE/P) before forest cover reduction, and SWE/P before and after forest cover reduction for MIROC5 in RCP 4.5 and HadGEM2-ES in RCP 8.5.

	Ensemble Mean					MIROC5 RCP 4.5				HadGEM2-ES RCP 8.5			
	Hist.	RCP 4.5		RCP 8.5		Pre-fire	Post-fire	Pre-fire	Post-fire	Pre-fire	Post-fire	Pre-fire	Post-fire
	1980s	2050s	2080s	2050s	2080s	1980s	1980s	2080s	2080s	1980s	1980s	2080s	2080s
μ SWE (cm)	6.6	1.97	1.37	1.43	0.72	8.40	8.04	1.33	1.15	6.38	4.42	0.08	0.05
μ P (cm)	75.7	79.23	78.19	79.46	81.08	77.20	77.20	79.74	79.74	76.96	77.14	79.89	79.89
μ SWE/P	0.08	0.02	0.015	0.016	0.0072	0.11	0.104	0.02	0.01	0.08	0.079	0.001	0.00626
% Δ		−75.0	−81.25	−80.0	−91.0			−81.81	−90.3			−98.75	−99.20
σ	0.059	0.027	0.026	0.024	0.029	0.069	0.069	0.029	0.028	0.054	0.056	0.003	0.002

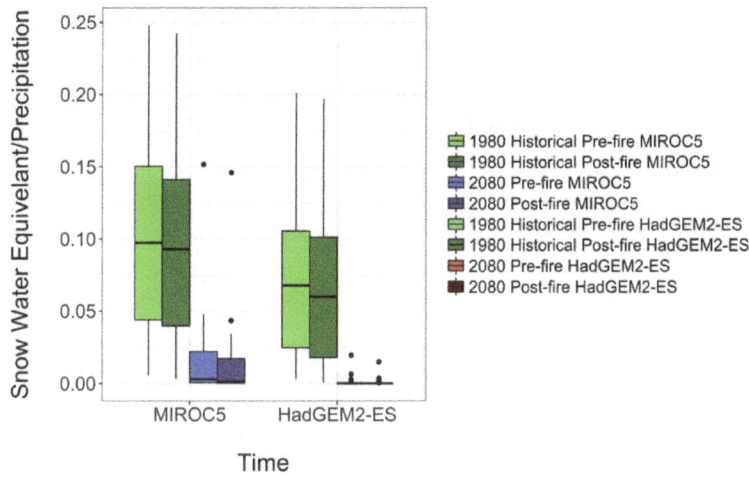

Figure 7. Ratio of snow water equivalent to precipitation for MIROC5 in RCP 4.5 and HadGEM2-ES in RCP 8.5 before and after forest cover reduction in comparison to historic conditions.

3.7. Potential Recharge after Fire-Burn

Cumulative basin-wide potential recharge after the fire-burn decreases in RCP 4.5 and RCP 8.5 by 14% and 26% in the 2080s in comparison to historical conditions before forest cover reduction (Figure 8; Table 7). After forest cover reduction, recharge decreases 13% and 22% in comparison to historical conditions in the RCP 4.5 and RCP 8.5 scenarios (Table 7). We see a decrease in recharge in the 2080s in both scenarios in comparison to historical conditions, but an increase in recharge after forest cover reduction in comparison to before forest reduction in the 1980s and 2080s (Table 7). Basin-wide recharge is not significantly different between any land cover conditions in RCP 4.5 and between historical pre-fire and historical post-fire, historical pre-fire and post-fire and between pre-fire and post-fire in RCP 8.5. Potential recharge is significantly different between historical pre-fire and future pre-fire, historical pre-fire and future post-fire and between future pre-fire and future post-fire in RCP 8.5. Ensemble mean basin recharge decreases 14% in the 2080s in RCP 8.5 and is exceeded in decrease by HadGEM2-ES before and after forest cover reduction (Table 7). We observe a slight increase in recharge after forest cover reduction in the 1980s and 2080s in both scenarios.

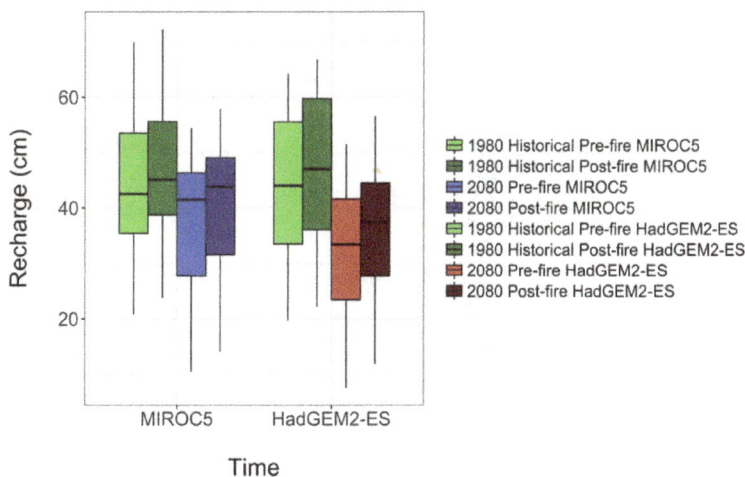

Figure 8. Potential mean basin recharge for MIROC5 in RCP 4.5 and HadGEM2-ES in RCP 8.5 before and after forest cover reduction in comparison to historic conditions.

Table 7. Ensemble mean of potential basin recharge before forest cover reduction, and potential basin recharge before and after forest cover reduction in RCP 4.5 and RCP 8.5 scenarios.

| | Ensemble Mean | | | | | MIROC5 | | | | HadGEM2-ES | | | |
| | Hist. | RCP 4.5 | | RCP 8.5 | | Pre-fire | Post-fire | Pre-fire | Post-fire | Pre-fire | Post-fire | Pre-fire | Post-fire |
	1980s	2050s	2080s	2050s	2080s	1980s	1980s	2080s	2080s	1980s	1980s	2080s	2080s
μ Recharge	42.77	40.71	38.36	39.60	36.46	44.55	47.20	37.89	40.87	43.73	46.54	32.14	36.06
Min	15.23	8.29	6.63	6.03	6.31	20.82	23.73	10.5	14.07	19.65	22.11	7.55	11.87
Max	72.79	75.62	77.60	70.63	75.09	69.88	72.20	54.36	57.88	64.26	66.79	51.48	56.65
% Δ		−4.82	−10.31	−7.41	−14.75			−14.95	−13.41			−26.50	−22.51
σ	11.19	12.07	11.34	11.75	11.44	12.96	12.75	11.11	10.86	13.75	13.79	11.65	11.59

4. Discussion

4.1. Caveats of Modeling

Model calibration may have been improved with a second stream gage located upstream on Meacham Creek, a tributary to the Umatilla River that provides a little over 50% of summer flows [60]. This would have required a second calibration. Consumptive use, which includes irrigation, municipal and domestic needs [61], was deemed negligible because diversions at the USGS stream gage, Umatilla River West Reservation Boundary near Pendleton, are minimal to total volume of streamflow [62]. The temporal and spatial behavior of groundwater in a heterogeneous geologic structure of the CRBGs could not be delineated with PRMS alone and is beyond the scope of this study. The CRBGs as a part of the Columbia Plateau Regional Aquifer System (CPRAS) cover 70,811 km^2 and warrant a regional approach in understanding groundwater behavior [63].

4.2. Temperature and Precipitation

A projected warming rate in the western U.S. is 0.1–0.6 °C per decade [64]. In the URB, there is high uncertainty and variability across the GCMs as can be seen in the wide variation of temperature and precipitation change throughout the 21st century (Figure 3). Mean temperature increases 3.3 °C by the end of the century in RCP 8.5, similar to a +3.2 °C increase by the 2080s predicted by Chang and Jung [8], in the Willamette River, OR. Dickerson-Lange and Mitchell [62], predicted a 1.8–3.5 °C mean increase in spring and summer temperatures by the 2050s in one scenario in northwestern Washington. Precipitation is variable in summer flows and increases as much as 11.3% in RCP 8.5 by the end of the century in the URB (Figure 3), where a 15%–21% increase is seen in northwestern WA in two models [64]. This is in agreement with Vynee et al. [65], who observed approximately a 10%–18% increase in precipitation by the mid and end of the century in the URB. With increased temperatures and less snow to hold increased precipitation, the frequency and magnitude of floods are predicted to increase [64].

4.3. Snow Water Equivalent and Precipitation

April 1 SWE is a function of winter accumulation and ablation. SWE/P substantially decreases with each time period, indicating a hydrologic regime shift from a snow-rain-dominated to a rain-dominated basin. This is consistent with predictions in the Pacific Northwest [5,14,64–68]. Vynee et al. [65] predicted SWE to decrease more than 50% by the 2080s in the URB. A considerable change in basin area-weighted SWE has been observed to affect mid-elevation areas in the rain and snow transition zone [15]. In post-fire conditions, there is a substantial decrease in SWE in the 2080s for both land cover conditions, a decrease of greater than 90%. This could be due to varying energy balances at the land and atmosphere interface, including radiative fluxes and changes in albedo, which can significantly influence the melting snow rate and the intensity of reflection by snow cover. Albedo was observed to be higher after a forest fire and lower after afforestation [68]. Further analysis of montane snowpacks that store winter precipitation and provide water for the rest of the year is required for climate adaptation planning in dam water releases and flood control [27].

4.4. Runoff Behavior

Precipitation and temperature are the main drivers of the magnitude and timing of streamflow [64]. At the end of the 21st century, after forest cover reduction, ensemble mean CT occurs earlier in the year by five weeks in RCP 4.5 and by 4.1 weeks in RCP 8.5. Post-fire parameters, including an 80% decrease in both summer and winter cover density and a 40% increase in the solar radiation transmission coefficient (Table 1), may have more effect on peak discharge during individual precipitation events than CT. Runoff trends, even if subtle, can be detrimental to fish habitat and growing seasons of wheat and green peas in the URB, for instance.

At the end of the 21st century, seasonal variability of ensemble winter flows is projected to increase (up to 98%) with decreases in summer flows (up to 65% reduction). Forest cover reduction is projected to amplify this variability further, with increases in winter flows by 85% in RCP 4.5 and 72% in RCP 8.5 in the UIR in particular (Table 4). An increase in the ratio of winter rainfall to winter snowfall is observed here, where precipitation is not being held in the snowpack due to warming temperatures as seen across the western United States. Jung and Chang [5] observed negative runoff trends in the spring and summer and positive trends in the fall and winter in the Willamette River Basin, OR. Similarly, Dickerson-Lange [64] observed increases in winter discharge from 34%–60% by midcentury and decreases in summer flows from −20% to −30% in Northwestern, WA. In the Deschutes Basin in central Oregon, winter flows are projected to increase 80%–115% in the Cascade Range [4].

4.5. Potential Basin Recharge and Base-Flow

A decrease in recharge after forest cover reduction in both scenarios in the 2080s, but not greater than historic conditions, may be due to decreased canopy interception and less evaporation occurring at the watershed surface with an increased potential for infiltration to occur, contributing to basin recharge. The ensemble mean of basin recharge is projected to remain within the range of historic levels before forest cover reduction with slight declines throughout the 21st century. It is most likely that fire-burned areas are relatively small compared to the whole basin area, and the recharge rate may vary over space with the shift in climate as reported in other Oregon watersheds in a semi-arid climate [4]. Historic mean basin recharge is 42 cm/year, within range of previous studies of 2.0 to 36.0 cm/year [21,69]. After forest cover reduction, mean recharge decreases by 1.9 cm in RCP 4.5 and decreases 6.71 cm in RCP 8.5 in comparison to historic conditions.

Quantifying groundwater is difficult due to the spatial and temporal variability of water below the subsurface [4]. The estimation of aquifer recharge and groundwater availability is critical to water management to meet domestic, municipal and ecological needs. The decline of groundwater levels in the URB has been addressed by The City of Pendleton, where the Aquifer Storage and Recovery program (ASR) lowered the city's dependence on groundwater from 62% down to 3%. Since then, groundwater declines were observed to be 340 cm/year and down to 200 cm/year after ASR was implemented in 2004 [70].

4.6. Future Work

PRMS files may be adapted to combine with the Modular Groundwater Flow Model (MODFLOW), a numerical groundwater model, to input to the Groundwater and Surface-Water Flow Model (GSFLOW), a coupled groundwater and surface-water flow model, to increase the understanding of the spatial and temporal behavior of groundwater. A dynamic global vegetation model may help identify parameters that account for regrowth, burn severity and intensity, to improve the understanding of the effects of fire-burns on a watershed system. Soil water repellence for example, has been found to last anywhere from one–six years, where a shorter temporal scale may best capture watershed response to fire [71,72]. Spatial analysis at a finer scale will only enhance localized efforts for the management of ecosystem services.

5. Conclusions

Increasing global mean temperature and changing precipitation are driving factors in runoff behavior. The uncertainty in the effects of climatic change and variability and anthropogenic influences on a hydrologic regime make it imperative to study their effects on natural resources. Using PRMS, a runoff model was calibrated for the upper URB, to characterize trends in runoff, snowpack, recharge and other components of the water budget to understand water availability in a changing climate and forest cover reduction. The effects of fire and climate shifts on runoff behavior are largely understudied in the URB, making this study unique.

A hydrologic regime shift is observed in the URB, from a snow-rain-dominated to a rain-dominated basin, as observed in SWE/P, as an important metric that shows increased sensitivity to climactic change in the URB throughout the 21st century before and after forest cover reduction. The ratio of SWE/P is shown to significantly decrease in both scenarios across the century before forest cover reduction. After forest cover reduction, similar trends in mean CT, seasonal flows and SWE/P are observed with a substantial decrease in SWE/P and an increase in winter flows in RCP 4.5 in the 2080s.

Mean basin recharge is sustained throughout the 21st century with slight declines in each subsequent time period before forest cover reduction, while after post-fire simulation, basin recharge is projected to increase. Due to the complexity of groundwater behavior in the CRBGs, basin recharge should be explored further with a numerical groundwater flow model.

This study provides further insight to secure freshwater resources for ecosystem function and cultural resources in the URB. Runoff modeling is a valuable tool to inform water and natural resources management to improve adaptive capacity, including flood control, dam releases and in-stream flow restoration practices.

Acknowledgments: The research was funded by the Geology Foundation at Portland State University and the Intertribal Timber Council. We appreciate John Risley of the U.S. Geological Survey and Professors Joseph Maser and Scott Burns of Portland State University for their careful reviews of the initial version of this manuscript. We thank Kate Ely, with Department of Natural Resources Water Resources Program at Confederated Tribes of the Umatilla Indian Reservation, for providing background materials on the study area. We also thank Mathew Dorfman, with the City of Portland, Environmental Services, for review of statistical analysis. The views expressed are our own and do not necessarily reflect those of sponsoring agencies.

Author Contributions: Kimberly Yazzie and Heejun Chang conceived of and designed the hydrologic modeling and analyses. Kimberly Yazzie performed the modeling and analyses. Kimberly Yazzie and Heejun Chang interpreted the results and wrote the paper.

Conflicts of Interest: The authors declare no conflict of interest. The founding sponsors had no role in the design of the study; in the collection, analyses or interpretation of data; in the writing of the manuscript; nor in the decision to publish the results.

Appendix A

Table A1. Global climate models (GCM) used in this study. The ensemble mean was taken of all ten GMCs. MIROC5 and HadGEM2-ES were used for forest cover reduction analysis.

Model Name	Model Agency	Country
CNRM-CM5	Natl. Centre of Meteorological Res.	France
HadGEM2-ES	Met Office Hadley Ctr.	UK
CanESM2	Canadian Ctr. for Climate Modeling & Analysis	Canada
MIROC5	Atmosphere & Ocean Res. Inst, Japan & Natl. Inst. for Env. Studies, Japan Agency for Marine-Earth Sci. and Tech.	Japan
NorESM1-M	Norwegian Climate Ctr.	Norway
CSIRO-Mk3.6.0	Commonwealth Sci. & Industrial Res. Org./Queensland Climate Change Ctr. of Excellence	Australia
MRI-CGCM3	Meteorological Res. Inst.	Japan
INM-CM4	Inst. for Numerical Mathematics	Russia
BCC-CSM1.1	Beijing Climate Ctr., China Meteorological Admin.	China
GFDL-ESM2M	NOAA Geophysical Fluid Dynamics Laboratory	USA

Table A2. Datasets, models and tools used for model parameters and data analysis.

Data	Resolution	Source
Historic Climate Data	4 km	Abatzaglou (2012)
Future Climate Data	4 km	Abatzaglou (2012)
Streamflow: U.S. Geological Survey Stream Gage 14020850		USGS (2013)
Soils: NRCS State Soils Geographic	30 m	STATSGO (2013)
Land Use and Land Cover: Nat'l Land Cover Data	30 m	USGS (2013)
DEM: National Elevation Dataset	30 m	USGS (2013)
Point data and acres burned in the URB		U.S. Forest Service, Umatilla Natl. Forest
Models and Tools	Version	
Precipitation Runoff Modeling System (PRMS)	3.0.5	USGS (2013)
Geo Data Portal (GDP)		USGS (2013)
Let Us Calibrate (LUCA)		USGS (2013)
Web-based Hydrograph Analysis Tool		Purdue University (2015)

Table A3. Final model parameters used after calibration.

Step	Calibration Dataset	Parameter Name	Final Value	Parameter Range
1	Water Balance	rain_cbh_adj_mo	1.128	0.6–1.4
		snow_cbh_adj_mo	1.4	0.6–1.4
2	Daily Flow Timing (all flows)	adjmix_rain_hru_mo	0.4–1.4	0.6–1.4
		cecn_coef	2.12	2.0–10.0
		emis_noppt	0.975	0.76–1.0
		freeh2o_cap	0.019	0.01–0.2
		K_coef	23.859	1–24.0
		potet_sublim	0.541	0.1–0.75
		slowcoef_lin	0.004	0.001–0.5
		soil_moist_max	2.14–12.537	2–10
		soil_rechr_max	1.643	1.5–5
		tmax_allrain_hru_mo	22–52	34–45
		tmax_allsnow_hru	37	30–40
3	Daily Flow Timing (high flows)	fastcoef_lin	0.005	0.001–0.8
		pref_flow_den	0.1	0–0.1
		sat_threshold	3.031–13.955	1.0–15.0
		smidx_coef	0.001	0.001–0.06
4	Daily Flow Timing (low flows)	gwflow_coef	0.024	0.001–0.1
		soil2gw_max	0.103	0–0.5
		ssr2gw_rate	0.582	0.05–0.8
		gwflow_coef	0.024	0.001–0.5
		gwsink_coef	0.02	0.0–0.05
		soil2gw_max	0.103	0–0.5
		ssr2gw_rate	0.582	0.05–0.8
		soil_moist_max	2.14–12.537	2–10
		slowcoef_sq	0.161	0.05–0.3

References

1. Oki, T.; Kanae, S. Global hydrological cycles and world water resources. *Science* **2006**, *313*, 1068–1072. [CrossRef] [PubMed]

2. Kundzewicz, Z.W.; Mata, L.J.; Arnell, N.W.; Doll, P.; Kabat, P.; Jimenez, B.; Miller, K.A.; Oki, T.; Sín, Z.; Shiklomanov, I.A. Freshwater resources and their management. In *Impacts, adaptation and vulnerability. Contribution of Working Group II to the Fourth Assessment Report of the Intergovernmental Panel on Climate Change*; Parry, M.L., Canziani, O.F., Palutikof, J.P., van der Linden, P.J., Hanson, C.E., Eds.; Cambridge University Press: Cambridge, UK, 2007; pp. 173–210.

3. Qi, S.; Sun, G.; Wang, Y.; Mcnulty, S.; Myers, J. Streamflow response to climate and landuse changes in a coastal watershed in North Carolina. *Trans. ASABE* **2009**, *52*, 739–749. [CrossRef]

4.   Waibel, M.S.; Gannett, M.W.; Chang, H.; Hulbe, C.L. Spatial variability of the response to climate change in regional groundwater systems—Examples from simulations in the Deschutes Basin, Oregon. *J. Hydrol.* **2013**, *486*, 187–201. [CrossRef]

5.   Jung, Il-W.; Chang, H. Assessment of future runoff trends under multiple climate change scenarios in the Willamette River Basin, Oregon, USA. *Hydrol. Process.* **2011**, *25*, 258–277. [CrossRef]

6.   Hamlet, A.F.; Lettenmaier, D.P. Effects of 20th century warming and climate variability on flood risk in the western U.S. *Water Resour. Res.* **2007**, *43*, 1–17. [CrossRef]

7.   Abatzoglou, J.T.; Rupp, D.; Mote, P. Understanding seasonal climate variability and change in the Pacific Northwest of the United States. *J. Clim.* **2014**, *27*, 2125–2142. [CrossRef]

8.   Chang, H.J.; Jung, I.W. Spatial and temporal changes in runoff caused by climate change in a complex large river basin in Oregon. *J. Hydrol.* **2010**, *388*, 186–207. [CrossRef]

9.   Elsner, M.; Cuo, L.; Voisin, N.; Deems, J.; Hamlet, A.; Vano, J.; Mickelson, K.; Lee, S.; Lettenmaier, D. Implications of 21st century climate change for the hydrology of Washington State. *Clim. Chang.* **2010**, *102*, 225–260. [CrossRef]

10. Hamlet, A.F.; Carrasco, P.; Deems, J.; Elsner, M.M.; Kamstra, T.; Lee, C.; Mauger, G.; Salathe, E.P.; Tohver, I.; Whitely Binder, L. Final Project Report for the Columbia Basin Climate Change Scenarios Project. 2010. Available online: http://warm.atmos.washington.edu/2860/report/ (accessed on 1 August 2015).

11. Surfleet, C.G.; Tullos, D.; Chang, H.; Jung, I.-W. Selection of hydrologic modeling approaches for climate change assessment: A comparison of model scale and structures. *J. Hydrol.* **2012**, *464–465*, 233–248. [CrossRef]

12. Knowles, N.; Dettinger, M.D.; Cayan, D.R. Trends in snowfall versus rainfall in the western United States. *J. Clim.* **2006**, *19*, 4545–4559. [CrossRef]

13. Abatzoglou, J.T. Influence of the PNA on declining mountain snowpack in the Western United States. *Int. J. Climatol.* **2011**, *31*, 1135–1142. [CrossRef]

14. Mote, P.W.; Hamlet, A.F.; Clark, M.P.; Lettenmaier, D.P. Declining mountain snowpack in western North America. *Bull. Am. Meteorol. Soc.* **2005**, *86*, 6. [CrossRef]

15. Mastin, M.C.; Chase, K.J.; Dudley, R.W. Changes in spring snowpack for selected basins in the United States for different climate-change scenarios. *Earth Interact.* **2011**, *15*, 1–18. [CrossRef]

16. Mote, P.W.; Parson, E.A.; Hamlet, A.F.; Keeton, W.S.; Lettenmaier, D.; Mantua, N.; Miles, E.L.; Peterson, D.W.; Peterson, D.L.; Slaughter, H.; et al. Preparing for climatic change: The water, salmon, and forests of the Pacific Northwest. *Clim. Chang.* **2003**, *61*, 45–88. [CrossRef]

17. Jung, I.; Chang, H. Climate change impacts on spatial patterns in drought risk in the Willamette River Basin, Oregon, USA. *Theor. Appl. Climatol.* **2011**, *108*, 355–371. [CrossRef]

18. Drost, B.; Whiteman, K.J. Washington Department of Ecology. In *Surficial Geology, Structure, and Thickness of Selected Geohydrologic Units in the Columbia Plateau, Washington*; U.S Geological Survey Water-Resources Investigations Report 84–4326; Washington Department of Ecology: Tacoma, DC, USA, 1986.

19. Vaccaro, J.J. *Plan of Study for the Regional Aquifer-System Analysis, Columbia Plateau, Washington, Northern Oregon, and Northwestern Idaho Water-Resources Investigations Report 85-4151*; U.S. Department of the Interior, U.S. Geological Survey: Tacoma, WA, USA, 1986.

20. Tolan, T.L.; Reidel, S.P.; Beeson, M.H.; Anderson, J.L.; Fecht, K.R.; Swanson, D.A. Revisions to the areal extent and volume of the Columbia River Basalt Group (CRBG). *Geol. Soc. Am.* **1987**, *19*, 458.

21. Bauer, H.H.; Vaccaro, J.J. *Estimates of Groundwater Recharge to the Columbia Plateau Regional Aquifer System, Washington, Oregon, and Idaho, for Predevelopment and Current Land-Use Conditions Water Resources Investigations Report 88–4108*; U.S. Geological Survey: Tacoma, WA, USA, 1990.

22. Chang, H.; Psaris, M. Local landscape predictors of maximum stream temperature and thermal sensitivity in the Columbia River Basin, USA. *Sci. Tot. Environ.* **2013**, *461–462*, 587–600. [CrossRef] [PubMed]

23. Burns, E.R.; Snyder, D.T.; Haynes, J.V.; Waibel, M.S. Groundwater status and trends for the Columbia Plateau Regional Aquifer System, Washington, Oregon, and Idaho Scientific Investigations Report 2012–5261. U.S. Department of the Interior, U.S. Geological Survey, 2012. Available online: http://pubs.er.usgs.gov/publications/sir20125261 (accessed on 1 September 2013).

24. Hansen, A.J.; Vaccaro, J.J.; Bauer, H.H. *Ground-Water Flow Simulation of the Columbia Plateau Regional Aquifer System, Washington, Oregon, and Idaho Water-Resources Investigations Report 91–4187*; U.S. Department of the Interior, U.S. Geological Survey: Tacoma, WA, USA, 1994.

25. Moody, J.A.; Martin, D.A. Post-fire, rainfall intensity—Peak discharge relations for three mountainous watersheds in the western USA. *Hydrol. Process.* **2001**, *15*, 2981–2993. [CrossRef]

26. Vieira, D.; Fernández, C.; Vega, J.; Keizer, J. Does soil burn severity affect the post-fire runoff and interrill erosion response? A review based on meta-analysis of field rainfall simulation data. *J. Hydrol.* **2015**, *523*, 452–464. [CrossRef]

27. Cerda, A. Post-fire dyamics of erosional processes under mediterranean climatic conditions. *Z. Feur Geomorphol. Neue Folge* **1998**, *42*, 373–398.

28. Martin, D.A.; Moody, J.A. Comparison of soil infiltration rates in burned and unburned mountainous watersheds. *Hydrol. Process.* **2001**, *15*, 2893–2903. [CrossRef]

29. Robichaud, P.R. Measurement of post-fire hillslope erosion to evaluate and model rehabilitation treatment effectiveness and recovery. *Int. J. Wildland. Fire* **2005**, *14*, 475–485. [CrossRef]

30. Turner, D.P.; Conklin, D.R.; Bolte, J.P. Projected climate change impacts on forest land cover and land use over the Willamette River Basin, Oregon, USA. *Clim. Chang.* **2015**, *133*, 335. [CrossRef]

31. Rulli, M.C.; Offeddu, L.; Santini, M. Modeling post-fire water erosion mitigation strategies. *Hydrol. Earth Syst. Sci.* **2013**, *17*, 2323–2337. [CrossRef]

32. Terranova, O.; Antronico, L.; Coscarelli, R.; Iaquinta, P. Soil erosion risk scenarios in the Mediterranean environment using RUSLE and GIS: An application model for Calabria (Southern Italy). *Geomorphology* **2009**, *112*, 228–245. [CrossRef]

33. Sheehan, T.; Bachelet, D.; Ferschweiler, K. Projected major fire and vegetation changes in the Pacific Northwest of the conterminous United States under selected CMIP5 climate futures. *Ecol. Model.* **2015**, *317*, 16–29. [CrossRef]

34. Oregon Water Resources Department. Available online: https://www.oregon.gov/owrd/Pages/law/integrated_water_supply_strategy.aspx (accessed on 15 October 2015).

35. Umatilla River Subbasin Local Advisory Committee, Oregon State Deptartment of Agriculture, & Umatilla Soil Water Conservation District. *Umatilla Agricultural Water Quality Management Area Plan*; The Umatilla Local Advisory Committee: Umatilla, FL, USA, 2012.

36. Ely, K. *Water Resources Status, A Study of the Water Resources Availability and Demand in the Umatilla River Basin, Oregon*; U.S. Bureau of Indian Affairs: Pendleton, OR, USA, 2001.

37. Jones, K.L.; Poole, G.C.; Quaempts, E.J.; O'Daniel, E.; Beechie, T. Umatilla River Vision Report. 2008. Available online: http://ctuir.org/DNRUmatillaRiverVision.pdf (accessed on 1 September 2013).

38. Hughes, M. Channel Change of the Upper Umatilla River during and between Flood Periods: Variability and Ecological Implications. Ph.D. Thesis, University of Oregon, Eugene, OR, USA, 2008.

39. Ely, K. Groundwater and surface water are they related? *Confed. Umatilla J.* **2012**, *16*, 31.

40. U.S. Forest Service. *Umatilla and Meacham Ecosystem Analysis*; USDA Forest Service Pacific Northwest Region Umatilla National Forest: Pendleton, OR, USA, 2001.

41. Confederated Tribes of the Umatilla Indian Reservation. *Forest Management Plan: An Ecological Approach to Forest Management*; Mason Bruce & Gerard, Inc. Publication: Bothell, WA, USA, 2010.

42. Homer, C.G.; Dewitz, J.A.; Yang, L.; Jin, S.; Danielson, P.; Xian, G.; Coulston, J.; Herold, N.D.; Wickham, J.D.; Megown, K. Completion of the 2011 national land cover database for the conterminous United States-Representing a decade of land cover change information. *Photogr. Eng. Rem. Sens.* **2015**, *81*, 345–354.

43. Markstrom, S.L.; Regan, R.S.; Hay, L.E.; Viger, R.J.; Webb, R.M.T.; Payn, R.A.; LaFontaine, J.H. *PRMS-IV, the Precipitation-Runoff Modeling System, Version 4: U.S. Geological Survey Techniques and Methods*; U.S. Geological Survey: Reston, VA, USA, 2015.

44. Abatzoglou, J.T. Development of gridded surface meteorological data for ecological applications and modelling. *Int. J. Climatol.* **2013**, *33*, 121–131. [CrossRef]

45. Mitchell, K.E.; Lohmann, D.; Houser, P.R.; Wood, E.F.; Schaake, J.C.; Robock, A.; Cosgrove, B.A.; Sheffield, J.; Duan, Q.; Luo, L.; et al. The multi-institution North American Land Data Assimilation System (NLDAS): Utilizing multiple GCIP products and partners in a continental distributed hydrological modeling system. *J. Geophys. Res.* **2004**, *109*, 1–32. [CrossRef]

46. Daly, C.; Halbleib, M.; Smith, J.; Gibson, W.; Doggett, M.; Taylor, G.; Curtis, J.; Pasteris, P. Physiographically sensitive mapping of climatological temperature and precipitation across the conterminous United States. *Int. J. Climatol.* **2008**, *28*, 2031–2064. [CrossRef]

47. Abatzoglou, J.T.; Brown, T.J. A comparison of statistical downscaling methods suited for wildfire applications. *Int. J. Climatol.* **2012**, *32*, 772–780. [CrossRef]

48. Taylor, K.E.; Stouffer, R.J.; Meehl, G.A. An overview of CMIP5 and the experiment design. *Bull. Am. Meteorol. Soc.* **2011**, *93*, 485–498. [CrossRef]

49. Van Vuuren, D.; Edmonds, J.; Kainuma, M.; Riahi, K.; Thomson, A.; Hibbard, K.; Rose, S. The representative concentration pathways: An overview. *Clim. Chang.* **2011**, *109*, 5–31. [CrossRef]

50. Blodgett, D.L.; Booth, N.L.; Kunicki, T.C.; Walker, J.L.; Viger, R.J. *Description and Testing of the Geo Data Portal: A Data Integration Framework and Web Processing Services for Environmental Science Collaboration*; Open-File Report 2011–1157; U.S. Deptartment of Interior, U.S. Geological Survey: Reston, VA, USA, 2011.

51. Rupp, D.; Abatzoglou, J.; Hegewisch, K.; Mote, P. Evaluation of CMIP5 20th century climate simulations for the Pacific Northwest USA. *J. Geophys. Res.* **2013**, *118*, 10884–10906.

52. Mote, P.; Brekke, L.; Duffy, P.; Maurer, E. Guidelines for constructing climate scenarios. *EOS Trans. Am. Geophys. Union* **2011**, *92*, 257–258. [CrossRef]

53. Leavesley, G.H.; Lichty, R.W.; Troutman, B.M.; Saindon, L.G. *Precipitation Runoff Modeling System: User's Manual -Water-Resources Investigations Report 83–4238*; US Geological Survey: Denver, CO, USA, 1983.

54. Konrad, C. *Simulated Water-Management Alternatives Using the Modular Modeling System for the Methow River Basin, Washington U.S*; Geological Survey Open-File Report 2004–1051; U.S. Department of the Interior, U.S. Geological Survey: Denver, CO, USA, 2004.

55. Ebert, B.U.S.; Forest Service, Pendleton, OR, USA. Personal Communication, 2015.

56. Hay, L.E.; Umemoto, M. *Multiple-Objective Stepwise Calibration Using LUCA: U.S. Geological Survey Open-File Report 2006–1323*; US Geological Survey Water-Resources Investigations Report: Washington, DC, USA, 2006.

57. Krause, P.; Boyle, D.P.; Bäse, F. Comparison of different efficiency criteria for hydrological model assessment. *Adv. Geosci.* **2005**, *5*, 89–97. [CrossRef]

58. Moriasi, D.N.; Arnold, J.G.; Van Liew, M.W.; Binger, R.L.; Harmel, R.D.; Veith, T.L. Model evaluation guidelines for systematic quantification of accuracy in watershed simulations. *Trans. ASABE.* **2007**, *50*, 885–900. [CrossRef]

59. Gupta, H.; Kling, H.; Yilmaz, K.; Martinez, G. Decomposition of the mean squared error and NSE performance criteria; implications for improving hydrological modelling. *J. Hydrol.* **2009**, *377*, 80–91. [CrossRef]

60. Confederated Tribes of the Umatilla Indian Reservation. *Meacham Creek Flood Restoration and In-stream Enhancement Project Completion Report*; Tetra Tech Inc. Publication: Portland, OR, USA, 2012.

61. Cooper, R.M. *Determining Surface Water Availability in Oregon, State of Oregon Water Resources Department, Open File Report SW 02-002*; State of Oregon, Water Resources Department: Salem, OR, USA, 2002.

62. Ely, K.; Department of Natural Resources, Confederated Tribes of the Umatilla Indian Reservation, Pendleton, OR, USA. Personal Communication, 2013.

63. Cherkauer, D. Quantifying ground water recharge at multiple scales using PRMS and GIS. *Ground Water* **2004**, *42*, 97–110. [CrossRef] [PubMed]

64. Dickerson-Lange, S.; Mitchell, R. Modeling the effects of climate change projections on streamflow in the Nooksack River Basin, Northwest Washington. *Hydrol. Process.* **2014**, *28*, 5236–5250. [CrossRef]

65. Stacy, V.; Reder, B.; Hamilton, R.; Doppelt, B.; Dello, K.; Sharp, D. *Projected Future Conditions in the Umatilla River Basin of Northeast Oregon*; Report: Oregon Climate Change Research Institute, Climate Leadership Initiative Institute for a Sustainable Environment; University of Oregon: Eugen, OR, USA, 2010.

66. Stewart, I.T.; Cayan, D.R.; Dettinger, M.D. Changes toward earlier streamflow timing across western North America. *J. Clim.* **2005**, *18*, 1136–1155. [CrossRef]

67. Hamlet, A.F. Assessing water resources adaptive capacity to climate change impacts in the Pacific Northwest Region of North America. *Hydrol. Earth Syst. Sci.* **2011**, *15*, 1427–1443. [CrossRef]

68. Safeeq, M.; Grant, G.; Lewis, S.; Tague, C. Coupling snowpack and groundwater dynamics to interpret historical streamflow trends in the western United States. *Hydrol. Process.* **2013**, *27*, 655–668. [CrossRef]

69. Spane, F.; Webber, W. *Hydrochemistry and Hydrogeologic Conditions within the Hanford Site Upper Basalt Confined Aquifer System*; PNNL-10817; U.S. Department of Energy: Richland, WA, USA, 1995.

70. Pendleton Public Works Water Division. *City of Pendleton Water Management Plan*; Pendleton Public Works Water Division: Pendleton, OR, USA, 2010.

71. MacDonald, L.; Huffman, E. Post-fire soil water repellency; persistence and soil moisture thresholds. *Soil Sci. Soc. Am. J.* **2004**, *68*, 1729–1734. [CrossRef]

72. Henderson, G.; Golding, D. The effect of slash burning on the water repellency of forest soils at Vancouver, British Columbia. *Can. J. For. Res.* **1983**, *13*, 353–355. [CrossRef]

Permissions

The contributors of this book come from diverse backgrounds, making this book a truly international effort. This book will bring forth new frontiers with its revolutionizing research information and detailed analysis of the nascent developments around the world.

We would like to thank all the contributing authors for lending their expertise to make the book truly unique. They have played a crucial role in the development of this book. Without their invaluable contributions this book wouldn't have been possible. They have made vital efforts to compile up to date information on the varied aspects of this subject to make this book a valuable addition to the collection of many professionals and students.

This book was conceptualized with the vision of imparting up-to-date information and advanced data in this field. To ensure the same, a matchless editorial board was set up. Every individual on the board went through rigorous rounds of assessment to prove their worth. After which they invested a large part of their time researching and compiling the most relevant data for our readers.

The editorial board has been involved in producing this book since its inception. They have spent rigorous hours researching and exploring the diverse topics which have resulted in the successful publishing of this book. They have passed on their knowledge of decades through this book. To expedite this challenging task, the publisher supported the team at every step. A small team of assistant editors was also appointed to further simplify the editing procedure and attain best results for the readers.

Apart from the editorial board, the designing team has also invested a significant amount of their time in understanding the subject and creating the most relevant covers. They scrutinized every image to scout for the most suitable representation of the subject and create an appropriate cover for the book.

The publishing team has been an ardent support to the editorial, designing and production team. Their endless efforts to recruit the best for this project, has resulted in the accomplishment of this book. They are a veteran in the field of academics and their pool of knowledge is as vast as their experience in printing. Their expertise and guidance has proved useful at every step. Their uncompromising quality standards have made this book an exceptional effort. Their encouragement from time to time has been an inspiration for everyone.

The publisher and the editorial board hope that this book will prove to be a valuable piece of knowledge for researchers, students, practitioners and scholars across the globe.

List of Contributors

Maha AlSabbagh
Environmental Management Programme, Arabian Gulf University, Manama 26671, Bahrain

Sonja Kivinen
Department of Geographical and Historical Studies, University of Eastern Finland, P. O. Box 111, FI-80101 Joensuu, Finland
Department of Geography and Geology, University of Turku, FI-20014 Turku, Finland

Sirpa Rasmus
Department of Biological and Environmental Science, University of Jyväskylä, P.O. Box 35, FI-40014 Jyväskylä, Finland

Kirsti Jylhä and Mikko Laapas
Finnish Meteorological Institute, P. O. Box 503, FI-00101 Helsinki, Finland

Singay Dorji
United Nations University, Institute for the Advanced Study of Sustainability (UNU-IAS), 5 Chome-53-70 Jingumae, Shibuya, Tokyo 150-8925, Japan

Srikantha Herath
Ministry of Megapolis and Western Development, Battaramulla 10120, Sri Lanka
UNU-IAS, IR3S-University of Tokyo, and University of Peradeniya, Tokyo 150-8925, Japan

Binaya Kumar Mishra
United Nations University, Institute for the Advanced Study of Sustainability (UNU-IAS), 5 Chome-53-70 Jingumae, Shibuya, Tokyo 150-8925, Japan
UNU-IAS, Tokyo 150-8925, Japan

Julia Hackenbruch, Sebastian Müller and Janus Willem Schipper
South German Climate Office, Institute of Meteorology and Climate Research, Karlsruhe Institute of Technology, Hermann-von-Helmholtz-Platz 1, 76344 Eggenstein-Leopoldshafen, Germany

Tina Kunz-Plapp
Institute of Meteorology and Climate Research, Karlsruhe Institute of Technology, Hermann-von-Helmholtz-Platz 1, 76344 Eggenstein-Leopoldshafen, Germany

Élise Lépy
Faculty of Humanities, University of Oulu, 90014 Oulu, Finland

Leena Pasanen
Research Unit of Mathematical Sciences, University of Oulu, 90014 Oulu, Finland

Norman C. Treloar
540 First Avenue West, Qualicum Beach, BC V9K 1J8, Canada

Shannan M. Little, H. Henry Janzen, Roland Kröbel and Karen A. Beauchemin
Agriculture and Agri-Food Canada, Lethbridge Research and Development Centre, Lethbridge, AB T1J 4B1, Canada; Henry.Janzen@agr.gc.ca (H.H.J.); Roland.Kroebel@agr.gc.ca (R.K.);

Chaouki Benchaar
Agriculture and Agri-Food Canada, Sherbrooke Research and Development Centre, Sherbrooke, QC J1M 0C8, Canada; Chaouki

Emma J. McGeough
Department of Animal Science, University of Manitoba, Winnipeg, MB R3T 2N2, Canada

Shannan M. Little, Karen A. Beauchemin, H. Henry Janzen and Roland Kröbel
Agriculture and Agri-Food Canada, Lethbridge Research and Development Centre, Lethbridge, AB T1J 4B1, Canada

Chaouki Benchaar
Agriculture and Agri-Food Canada, Sherbrooke Research and Development Centre, Sherbrooke, QC J1M 0C8, Canada

Emma J. McGeough
Department of Animal Science, University of Manitoba, Winnipeg, MB R3T 2N2, Canada

Renata dos Santos Cardoso, Larissa Piffer Dorigon, Danielle Cardozo Frasca Teixeira and Margarete Cristiane de Costa Trindade Amorim
Department of Geography, São Paulo State University (UNESP), 305 Roberto Simonsen Street, Presidente Prudente, São Paulo 19060-900, Brazil

Muna Neupane
Central Department of Environmental Science, Tribhuvan University, Kathmandu 44600, Nepal

Subodh Dhakal
Department of Geology, Tri-Chandra Campus, Tribhuvan University, Kathmandu 44600, Nepal

Felix Dietzsch, Axel Andersson, Markus Ziese, Marc Schröder, Kristin Raykova,
Kirstin Schamm and Andreas Becker
Deutscher Wetterdienst, Frankfurter Straße 135, 63067 Offenbach (Main), Germany
Deutscher Wetterdienst, Bernhard-Nocht-Straße 76, 20359 Hamburg, Germany

Kimberly Yazzie
Department of Environmental Science and Management, Portland State University, Portland, OR 97207, USA

Heejun Chang
Department of Geography, Portland State University, Portland, OR 97207, USA

Index

www.ingramcontent.com/pod-product-compliance
Lightning Source LLC
Chambersburg PA
CBHW080257230326
41458CB00097B/5092